NOx EMISSION CONTROL TECHNOLOGIES IN
STATIONARY AND AUTOMOTIVE INTERNAL COMBUSTION ENGINES

NOx EMISSION CONTROL TECHNOLOGIES IN STATIONARY AND AUTOMOTIVE INTERNAL COMBUSTION ENGINES

Approaches Toward NOx Free Automobiles

Edited by

B. ASHOK

Associate Professor, Vellore Institute of Technology,
Vellore, Tamil Nadu, India

ELSEVIER

Elsevier
Radarweg 29, PO Box 211, 1000 AE Amsterdam, Netherlands
The Boulevard, Langford Lane, Kidlington, Oxford OX5 1GB, United Kingdom
50 Hampshire Street, 5th Floor, Cambridge, MA 02139, United States

Notices
Knowledge and best practice in this field are constantly changing. As new research and experience broaden our understanding, changes in research methods, professional practices, or medical treatment may become necessary.

Practitioners and researchers must always rely on their own experience and knowledge in evaluating and using any information, methods, compounds, or experiments described herein. In using such information or methods they should be mindful of their own safety and the safety of others, including parties for whom they have a professional responsibility.

To the fullest extent of the law, neither the Publisher nor the authors, contributors, or editors, assume any liability for any injury and/or damage to persons or property as a matter of products liability, negligence or otherwise, or from any use or operation of any methods, products, instructions, or ideas contained in the material herein.

Library of Congress Cataloging-in-Publication Data
A catalog record for this book is available from the Library of Congress

British Library Cataloguing-in-Publication Data
A catalogue record for this book is available from the British Library

ISBN: 978-0-12-823955-1

Publisher: Matthew Deans
Acquisitions Editor: Carrie Bolger
Editorial Project Manager: Joshua Mearns
Production Project Manager: Poulouse Joseph
Cover Designer: Mark Rogers

Typeset by STRAIVE, India

This book is dedicated to all my research team members, students, and collaborators.

Contents

Contributors

Avinash Alagumalai
Department of Mechanical Engineering, GMR Institute of Technology, Rajam, Andhra Pradesh, India

Rajarajan Amudhan
Department of Mechanical Engineering, Mepco Schlenk Engineering College, Sivakasi; Department of Mechanical Engineering, CK College of Engineering and Technology, Cuddalore, India

B. Ashok
Engine Testing Laboratory, School of Mechanical Engineering, Vellore Institute of Technology, Vellore, Tamil Nadu, India

Dhinesh Balasubramanian
Department of Mechanical Engineering, Mepco Schlenk Engineering College, Sivakasi, India; Mechanical Engineering, Faculty of Engineering; Center for Alternative Energy Research and Development, Khon Kaen University, Khon Kaen, Thailand

Saravanan Balusamy
Department of Mechanical and Aerospace Engineering, Indian Institute of Technology Hyderabad, Hyderabad, India

Madhu Sudan Reddy Dandu
Department of Mechanical Engineering, Sree Vidyanikethan Engineering College, Tirupati, Andhra Pradesh, India

P. Dhamodharan
SSN College of Engineering, Chennai, Tamil Nadu, India

P. Ganesh Kumar
Alstom Transport India Ltd, Chennai, India

Ashwin Jacob
Engine Testing Laboratory, School of Mechanical Engineering, Vellore Institute of Technology, Vellore, Tamil Nadu, India

P. Jayalakshmi
Hindustan College of Engineering & Technology, Coimbatore, Tamil Nadu, India

Amin Jodat
Department of Mechanical Engineering, University of Bojnord, Bojnord, North Khorasan, Iran

C. Kannan
Department of Automotive Engineering, School of Mechanical Engineering, Vellore Institute of Technology, Vellore, Tamil Nadu, India

C. Karthick
Engine Testing Laboratory, School of Mechanical Engineering, Vellore Institute of Technology, Vellore, Tamil Nadu, India

A. Naresh Kumar
Department of Mechanical Engineering, Lakireddy Bali Reddy College of Engineering, Mylavaram, Andhra Pradesh, India

Omid Mahian
School of Chemical Engineering and Technology, Xi'an Jiaotong University, Xi'an, Shaanxi, China

M. Mohamed Ibrahim
Automotive Research Centre, School of Mechanical Engineering, Vellore Institute of Technology, Vellore, Tamil Nadu, India

T. Mohanraj
Department of Mechanical Engineering, Amrita School of Engineering, Amrita Vishwa Vidyapeetham, Coimbatore, Tamil Nadu, India

Kasianantham Nanthagopal
Engine Testing Laboratory, School of Mechanical Engineering, Vellore Institute of Technology, Vellore, Tamil Nadu, India

Inbanaathan Papla Venugopal
Department of Mechanical Engineering, Mepco Schlenk Engineering College, Sivakasi, India

R. Prakash
School of Mechanical Engineering, Vellore Institute of Technology, Vellore, India

Pajarla Saiteja
Engine Testing Laboratory, School of Mechanical Engineering, Vellore Institute of Technology, Vellore, Tamil Nadu, India

Pemmareddy Saiteja
Engine Testing Laboratory, School of Mechanical Engineering, Vellore Institute of Technology, Vellore, Tamil Nadu, India

R. Sakthivel
Department of Mechanical Engineering, Amrita School of Engineering, Amrita Vishwa Vidyapeetham, Coimbatore, Tamil Nadu, India

S.V. Saravanan
Department of Mechanical Engineering, Asian College of Engineering and Technology, Coimbatore, India

G. Sathish Sharma
School of Mechanical Engineering, Vellore Institute of Technology, Vellore, India

S. Sathishkumar
Automotive Research Centre, School of Mechanical Engineering, Vellore Institute of Technology, Vellore, Tamil Nadu, India

S. Sidharth
Robert Bosch Engineering and Business Solutions Pvt Ltd, Coimbatore, India

M. Sugavaneswaran
School of Mechanical Engineering, Vellore Institute of Technology, Vellore, India

A. Tamilvanan
Department of Mechanical Engineering, Kongu Engineering College, Erode, Tamil Nadu, India

R. Vignesh
Engine Testing Laboratory, School of Mechanical Engineering, Vellore Institute of Technology, Vellore, Tamil Nadu, India

T. Vijayakumar
Department of Automotive Engineering, School of Mechanical Engineering, Vellore Institute of Technology, Vellore, Tamil Nadu, India

Tanakorn Wongwuttanasatian
Mechanical Engineering, Faculty of Engineering; Center for Alternative Energy Research and Development, Khon Kaen University, Khon Kaen, Thailand

Preface

In recent years, increased awareness of climate change and environmental deterioration due to vehicular pollutants has sparked governments to establish control policies to manage these crises. Modern engines are not only focused on engine performance, but also on emission aspects. Recent emission legislation has particularly focused on limits on nitrogen oxides (NO_x). Most countries in the world try to reduce pollutions by formulating various emission norms, recent focusing on nitrogen oxides. In this context, extensive research has been carried out on internal combustion (IC) engines to mitigate harmful emissions, mainly NO_x emissions. NO_x emissions generated from vehicle tailpipes have detrimental effects on human health and the environment while causing imbalances to the ecosystem. Hence, the authors identified that there is a need to understand the current stance of NO_x emission formation, control techniques, and the corresponding technologies employed in IC engines. To facilitate this, the authors decided to formulate an in-depth literary correlation between the theory and practical aspects of NO_x formation and control techniques in IC engines, thereby educating and providing the scientific research community with a comprehensive research work.

The primary motivation for writing this book is to provide a possible solution to reduce global NO_x emissions generated from automobiles by establishing control techniques that are efficient in reducing harmful emissions. The authors also intend to provide researchers working in the fields of automotive, mechanical, mechatronics, and chemical engineering a concise work covering the theoretical and experimental aspects of NO_x emission formation and control techniques for IC engines. Furthermore, the authors aim to provide advanced NO_x control techniques that have the potential to be commercially viable without major modification to the existing engine design as per the legislative emission norms. Within this context, this book covers NO_x and particulate matter formation, NO_x control, after-treatment devices, and fuel modification techniques. Also, related information for IC engines is discussed in a very extensive manner. The results observed in the literature are scrutinized and deconstructed to understand every aspect that led to NO_x formation and control. In some cases, controversial results are cross-examined by experimental testing to understand the NO_x formation concepts in a practical manner.

Hence, to provide an in-depth understanding related to NO_x emission formation and control aspects in IC engines, this book will be an essential literary tool for academics, researchers, and industrial experts working in the automotive sector. The whole idea of this book was initiated and carried by Dr. B. Ashok and his research team from the Vellore Institute of Technology, India. The contributors for this work are selected in such a way to provide a better outcome in the proposed area. Further, as editor and author, Ashok would like to express his heartfelt gratitude to all the contributors for being a part of this exceptional work. I am humbled by their sincerity and commitment throughout this journey. More importantly, I would like to specially thank the research scholars and collaborators who are working with me for helping to organize and evaluate the chapters for any inconsistencies in the structure and content as a whole. Moreover, Ashok and his research team would like to thank the Vellore Institute of Technology, India, for providing the research facilities and a wealth of knowledge in the form of literary subscriptions for e-resources and books from the library for reference.

About the editor

B. Ashok started his professional career as specialist engineer in vehicle electrical and electronics at M/S. Force Motors India Ltd, Pune after his M.Tech degree. During his tenure, he played a lead role in the implementation of electronic control for the BS IV engine and the antilock braking system in the vehicle portfolio. After the industrial experience, he joined the automotive department at Vellore Institute of Technology, Vellore, India. During his doctoral research work, he developed a low-cost virtual sensor for the throttle position sensing application for IC engines using soft computing techniques. Ashok has developed various techniques for the effective implementation of biofuels in IC engines, including fuel reformulation, engine parameter optimization, and design modification.

Furthermore, his expertise in mechatronics engineering helps to evolve novel interdisciplinary research in the areas of automotive electronics and automotive engineering systems. Those outcomes of his research have led to the publication of eight book chapters and 110 research papers as well as two patents, with a cumulative SCI Impact Factor of 456. Furthermore, his research is in the area of IC engines, emission control, and alternate mobility for the transport sector. The outcome of that research is acknowledged strongly by the international research community through 2500 citations, an H-Index of 30, and an I-10 Index of 50, as per the research database. He secured an amount of 1.4 Crores (INR) worth from various funding agencies such as DST, ASEAN and the Royal Academy, UK, in the areas of biofuels, flex-fuel engines and electric vehicle development. Ashok has been named as a "Top 2% scientist in the world" in a recent study conducted by researchers at Stanford University. His current research is focused on the establishment of a hybrid strategy for IC engines, flex-fuel engine development, and effective control strategies for electric powertrain configuration.

CHAPTER 1

Emission formation in IC engines

B. Ashok[a], A. Naresh Kumar[b], Ashwin Jacob[a], and R. Vignesh[a]
[a]Engine Testing Laboratory, School of Mechanical Engineering, Vellore Institute of Technology, Vellore, Tamil Nadu, India
[b]Department of Mechanical Engineering, Lakireddy Bali Reddy College of Engineering, Mylavaram, Andhra Pradesh, India

1.1 Introduction

The emissions generated by both transportation and nontransportation sector internal-combustion (IC) engines are considerable and are a major challenge for both the research community and governments. Initially, toxic pollutants generated from IC engines were lower. Also, there were fewer vehicles. However, in the mid-20th century, improvements in living standards have led to increasing energy demands, resulting in a huge number of automobiles on roads. Heavy vehicles used for carrying different goods also contributed to this increase. As a result, fossil fuel consumption, both in liquid and gaseous form [1,2], severely increased from 1960, causing air pollution. On the other side, oils obtained from waste vegetable seeds [3] and peels no doubt have fewer hydrocarbon chains but generate less brakepower. Due to the increase in the number of automobiles, extreme levels of pollutants are released, causing serious effects. The use of petroleum-based fossil fuels is the primary reason behind these adverse effects. These pollutants appear in solid or gaseous states. But the chemical substances containing carbon as a constituent are of major influence. Some of the pollutants from vehicles are hydrocarbons (HC), carbon monoxide (CO), nitrogen oxides (NOx), and particulate matter (PM). Out of these, gases such as CO_2 and H_2O are released due to complete combustion while a majority of toxic gases are released due to various reasons such as incomplete combustion, heterogeneity of the air-fuel mixture, and the nonavailability of oxygen. Meanwhile, the development of high temperatures due to complete combustion also generates NOx caused by inbuilt diatomic nitrogen present in air and fuel. All the above-mentioned emissions such as HC, CO, NOx, CO_2, and PM are primary pollutants released from IC engines and are anthropogenic in nature. The various toxic emissions formed in SI and CI engines are shown in Figs. 1.1 and 1.2. CO is generated due to a

NOx Emission Control Technologies in Stationary and Automotive Internal Combustion Engines
https://doi.org/10.1016/B978-0-12-823955-1.00001-2

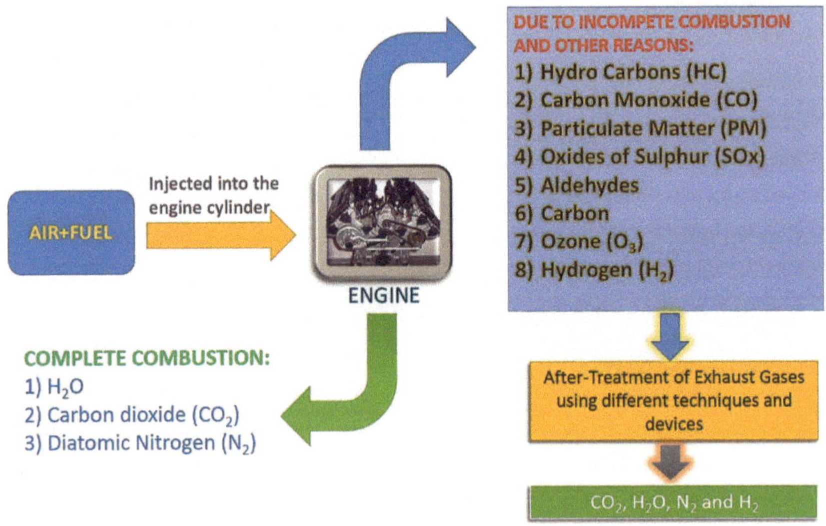

Fig. 1.1 Emissions resulting from complete and incomplete burning.

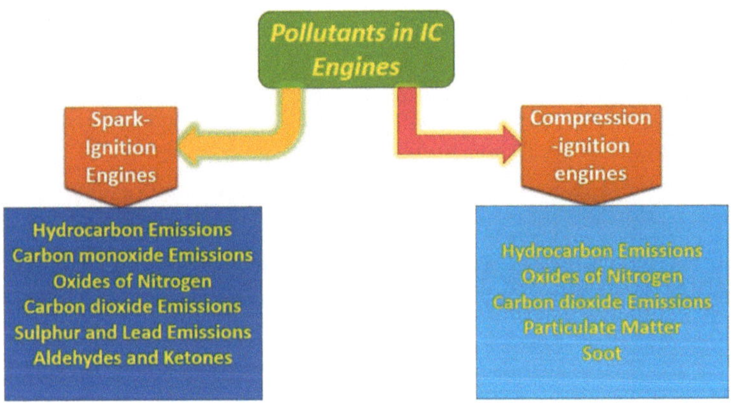

Fig. 1.2 Pollutants formed in spark-ignition and compression-ignition engines.

deficiency of oxygen and accounts for 50% of the total emissions. The other pollutant, HC, is generated due to incomplete combustion and evaporation of the fuel, which is highly carcinogenic in nature. Particulate matter formed inside the engine cylinder is much less in diameter and contains solid carbon particles. NOx generated inside the engine cylinder reacts with atmospheric gases and forms toxic substances such as nitric acid. But the reaction of the primary pollutants with the chemical constituents present in the atmosphere generates secondary pollutants such as ozone (O_3). The presence of ozone at higher altitudes protects the Earth from ultraviolet radiation, but its presence in the lower atmosphere is harmful. This can damage vegetation and cause lung disorders in human beings. Not only automobiles but pollutants from power plants can also cause adverse effects such as an increase in the temperature of the Earth's atmosphere, respiratory disorders, cancerous diseases, and greenhouse effects such as depletion of the ozone layer. All these factors have forced governments to impose stricter regulations on various pollutants from automobiles. These regulations were initially imposed in developed countries but were later imposed by governments of developing nations as well. A detailed discussion regarding the emission regulations is illustrated in the next section. The types of pollutants formed due to different reasons and their characteristics are presented in Table 1.1.

1.2 Emission standards

Pollutants emitted from automobiles have been a matter of concern for many years. Compared to SI engines, pollutants from CI engines, particularly from heavy-duty vehicles, cause considerable damage to human health. Additionally, the CI engine application in view of various advantages is comparatively broader than SI engines. Different pollutants emerging from the transportation sector cause significant damage to human health and the atmosphere. In view of the damage caused to air quality, many nations have set regulations for emissions released from the tailpipe [4]. These regulations were set for highly toxic gases such as carbon monoxide, particulate matter, hydrocarbons, and oxides of nitrogen. Emission regulations represent the maximum permissible toxic effluents that can be released from various categories of vehicles. A step forward in this direction is initially done by European nations, the United States, and Japan. The standards were set separately by these nations with other developing nations following one of the above-mentioned emission standards. The developing country India also initiated a mild emission regulation program in 1996 and followed European emission regulations with differences in testing conditions such as speed and temperature. Emission standards were represented as Bharat Stage Emissions Standards (BS) in India. In spite of differences in testing conditions, the maximum permissible pollutant limits set by India were the same as European emission standards (EURO). According to the directions of the supreme court, EURO-I emission norms were made mandatory for all private vehicles in India in 1999 and EURO-II emission norms in 2000. All the newer vehicles manufactured have to be compliant with these regulations.

Table 1.1 Pollutants formed in IC engines and their characteristics.

S. No.	Type of pollutant	Characteristic of pollutant
1	Particulate matter	Tiny particles with diameters less than 2.5 μm that primarily contain carbon
2	Hydrocarbons	Compounds containing carbon and hydrogen atoms
3	Carbon dioxide	Odorless at low concentrations but exhibits acidic color at high concentrations. Contains a carbon atom bonded to two oxygen atoms
4	Carbon monoxide	Contains each atom of carbon and oxygen with bond length of 112.8 pm. It is a colorless, tasteless, and odorless gas
5	Oxides of nitrogen emissions (NO and NO_2)	NO is a colorless, flammable gas having little odor. NO has one unpaired electron (free radical). NO2 is a nonflammable gas but poisonous and has a deep orange-red color
6	Sulfur oxides (SO_2) and Lead oxides (PbO_2)	SO_2 is colorless, toxic, and has a suffocating odor. Furthermore, SO_2 leads to the formation of sulfuric acid and results in acid rains. PbO_2 is dark brown in color, nonflammable, and insoluble in water
7	Ozone (O_3)	Ozone (O_3) is bluish in color and a strong oxidizer with good solubility. Ozone can be explosive if its concentration exceeds 20% in a mixture
8	Aldehydes and ketones	They are organic compounds that are soluble in water, but their solubility decreases with an increase in carbon length. With an increase in molecular weight, their boiling point increases

Emission standards have been revised from time to time in view of the threat to the atmosphere. BS-III and BS-IV emission regulations were enforced in India in 2010 and 2017. Further, BS-VI emission regulations were imposed in New Delhi in 2018 in view of the heavy air pollution in the nation's capital. However, BS-VI regulations were made mandatory for the entire country on April 1, 2020. The overview of the EURO emission regulations for light-duty, heavy-duty, and off-road vehicles is shown in Tables 1.2, 1.3, and 1.4. To meet these stricter emission regulations, particularly in diesel engines, after-treatment devices such as diesel particulate filters and

Table 1.2 EURO emission regulations for heavy-duty diesel (CI) engines.

Emission	Units	EURO-I	EURO-II	EURO-III	EURO-IV	EURO-V	EURO-VI
Year of implementation	g/kWh	1992	1998	2000	2005	2008	2013
CO		4.5	4	2.1	1.5	1.5	1.5
HC		1.1	1.1	0.66	0.46	0.46	0.13
NOx		8	7	5	3.5	2	0.4
PM		0.36	0.15	0.1	0.02	0.02	0.01

Table 1.3 EURO emission regulations for passenger cars (diesel engines).

Emission	Units	EURO-I	EURO-II	EURO-III	EURO-IV	EURO-V	EURO-VI
Year of Implementation	g/km	1992	1996 (DI)	2000	2005	2011	2014
CO		2.72	1	0.64	0.5	0.5	0.5
HC		–	–	–	–	–	–
HC + NOx		0.97	0.9	0.56	0.3	0.23	0.17
NOx		–	–	0.5	0.25	0.18	0.08
PM		0.14	0.1	0.05	0.025	0.005	0.005

Table 1.4 EURO emission regulations for light commercial vehicles (diesel engines).

Emission	Units	EURO-I	EURO-II	EURO-III	EURO-IV	EURO-V	EURO-VI
CLASS-I for ≤ 1305 kg							
Year of implementation	g/km	1994	1997 (DI)	2000	2005	2011	2014
CO		2.72	1	0.64	0.5	0.5	0.5
HC		–	–	–	–	–	–
HC + NOx		0.97	0.9	0.56	0.3	0.23	0.17
NOx		–	–	0.5	0.25	0.18	0.08
PM		0.14	0.1	0.05	0.025	0.005	0.005
CLASS-II for 1305–1760 kg							
Year of implementation	g/km	1994	1998 (DI)	2001	2006	2011	2015
CO		5.17	1.25	0.8	0.63	0.63	0.63
HC		–	–	–	–	–	–
HC + NOx		1.4	1.3	0.72	0.39	0.295	0.195
NOx		–	–	0.65	0.33	0.235	0.105
PM		0.19	0.14	0.07	0.04	0.005	0.005
CLASS-III for > 1760 kg							
Year of implementation	g/km	1994	1998 (DI)	2001	2006	2011	2015
CO		6.9	1.5	0.95	0.74	0.74	0.74
HC		–	–	–	–	–	–
HC + NOx		1.7	1.6	0.86	0.46	0.35	0.215
NOx		–	–	0.78	0.39	0.28	0.125
PM		0.25	0.2	0.1	0.06	0.005	0.005

techniques such as exhaust gas recirculation and selective catalytic reductions were essential. On the other side, three-way catalytic converters are needed in petrol engines.

1.3 Exhaust pollutants from spark ignition engines

The word "pollutant" is a combined term used to represent the undesirable gaseous emissions and solid particles released from both SI and CI engines. Pollutants from SI engines are the major issue because their harmful impact on the atmosphere and human health. Different regulated and unregulated emissions are released from SI engines. The reasons for the formation of these toxic pollutants include incomplete combustion, valve overlap, non-availability of oxygen, high temperatures, and the heterogeneity of the mixture. The formation of each pollutant in SI engines is extensively discussed in the following subsections.

1.3.1 Regulated emissions

The emissions released from SI engines are not only harmful to the atmosphere but also influence humankind negatively to a major extent. Regulated harmful pollutants from SI engines include HC, NOx, CO_2, PM, and CO.

1.3.1.1 Hydrocarbon emissions

Hydrocarbon pollutants are highly toxic in nature and their group composition typically varies from methane to 4-hydroxybiphenyl. They are generally represented by C_XH_Y. This group of pollutants has a significant effect on both human health and the environment. Hydrocarbons represent the chemical energy lost as well as the effect on the environment due to their toxic nature. HC emissions are generated in SI as well as CI engines. However, certain differences exist in the mechanism of HC formation between SI and CI engines. HC emissions are mainly formed during the scavenging process in two-stroke engines and due to the nonstoichiometric mixture in four-stroke engines. The main sources of hydrocarbon emissions in SI engines are an overrich mixture, flame quenching near the combustion chamber walls, deposits on the combustion chamber walls at high temperatures, the flow of the air-fuel mixture into the crevices, and improper mixing of air and fuel.

HC emission formation strongly depends on the type of air-fuel mixture and is found to be minimum near stoichiometric conditions. When the air-fuel mixture is too rich, part of the air-fuel mixture cannot find oxygen, hence resulting in HC emissions. Combustion in rich mixture conditions also represents the chemical energy lost that is otherwise converted into useful brake power. During the combustion process, part of the flame comes near the combustion chamber walls and heat transfer occurs from the flame to the combustion chamber walls. This results in the loss of heat from the flame to the walls. Thus the flame cannot be sustained near the combustion chamber walls and that part of the flame near the walls comes out as HC emissions. The other reason for flame quenching in SI engines is a decrease in the cylinder temperature during the expansion stroke. Because of this, the temperature of the end charge decreases suddenly before the flame burns it. This results in faster cooling of the unburnt charge or sometimes the flame can be extinguished. The unburnt charge that had undergone quenching burns poorly and is exhausted as HC emissions. However, the flame quenching effect can be reduced by providing additional turbulence in the combustion chamber. During combustion, a certain portion of the air-fuel vapor is forced into the crevice volume. The crevice volume is the small volume between the piston and rings where only vapors can enter. After the completion of the combustion process, the pressure inside the cylinder reduces more than that of the crevice volume and the air-fuel vapor stored in the crevice volume enters the combustion chamber, resulting in HC emissions. Even though the correct air-fuel mixture enters the engine cylinder, a minor quantity of HC emissions would be formed due to the undermixing of the fuel and air. This may be due to faulty combustion chamber design, low turbulence, and aging of the engine.

1.3.1.2 Carbon monoxide emissions

Carbon monoxide is a highly poisonous gas that can have a significant effect on the human respiratory system when inhaled in large quantities. CO emissions are formed during rich air-fuel mixture conditions resulting in incomplete combustion of carbon-rich fuels such as gasoline and natural gas. The major reason for the formation of CO emissions in SI engines is due to nonstoichiometric operation. The equivalence ratio increases during high load and accelerating conditions, which means that the fuel quantity is greater and the oxygen quantity is decreased more than the required amount for the complete burning of fuel. Due to the nonavailability of oxygen in the rich mixture, part of the air-fuel mixture that must be converted to CO_2 comes out as CO. Low outside temperature can also lead to significant amounts of CO emissions. When the temperature is low, the combustion efficiency will be less. Also, the after-treatment devices incorporated to regulate emissions would become fully operational only at higher temperatures. As a result, large concentrations of CO are formed under those conditions. Carbon monoxide pollutant formation also gets exaggerated where nearby hills are present. Any sources that limit the airflow can drastically increase the emissions. Hilly areas limit pollutant dispersions and thereby cause an emission increase. Moreover, HC emissions that were partially oxidized during the late expansion stroke also emerge as CO emissions. In addition to the rich mixture, the heterogeneous nature of the mixture as well as incomplete combustion can also cause CO emissions in CI engines. Therefore, idling conditions, high load conditions, and high acceleration conditions are favorable situations for CO formation. Additionally, at low temperatures, the accumulation of more carbon monoxide in a particular region of the CI engine combustion chamber can also lead to the formation of PM and secondary pollutants such as O_3. CO when inhaled reacts with the hemoglobin present in the blood and forms carboxy-hemoglobin (COHB). The presence of COHB in the blood is an indication of CO in the blood. The presence of COHB at 10% causes a headache, hearing and vision problems at 20%, vomiting and weakness at 30%, and death at concentrations of 50%. Furthermore, the inhalation of more CO can also result in the replacement of oxygen from hemoglobin. Prolonged inhalation of CO results in decreased oxygen transportation efficiency to key parts of the human system. This can affect listening and speech and is more often fatal for people with heart diseases. Thus, the presence of CO not only reduces the oxygen-carrying capacity but also decreases the absorption of oxygen by key parts of the body.

1.3.1.3 Oxides of nitrogen emissions

Nitrogen enters the engine cylinder through atmospheric air and minor traces of nitrogen are also present in the fuel. The formation of NOx is a strong function of in-cylinder temperature. The NO and NO_2 group of compounds are generally referred to as NOx. NOx emissions are produced according to a chain of reactions. Thus, it is evident that NOx emissions are greater at slightly leaner conditions due to additional oxygen availability. This does not, however, mean that NOx is formed at lean mixtures only. In rich mixture conditions, NOx would be formed due to high temperatures developed inside the engine cylinder. During the normal combustion process, the fuel injected before the end of the compression stroke reacts with oxygen present in the air, thus liberating heat. When the temperatures during combustion are less than 2000 K, the nitrogen entered into the engine cylinder does not react with oxygen and is emitted from the engine as diatomic nitrogen. Hence, nitrogen exists in a diatomic state and is stable when temperatures are less than 2000 K. But a considerable amount of NOx would be generated between the cylinder temperatures of 2000–3000 K. At high temperatures above 2500 K, the nitrogen present in the diatomic state gets dissociated into monoatomic nitrogen, which is an unstable and highly reactive compound. This monoatomic nitrogen reacts with oxygen present in the air to form NO and NO_2, according to the following reactions:

$$N_2 \rightarrow 2N$$
$$O + N_2 \rightarrow NO + N$$
$$N + O_2 \rightarrow NO + O$$
$$N + OH \rightarrow NO + H$$

1.3.1.4 Sulfur and lead emissions

Sulfur emissions (SOx) are released from the engine because the sulfur content in the fuel. Nowadays, the sulfur content in fuel is filtered in the refining stage. But due to the increasing number of automobiles, sulfur emissions pollute the ambient air quality. Sulfur emissions are particularly high when the engine is consuming a large amount of fuel, that is, during accelerating and idling. Sulfur emissions in the exhaust vary from 500 to 6000 ppm. Sulfur in the fuel can react with hydrogen and oxygen to form H_2S and SO_2. Further, available oxygen in the exhaust can lead to sulfur trioxide formation. SO_2 and SO_3 in the atmosphere are highly toxic, as they can react with water vapor in the atmosphere to form sulfuric acid. These acid substances again reach the surface of the Earth in the form of acid rain. Apart from this effect, SOx in the atmosphere reacts with other substances to form solid particles. These solid particles further reduce visibility during fog. Higher concentrations of SOx in the atmosphere damage plant growth and trees. Regarding the health effects of SOx emissions, even exposure to SOx in minor concentrations can affect the respiratory system, particularly with children and those with a poor respiratory background. Minute fine particles formed in the atmosphere by the reaction of SOx with atmospheric components can deeply penetrate into the lungs and cause cancerous diseases such as lung cancer and other serious respiratory disorders. The reactions involving the above-illustrated process are presented below.

$$S + H_2 \rightarrow H_2S$$
$$S + O_2 \rightarrow SO_2$$
$$SO_2 + O \rightarrow SO_3$$
$$2SO_2 + O_2 \rightarrow 2SO_3$$
$$SO_3 + H_2O \rightarrow H_2SO_4$$

The presence of lead in gasoline leads to serious pollutants that cause damage the engine and also significantly affect human health. Lead oxide in the atmosphere causes gastrogenic problems in humans. Moreover, lead oxide is a carcinogenic agent that affects reproduction in humans and causes various cancers such as lung cancer, kidney cancer, and blood cancer. Lead and lead oxide released from the exhaust mixes with soil and remains at the ground level for many years, thus affecting the fertility of the soil and in turn inhibiting plant growth. In spite of all the above-mentioned disadvantages, lead is used in gasoline before the 1990s because lead increases the octane number of the fuel. An increase in the octane number permits increasing the compression ratio and brake thermal efficiency as well. Furthermore, lead contact with cylinder surfaces hardens the walls engine, and therefore lead is a strong surface hardening agent. At the same time, the absorption of air-fuel vapor by the cylinder walls is reduced due to the presence of lead, and hence HC emissions were lower when leaded gasoline is used before the 1990s. Hence, phasing out lead is not immediately possible during those times. But due to serious effects on health and the atmosphere, lead is phased out in stages after the 1990s and unleaded gasoline is established for SI engines.

1.3.2 Unregulated emissions

Among the emissions released from SI engines, certain portions of pollutants such as hydrocarbons, carbon monoxide, nitrogen oxides, and carbon dioxide are regulated because of their environmental effects. Some percentage of the pollutants remains unregulated, yet they are potentially hazardous substances and negative contributors to air quality.

1.3.2.1 Aldehydes and ketones

Aldehydes and ketones are carbonyl compounds that are subject to intensive research as they are carcinogenic agents and can react with multiple substances easily. Carbonyl compounds released from IC engines can result in the formation of ozone at ground level. Aldehydes and ketones present neither in fuel nor air are partially oxygenated compounds released from the exhaust of the IC engine. Aldehydes and ketones are principally formed because of the sudden termination of oxidation reactions between the air-fuel due to multiple reasons such as a decrease in temperature, more viscosity, poor oxygen content, and poor vaporization. Some of the carbonyl compounds formed in IC engines include formaldehyde, acetaldehyde, propionaldehyde, butyr-aldehyde, valer-aldehyde, hexanol, oenanth-aldehyde, and aromatic compounds such as benzaldehyde and aceto-phenone. Of these, formaldehyde is a highly reactive compound that forms ozone at ground level by photochemical oxidation, resulting in photochemical smog. Aldehyde formation begins with a reaction of alkyl radicals with oxygen or hydrogen. Alkyl radicals are molecules that have unpaired valence electrons and are principally formed during the oxidation of hydrocarbons. Initially, aldehyde formation starts with formaldehyde (CH_2O) generation, which is formed by the reaction of a methyl radical with an oxygen radical. The formation of formaldehyde is the initial reaction and responsible for the formation of different carbonyl compounds. Not only higher carbonyl compounds, but the reaction of formaldehyde with hydrogen atoms, hydrogen radicals, an oxygen atom, an oxygen radical, and an OH radical can also lead to the formation of carbon monoxide, which is toxic as explained in the earlier sections. Aldehydes can form at all temperature ranges, but most of them have the probability to form at low temperatures. On the whole, aldehyde emissions are formed during the oxidation process of hydrocarbons. The oxidation process of hydrocarbons continues after leaving the cylinder, such as in the exhaust manifold and after-treatment devices. Therefore, the total amount of produced hydrocarbons and aldehydes leaving the tailpipe will be less than that formed inside the cylinder. Any factor that increases the oxidation rate and exhaust temperature of hydrocarbons decreases aldehyde emissions. Various factors such as spark timing, load, speed, compression ratio, and air-fuel ratio can affect the formation of formaldehyde emissions significantly.

1.4 Exhaust pollutants from compression ignition engines

Compared to SI engines, the emissions released from CI engines are harmful to the atmosphere and human health; this is attributed to nonpremixed combustion in CI engines. Even though air availability is abundant in CI engines, different pollutants are formed; the reasons for this are extensively illustrated in the following subsections.

1.4.1 Regulated emissions

Different regulated pollutants in CI engines include hydrocarbons, carbon monoxide, nitrogen oxides, particulate matter, and soot. Poor mixing, the heterogeneity of the mixture, high temperatures, and abundant oxygen availability are the principal reasons for the above pollutants to form.

1.4.1.1 Hydrocarbons emissions

Compared to SI engines, CI engines release a considerable amount of HC emissions, as the molecular weight of diesel is higher. Also, the heterogeneous nature of combustion and less time for air-fuel mixing contribute to HC emissions in CI engines. In CI engines, due to the separate induction of air and fuel, there are numerous lean zones available where the equivalence ratio will be less than 1. The formation of lean zones is more predominant toward the walls of the combustion chamber, as the turbulence generated near the walls is slightly less. Therefore, this lean air-fuel mixture cannot burn itself and can only burn by absorbing heat from the combustion products. But unfortunately during certain operating conditions, this process cannot begin before the expansion and cooling of the cylinder starts in the expansion stroke. Hence, the fuel present in the lean and overlean mixture remains unburnt or partially burnt and is released as HC emissions. Due to the high compression ratio in CI engines, the pressure and temperature increase would be higher in CI engines. During combustion, part of the fuel vapor gets absorbed onto the metallic walls and is released back again into the combustion chamber only when the temperature of the walls decreases. As the decrease in temperature occurs in the late expansion stroke, the released fuel cannot undergo efficient combustion. Therefore, the released fuel in the late expansion stroke comes out into the tailpipe as HC emissions. The other source of HC emissions in CI engines is a small portion of the fuel left in the injector nozzle holes. The fuel left in the nozzle gets evaporated and slowly enters the combustion chamber. This happens only after the temperature increase occurs due to combustion. But then again, as the expansion process and cooling have started, this low-velocity fuel cannot mix with the air efficiently. This phenomenon is called the undermixing of fuel and air, resulting in a considerable amount of HC emissions. Flame quenching, which is a reason for HC formation in SI engines, is present in CI engines as well. On the whole, HC formation in CI engines is due to a too lean mixture, undermixing of the air-fuel mixture, flame quenching, and a small portion of the fuel left in the injector nozzle holes.

1.4.1.2 Particulate matter

PM is the most problematic emission of all pollutants in diesel engines, as it contains polyaromatic hydrocarbons. Particulate matter is categorized into

PM2.5 and PM10, as 2.5 and 10 represent the diameter of the particles in microns. The fine particles are far more dangerous and can enter key parts of the human system, causing carcinogenic diseases, when compared with bigger diameter carbon particles. PM is mainly composed of elementary carbon (soot), sulfur, unburnt hydrocarbons, and unburnt lubricating oil particles. However, the composition of carbon particles referred to as soot and partially burnt oil is more than the remaining substances. Particulate matter emitted from the engine is visible as black smoke. Therefore, the measurement of the intensity of smoke is directly proportional to the PM measurement. In view of the threat caused to the environment and human health, emission regulations for particulate emissions are stricter compared to the remaining pollutants. The main cause of particulate formation in CI engines is the heterogeneous nature of the air-fuel mixture. In CI engines, air is sent during the suction stroke and fuel is injected into the engine cylinder before the end of the compression stroke. Therefore, fuel and air are sent into the engine cylinder during separate periods. As fuel is injected before the end of the compression stroke, less time is available for the air and fuel to mix uniformly. Because of this, there are always local zones in CI engines. The air-fuel ratio in the local zones varies from too rich to too lean. In the too-rich zones, elementary carbon particles of different diameters will be formed because of poor oxygen availability in those zones. Particulate emissions are also formed in the too-lean zones because combustion cannot be initiated, which ends up in flame quenching. Moreover, the fuel injected initially in CI engines consumes most of the oxygen, resulting in a deficiency of oxygen for the later injected fuel in the same thermodynamic cycle. As a result, a large number of particles containing carbon atoms is formed by the later injected fuel in the cycle. The carbon atoms in the PM are in a hexagonal face-centered structure. Soot formation proceeds in four stages: pyrolysis, nucleation, surface growth, and agglomeration. During pyrolysis, hydrocarbons undergo decomposition. This occurs before the formation of soot. The carbonaceous soot particles are formed during the nucleation process, and during this process, molecular to particle conversion occurs. During nucleation, aromatics and aliphatics formed in pyrolysis undergo fragmentation, resulting in the formation of smaller size particles. In surface growth, unburnt hydrocarbons released from the partial combustion of fuel and lubricating oil get deposited on the particles formed during nucleation, resulting in the formation of primary particles. In the next stage, particle to particle collisions result in the formation of larger particles, and this stage in soot formation is referred to as agglomeration. The continuous joining of particles also results in the formation of the chain structure of particulates, which is referred to as aggregation. On the whole, PM is formed due to the inhomogeneity of the air-fuel mixture and the nonavailability of oxygen for the last injected fuel in CI engines.

1.4.1.3 Nitrogen oxides emissions

NOx formation in CI engines is higher than SI engines due to the large temperatures and pressures generated because of the high compression ratio. An increase in the compression ratio increases the temperature at the end of compression and thus NOx formation is higher during the initial stages of the expansion stroke. Compared to SI engines, NOx formation is higher in CI engines due to the abundant air availability. High temperatures, the oxygen content, and the residence time of the flame are the key drivers for the formation of NOx. In addition, a lower flame speed gives more time for nitrogen-oxygen reactions and hence the opportunity for NOx to form in CI engines. NOx formation, however, is not equal everywhere inside the combustion chamber because of the heterogeneous nature of the air-fuel mixture in diesel engines. But its formation is slightly higher where the local air-fuel mixture is lean. Varying engine parameters such as retarding the injection timing, decreasing the compression ratio, decreasing the injection pressure, and using EGR can reduce NOx by reducing in-cylinder temperatures. Of these techniques, circulating part of the exhaust gases into the inlet is an effective technique to reduce NOx without causing much damage to the efficiency and other emissions. Additionally, the trade-off between NOx and fuel consumption must be balanced while attempting to reduce NOx. The effect on particulate emissions while reducing NOx must also be taken into account.

1.4.1.4 Carbon monoxide emissions

CO is an intermediate product in the chemical reaction pathway formed mainly due to the nonavailability of oxygen. However, many carbon monoxide emissions are not formed in CI engines as they operate at lean mixtures. Additionally, constant air quantity entering at all operating conditions results in surplus oxygen and consequently forms fewer CO emissions in CI engines. Nevertheless, separate injection periods of air and fuel in CI engines make certain areas of the combustion chamber oxygen-deficient, thereby forming large quantities of CO and PM. Additional turbulence created after the initiation of combustion, however, will make oxygen accessible to the rich fuel, thereby decreasing the CO emissions slightly. Furthermore, a large decrease in temperatures during the power stroke is also a possible reason for CO formation. Carbon monoxide is a highly toxic and poisonous gas and the major effect on human beings is the blockage of oxygen to key parts of the body.

1.5 Environmental and health effects of engine emissions

Internal combustion engine emissions cause more harm to the atmosphere compared to other stationary and industrial sources [5]. Of all the pollutant sources, fixed and industrial sources cause 30%–40% of the damage to the atmosphere, whereas the remaining 60%–70% of harm will be done by the transportation sector. This is mainly because the fuel sources used in the IC engines are composed of hydrocarbon chains. Initially, the chemical energy in the fuel is converted into heat energy, and this heat energy in turn is used to generate rotary moment from the crankshaft. During the ideal conversion of chemical to heat energy (combustion) in IC engines, CO_2 and H_2O gases will be released. However, ideal combustion is never possible, thereby resulting in the release of toxic gases and hazardous solid particles from the engine exhaust into the atmosphere. The gases and solid particles include CO, HC, sulfur oxides (SOx), NOx, PM, and carbon particles. The damage caused by certain emissions to human beings and the atmosphere is severe, including cancerous diseases and acid rains. The impact of each pollutant, however, depends on the number of pollutants to which human beings are exposed and also on the atmospheric conditions. The strong winds in the atmosphere create additional turbulence for mixing emissions with atmospheric compounds, thereby creating secondary pollutants such as O_3 and acid rains. Acid rains are caused by the release of NOx and SOx into the atmosphere from IC engines. These gases are carried up to 2000–3000 km by strong winds. The carried gases (NOx and SOx) react with other components in the atmosphere to form nitric acid and sulfuric acid. These acids mix with water during rains and are brought to the ground level. The effect of pollutants (primary pollutants) emitted directly from IC engines or pollutants (secondary pollutants) formed indirectly because of the reaction of IC engine emissions with the atmosphere depends on the atmosphere where they are present. Pollutants at ground level are far more dangerous than at the upper atmospheric level [6]. Certain pollutants such as O_3 and PM cause adverse health effects. These risks include reduced functioning of the lungs and respiratory system, thickening of heart valves, nervous disorders, vision problems, headaches, speech problems, and even dangerous diseases such as cancer. The pollutants that enter the atmosphere directly or indirectly affect vegetation as well. The pollutants deposited on the leaves of plants reduce growth and resistance to different diseases. The increased temperature of the Earth's surface and the melting of glaciers are other adverse effects due to engine emissions. Although certain pollutants such as CO_2 increase the Earth's temperature, they are still essential for plant life in photosynthesis reactions. In view of the adverse effect caused by IC engine pollutants on the atmosphere and human health, stricter emission regulations have been enacted. The cause of pollutant formation in SI and CI engines and the effects of pollutants on the atmosphere and human health are illustrated in Tables 1.5 and 1.6.

1.5.1 Primary pollutants

CO, HC, PM, NOx, SOx, and lead are the primary pollutants formed from IC engines.

Hydrocarbons are formed as a result of incomplete or partial combustion; benzene, toluene, aldehydes, alkanes, and ethers are the various forms of hydrocarbons. All these compounds are carcinogenic agents resulting in leukemia, and they can also cause reduced vision, headaches, and dizziness. The effect of HC emissions on the atmosphere is also severe. Hydrocarbons reacting with nitrogen oxides result in the formation of ground-level O_3. HC emissions reduce the functioning of the immune system, reduce the body's ability to resist infections, and can also cause genetic disorders in animals.

Carbon monoxide in IC engines is formed due to the nonavailability of oxygen. CO, when inhaled in large quantities, causes headaches, reduced vision, and reduced breathing and learning abilities. CO reacts with hemoglobin in the blood to form carboxyhemoglobin. This replaces the hemoglobin in the blood. Hemoglobin is the key component that helps in the transport of oxygen to various vital organs of the body such as the heart and brain. CO reacts with oxygen atoms in the atmosphere to form CO_2, which causes the greenhouse effect.

NOx is formed as a result of higher temperatures developed inside the engine cylinder. Increased exposure to NOx causes genetic disorders, breathing difficulties, tiredness, and the formation of fluid in the lungs as well as reduces fertility in females. Similar to CO, NOx reacts with hemoglobin in the blood to form methemoglobin, which reduces the blood's oxygen-carrying capacity. From an atmospheric viewpoint, NOx reacts with water during rain and enters the ground level in the form of acid rain.

Table 1.5 SI engine pollutant formation root cause and effects on health and the environment.

S. No	Type of pollutant	Reasons for emission formation in SI engines	Environmental and health effects
1	Hydrocarbons	1. Flame quenching 2. Crevices in combustion chamber walls 3. Absorption of air-fuel mixture by lubricating oil 4. Poor air-fuel mixing	Leukemia, reduced eye vision, dizziness, and formation of ground-level ozone
2	Carbon monoxide	Deficiency of oxygen and improper combustion	Respiratory problems, breathing problems, causes greenhouse effect
3	Nitrogen Oxide Emissions	Stoichiometric burning, higher temperatures	Genetic disorders, reduced fertility, tiredness, forms methemoglobin by reacting with hemoglobin in the blood, acid rain
4	Sulfur and Lead Emissions	Presence of sulfur and lead in fuel samples	Formation of sulfuric acid in atmosphere, which is highly toxic, damages plant growth and causes acid rain, lung cancer and respiratory disorders, genetic disorders
5	Aldehydes and Ketones	Sudden termination of oxidation reactions, more viscosity, poor vaporization	Formation of ground-level ozone, photochemical smog, leads to formation of carbon monoxide
6	Carbon dioxide	Complete burning of air and fuel, oxidation of carbon monoxide	Causes greenhouse effect

PM is majorly formed in CI engines due to the inhomogeneity of the air and fuel mixture. PM predominantly affects the respiratory system and leads to reduced functioning of the lungs. Compared to bigger diameter particles, fine particles are a lot more dangerous to human health. CI engine-operated commercial vehicles and off-road vehicles are the main PM sources [7].

Table 1.6 CI engine pollutant formation root cause and effects on health and the environment.

S. No.	Type of pollutant	Reasons for emission formation in CI engines	Environmental and health effects
1	Hydrocarbons	1. Too lean mixture 2. Improper mixing of air and fuel 3. Flame quenching 4. Fuel left in the injector nozzle holes	Formation of ground-level ozone, reduced ability of photosynthesis in plants, genetic disorders in animals, causes various cancers
2	Oxides of Nitrogen Emissions	Surplus oxygen availability, lean burning, more compression ratio, high temperatures	Breathing problems, acid rain, reduced vision, decreased crop yield
3	Particulate Matter	Heterogeneous nature of the mixture, consumption of most oxygen by initially injected fuel	Reduced functioning of lungs, small particulates acts as carcinogens, premature death, asthma, and bronchitis
4	Carbon monoxide	Deficiency of oxygen	CO mixes with hemoglobin in blood and decreases its ability to transport oxygen, reduces heart functioning, causes respiratory problems
5	Carbon dioxide	Complete burning of air and fuel, oxidation of carbon monoxide	Global warming

1.5.2 Secondary pollutants

Secondary pollutants are formed due to the reaction of primary pollutants with compounds in the atmosphere. O_3 and Peroxyacetyl nitrate (PAN) are the two secondary pollutants formed by the influence of the primary pollutants NOx and HC. Diatomic nitrogen at high temperatures reacts with oxygen to form NO and NO_2. NO_2 in the presence of sunlight dissociates into NO and O. The reaction between O_2 and O results in O_3 at ground level. O_3 at ground level results in blurred vision, breathing problems, and increased consumption of oxygen. Furthermore, ground-level ozone badly affects those with respiratory problems such as asthma. Ozone near the Earth's surface also effects vegetation and plant growth, thus reducing their ability to resist diseases. Ozone above the Earth's surface is beneficial because it disintegrates and is reformed under the action of ultraviolet (UV) rays, thus preventing harmful UV rays from reaching the Earth's surface. UV rays reaching the Earth's surface result in skin diseases and skin cancer.

$$N_2 \rightarrow 2N$$
$$O_2 + N_2 \rightarrow 2NO$$
$$NO + HC + O_2 \rightarrow \text{Peroxyacetyl Nitrate (PAN)} + NO_2$$
$$2NO + O_2 \rightarrow 2NO_2$$
$$NO_2 \rightarrow NO + O$$
$$O + O_2 \rightarrow O_3$$

1.6 SI engine emission formation and its root cause

The use of SI engines has increased in the past decade. Even though there has been an improvement in combustion systems, ambient air quality has deteriorated due to an increase in the number of vehicles and an increase in distance each vehicle has travelled. Because of this, emissions—particularly carbon monoxide emissions—from SI engines have increased. This section describes the root cause of in-cylinder emissions in SI engines and techniques available to reduce pollutants before exhausting them into the atmosphere. The primary pollutants in SI engines are nitrogen oxide, hydrocarbons, carbon monoxide, and particulate matter. However, when compared to CI engines, particulate matter formed in SI engines is less. The pollutants formed in SI engines and their root cause are shown in Fig. 1.3. In addition to the environmental effect due to emissions in SI engines, pollutants such as NO and CO can also cause serious damage to human health. Emissions in SI engines are not only generated inside the engine cylinder, but are also generated due to the continuous evaporation of fuel. During certain conditions, evaporative pollutants such as hydrocarbons are equal to HC emissions released from the engine cylinder. Reducing exhaust emissions by using aftertreatment devices such as catalytic converters, diesel particulate filters, and selective catalytic reduction is explained in subsequent sections. A nonevaporative emission from the SI engine is NO, which is formed due to the dissociation of monoatomic nitrogen and CO that is formed due to oxygen

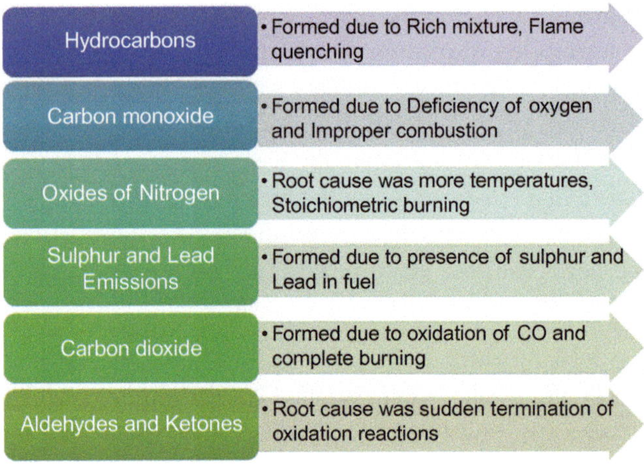

Fig. 1.3 Pollutants released in SI engines and root cause for formation.

nonavailability. On the other side, HC emissions are generated due to flame quenching, fuel evaporation, and the absorption of fuel in walls/ lubricating oil.

With respect to SI engines, NO formation is higher than nitrogen dioxide. NO_2 is higher in CI engines due to abundant oxygen availability. The formation of NO, though initially slower, drastically increases at higher temperatures. This is indicated as a thermal route for the formation of NOx emissions.

$$N2 + O \rightarrow NO + N$$
$$N + O2 \rightarrow NO + O$$
$$N + OH \rightarrow NO + H$$

However, a part of the NO emissions in the SI engine is also formed due to the reaction of formed hydrocarbons in the engine cylinder with diatomic nitrogen at high temperatures. Because of its high stability, diatomic nitrogen can only dissociate at higher temperatures. As temperatures generated inside the engine cylinder of the SI engine are less than those of the CI engine, NO formation through this mechanism is less. Other sources of NO emissions include the recombination of diatomic nitrogen and oxygen, as indicated by the following reactions. In contrast with the conventional prediction that NO emissions would be higher nearer the spark plug, various investigators have revealed that NO emissions are largely dependent on the temperature factor. At least 5–10 temperature zones in the SI engine combustion chamber contribute to the formation of NO.

$$CH + N_2 \rightarrow HCN + N$$
$$N + O2 \rightarrow NO + O$$
$$N_2 + O \rightarrow N_2O$$
$$N_2O + O \rightarrow NO + NO$$

Therefore, temperature is the main parameter affecting NOx emissions and any factor that increases temperature increases nitrogen oxides. The factors that affect the temperature are mainly the air quantity, the engine speed, the compression ratio, the valve overlap, and the spark timing. An increase in the compression ratio increases temperature and thus NOx emissions. An increase in the overlap increases the number of residual gases inside the cylinder. This contamination of residual gases with fresh air decreases the temperatures and thus NOx emissions. NOx emissions can also depend on the time at which combustion occurs. The crank angle at which combustion occurs strongly depends on spark timing. Advancing spark timing increases NO emissions because the combusted fuel will be again compressed with the advancement of spark timing, thus giving more opportunity for the peak temperatures and NO emissions to increase. On the other hand, retarding the spark timing leads to a reduction in maximum temperatures, avoiding N_2 dissociation and lowering NO emissions. An increase in engine speed results in decreased breathing ability because the number of cycles executed per minute is greater. This results in more quantity of residual gases in the cylinder and decreases NOx emissions. Another reason for decreased NOx emissions at higher speeds is the high combustion chamber temperatures generated during combustion will not prevail for a longer period of time. Therefore, there will be less time for nitrogen and oxygen to react and hence fewer NO emissions. CO emissions are formed in the SI engine due to a lack of oxygen, which results in poor conversion efficiency from CO to CO_2. When fuel undergoes a reaction with oxygen in the air, CO is formed initially. However, the conversion of CO into CO_2 strongly depends on oxygen availability. There will be a deficiency of oxygen in fuel-rich conditions when the engine is idling, accelerating, or high load conditions. In addition, the faster expansion of gases also hinders the oxidation reaction of CO and oxygen.

The combustion process in SI engines after spark initiation is not ideal. Different factors contribute to the deviation of the actual and ideal combustion processes. The escape of part of the unburnt charge from the combustion process is the main reason for this deviation. This escape is due to the main flame coming into contact with cold surfaces. The cold surfaces where combustion cannot continue absorb part of the flame and release it back into the combustion chamber as unburnt hydrocarbons during the expansion stroke. These released hydrocarbons partly undergo oxidation during expansion and the remaining comes out in the exhaust as unburnt hydrocarbons. Different researchers have conducted experiments on sources of HC emissions in SI engines. From the extensive literature observed, bad fuel atomization and propagation of flame through crevices were the major causes of HC formation. The absorption of flame on lubricating oil and leakage past the exhaust valve are also sources of HC formation. Surprisingly, the literature review also revealed that 10%–15% of the raw fuel combustion does not occur during combustion but occurs late in the expansion stroke or during exhaust. Fuel combustion during exhaust does not contribute to useful power output. The air-fuel ratio, spark timing, temperatures, load, and speed are the main factors governing HC formation. HC emissions show a decreasing trend at lean conditions, although temperatures inside the engine cylinder would be higher during stoichiometric and rich mixture conditions. The primary reason for this behavior is the availability of enough oxygen for the oxidation of fuel during lean mixture conditions. Increased coolant wall temperatures can also decrease HC emissions by 0.7%–1%. Retarding spark timing decreases the power output of the engine. However, it also leads to a reduction in hydrocarbon emissions due to combustion occurring in the exhaust. HC emissions decrease with an increase in engine speed due to less time being available for heat dissipation.

1.7 CI engine emission formation and its root cause

Various governments and environmental agencies have stressed the environmental impact from pollutants of both SI and CI engines. But the toxic substances produced by diesel engines have a considerable effect than SI engines. Different pollutants released from CI engines and the root cause for their formation are presented in Fig. 1.4. The origin of CI emissions will be illustrated in this section. Nitrogen oxides, PM, HC, and CO_2 are the primary CI engine emissions. Carbon monoxide is not formed to a considerable extent in CI engines because of more oxygen availability. Unlike SI engines, NO formation in CI engines is greater due to abundant oxygen availability. The reasons for the formation of NO emissions in CI engines are similar to those of SI engines. NO_2 is not in CI engines, however, due to less time available for the oxidation of NO to NO_2. NO formation is high in all mixture conditions in CI engines. In rich mixture and stoichiometric mixture

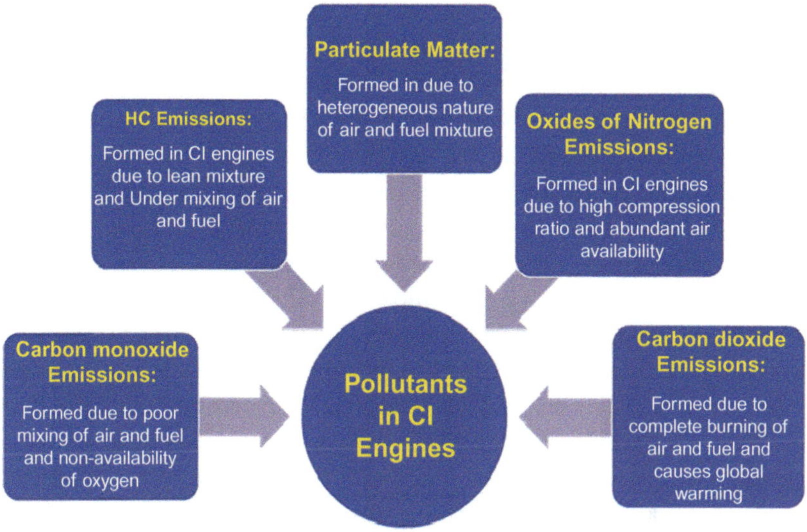

Fig. 1.4 Pollutants released in CI engines and the root cause for formation.

conditions, NO emissions are greater due to the high temperatures generated. In lean mixtures, NO emissions are high in CI engines due to low flame speeds. Low speeds result in enough time for the dissociation of monoatomic nitrogen and also for the reaction between monoatomic nitrogen and oxygen. NOx in CI engines is not formed equally in every area of the combustion chamber as in SI engines. In CI engines, fuel at high pressure is injected into the compressed air. However, due to less time available for the mixing of air and fuel, there are always local zones where the mixture type varies from too rich to too lean. Hence, NOx formation is different in different areas of the combustion chamber in CI engines. But, NOx formation is greater in the CI engine combustion chamber where the air-fuel ratio is lean. The most successful method many researchers have found for the reduction of nitrogen oxides is the retardation of injection timing and the recirculation of exhaust gases (EGR). In the EGR technique, part of the exhaust gases are recirculated with fresh air. The dilution of exhaust gases with fresh air decreases the temperature of the fresh air, thereby hindering the dissociation of diatomic nitrogen. Thus NOx emissions are reduced with the usage of EGR. Both techniques have influences such as an increase in fuel consumption as well as an increase in HC and PM emissions. But slight injection timing retardation and the use of EGR will cause very little damage to BSFC. Hence, injection timing and the EGR rate should be carefully selected based on BSFC and emission parameters.

The reasons for the formation of hydrocarbon emissions in SI and CI engines are different. However, certain similarities exist in HC formation between SI and CI engines. The primary reasons for HC formation in CI engines are overmixing and undermixing of the air-fuel mixture, overpenetration of fuel spray and fuel leakage past the injector nozzle holes, and flame quenching. The formation of HC emissions due to flame quenching will occur in SI engines also. Overmixing of the air and fuel makes too lean a mixture, resulting in difficult burning and incomplete combustion. On the other hand, undermixing of air and fuel results in a lack of air availability to certain areas of the combustion chamber. The air deficiency always causes incomplete combustion and thus HC emissions. During injection at the end of the compression stroke, part of the fuel gets trapped in the injector nozzles. The trapped fuel enters the engine cylinder later in the cycle during late expansion or the exhaust stroke. Fuel entering the cylinder during the late expansion stroke not only results in incomplete combustion, but can also lead to negative output from the cylinder. The fuel trapped in the injector nozzle holes can also enter with the fuel in the next cycle, thus making the air-fuel mixture too rich for the next thermodynamic cycle. The too rich air-fuel mixture cannot undergo combustion completely, resulting in a drastic rise in HC emissions. PM is the most dangerous pollutant in CI engines. PM consists primarily of unburnt oil, unburnt fuel, and carbon. Sulfur also gets attached to carbon nuclei and contributes to PM emissions. Fine particulates of less than or equal to 2.5 μm in diameter (PM2.5) are possibly the most harmful. PM2.5 particles can travel deep into the respiratory system and can carry carcinogenic agents into the lungs.

1.8 Concept of emission mitigation technologies for NOx emissions

NOx emissions can be reduced by different after-treatment devices such as three-way catalytic converters and selective catalytic reduction. However, different engine parameters can be altered to reduce NOx emissions at less cost. The following sections provide a brief overview of the reduction of NOx emissions by changing engine parameters and after-treatment devices.

1.8.1 Engine design and operation parameter-based NOx emission control

Changes in the bore-to-stroke ratio, combustion chamber geometry, advancing/retarding the injection timing, increasing/decreasing the injection pressure, and exhaust gas recirculation are the various options available to decrease NOx emissions without using after-treatment devices. A brief overview of the above methods is discussed in the following sections.

1.8.1.1 Alteration of injection timing

Out of various parameters in injector design, the timing of the fuel is the important parameter affecting the engine performance. The timing at which the fuel is injected into the engine cylinder before the end of the compression stroke is represented as injection timing. The alteration of injection timing can be in two ways– advancing or retarding the injection timing. Both have their own influence on performance and emissions. Generally, retarding the injection timing will be employed to reduce NOx emissions in IC engines. When the timing of the injection is advanced, more time will be available for air-fuel mixing before the combustion takes place, thus resulting in peak pressures to be developed close to TDC. Furthermore, maximum temperatures will also be greater because of efficient combustion. Greater temperatures always result in the dissociation of diatomic nitrogen and generate more NOx emissions. On the other hand, when the injection timing is retarded, the ignition delay period is less, which results in less time for fuel-air mixing. This leads to the formation of a heterogeneous mixture and decreases the temperature inside the engine cylinder. Hence, poorer combustion and lower temperatures hinder NOx formation, but at the cost of reduced performance from the engine.

1.8.1.2 Technique of exhaust gas recirculation

Exhaust gas recirculation is a widely accepted method to reduce NOx emissions at a lower cost. In this method, part of the exhaust gases are recirculated with fresh air during the suction stroke. EGR produces three different kinds of effects in the engine cylinder. The first is the dilution effect, resulting in decreased oxygen content to the engine cylinder and hence a reduction in volumetric efficiency. Therefore, less oxygen will be accessible to react with monoatomic nitrogen. Second, the chemical effect of the EGR technique results in the entry of CO_2 and H_2O gases into the exhaust to mix with fresh air. Finally, the thermal effect results in an increment of the specific heat capacity of the mixture. Hence, more heat is required to vaporize the fuel and the fuel takes more time to reach its auto-ignition temperature, resulting in an increase in the delay period. The layout of the EGR system is shown in Fig. 1.5.

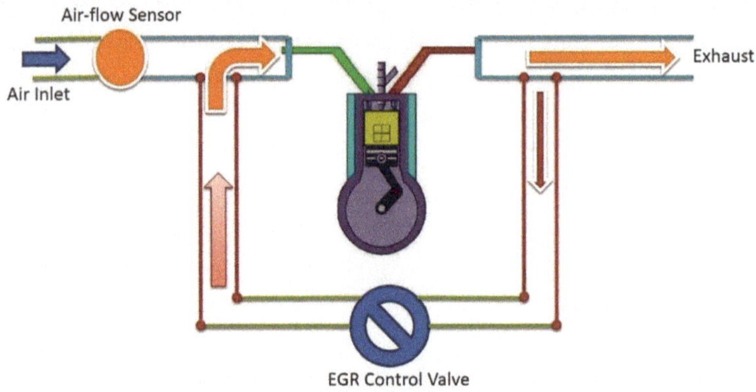

Fig. 1.5 Layout of the EGR system.

1.8.1.3 Usage of alcohols

NOx emissions are always a major hurdle in CI engines due to their operation at high compression ratios and high temperatures. Various after-treatment devices such as catalytic converters, selective catalytic reduction, etc., are available to reduce NOx emissions, but the use of these devices increases the cost and complexity of the system. Mixing diesel with alcohol additives or direct use of higher alcohols such as propanol, pentanol, hexanol, octanol, and decanol will be a viable option according to the recommendations of different investigators. Additional oxygen molecules present in the alcohols help to increase brake thermal efficiency and reduce toxic emissions such as HC, CO, and PM. Interestingly, NOx emissions would also reduce with alcohol use in CI engines. This is due to the quenching effect caused by the high latent heat of evaporation. The quenching effect reduces the in-cylinder gas temperatures and as a result, NOx emissions decrease. However, certain issues such as poor miscibility with diesel, inferior lubrication, and phase separation when mixed with diesel and biodiesel are some of the issues to be addressed for the large-scale use of alcoholic additives commercially.

1.8.1.4 Alteration of injection pressure

In diesel engines, the time available for mixture preparation is very limited, and additionally, the mixture is heterogeneous because the air and fuel mixture is not premixed. Moreover, if the diameter of the fuel droplets is higher, the fuel cannot mix with the air properly, leading to incomplete combustion. Hence, a finely atomized spray of fuel is necessary to ensure the efficient mixing of air and fuel. The efficient mixing of fuel with air helps the fuel to reach its self-ignition temperature quickly. Hence, the pressure of the fuel at the time of injection must be very high to ensure the atomization and vaporization of the fuel. A higher injection pressure improves the fuel spray pattern and increases the premixed combustion phase. This results in efficient combustion and maximum temperatures too. Higher temperatures inside the engine cylinder result in the formation of monoatomic nitrogen which can readily react with oxygen to form oxides of nitrogen. Contrarily, decreased injection pressure reduces the ignition delay period and causes poor penetration of the fuel spray, resulting in improper mixing of the air-fuel mixture. The premixed combustion rate would also become slower at reduced injection pressures. Because of all these circumstances, fuel takes more time to reach the self-ignition temperature and proceeds the combustion to late expansion stroke. As temperature decreases during the expansion stroke, lesser NOx will be formed according to Zeldovich mechanism.

1.8.2 After treatment-based NOx emission control

Pollutants such as HC, CO, NOx, and particulate matter released from internal combustion engines are undesirable to both the environment and human beings. As a result, stricter emission regulations have been imposed by governments of different nations. Therefore, different pollutant mitigation technologies have been created to reduce tail-pipe emissions. The available emission reduction technologies are presented in detail in Fig. 1.6. Various after-treatment devices developed for reducing NOx emissions are discussed in the following sections.

1.8.2.1 Three-way catalytic converter

The catalytic converter is the most widely used method to reduce harmful emissions such as HC, CO, and NOx. The inlet of the catalytic converter unit is interfaced to the exhaust of the engine and the outlet is connected to the silencer of the engine [8]. Reactions essential to reduce the toxic pollutants are made to occur in the catalytic converter, even at lower temperatures. This happens because of the action of the catalyst. A catalyst speeds up the oxidation and reduction reactions required to minimize the emissions. Platinum and palladium are used as oxidation catalysts to perform oxidation reactions. During these reactions, carbon monoxide and hydrocarbons are oxidized into a lesser amount of toxic products. Also, rhodium is used as a reduction catalyst and the reduction reactions reduce the nitrogen oxide emissions.

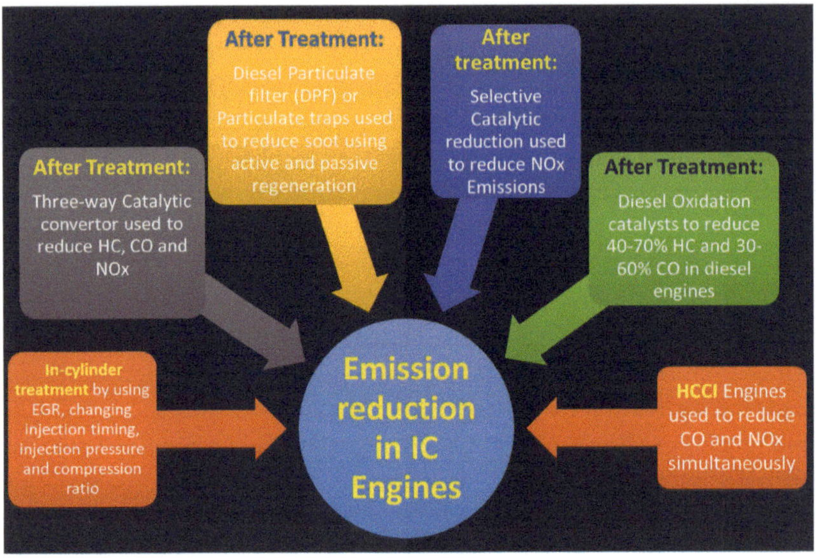

Fig. 1.6 Various emission mitigation technologies currently used in IC engines.

The essential components of the catalytic converter are a honeycomb ceramic substrate, a washcoat, a catalyst, an interface, and canning. The washcoat, basically alumina material, is the intermediate layer coated on the substrate. Catalysts are the materials placed on the washcoat to fasten the reactions. Lead in petroleum fuels, however, poisons the catalyst and therefore will reduce the efficiency of the catalyst [9]. Interface mats in the catalytic converter are used to protect the whole unit from vibration and expansion due to high temperatures. The oxidation catalyst speeds up the reaction of carbon monoxide and hydrocarbons with oxygen, thus converting them into less-toxic products such as H_2O and CO_2. The reduction catalysts, on the other hand, attack nitrogen oxides, thus breaking them into nitrogen and oxygen. The catalyst action can be better understood as follows. Toxic pollutants such as hydrocarbons, carbon monoxide, and nitrogen oxides are absorbed onto the catalyst material [10]. The catalytic action weakens the bonds of HC, CO, and NOx, thus causing the different constituents in the bonds to readily react with other components. Oxygen is the main atom available for the weakened bonds to react. Therefore, hydrogen and carbon atoms in HC and CO readily react with oxygen to form H_2O and CO, which are relatively less-toxic components for the atmosphere and humankind. However, in diesel engines, only oxidation catalysts are employed and nitrogen oxides are removed by other methods in CI engines.

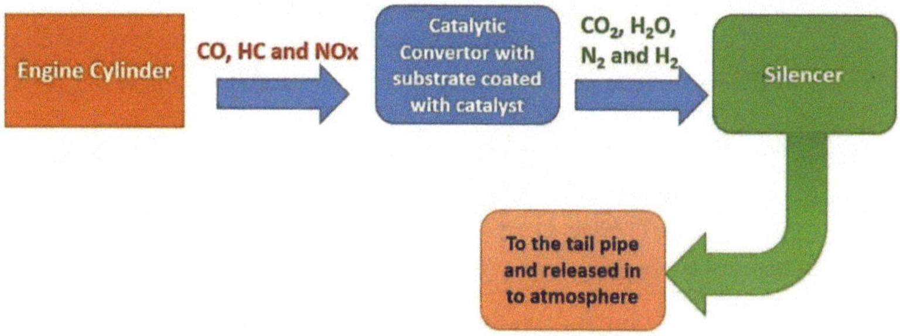

Fig. 1.7 Illustration of the workings of a catalytic converter.

The reactions occurring inside the catalytic converter are presented below with a diagrammatic representation, as shown in Fig. 1.7.

Oxidation reactions in the catalytic converter:

$$2H_2 + O_2 \rightarrow 2H_2O$$
$$HC + O_2 \rightarrow H_2O + CO_2$$
$$2CO + O_2 \rightarrow 2CO_2$$

Reduction reactions in the catalytic converter:

$$NO + HC \rightarrow N_2 + CO_2 + H_2O$$
$$2NO + 2CO \rightarrow N_2 + 2CO_2$$
$$2NO + 5H_2 \rightarrow 2NH_3 + 2H_2O$$
$$2NO + 2H_2 \rightarrow N_2 + H_2O$$

1.8.2.2 Selective catalytic reduction

A large number of nitrogen oxides are formed in IC engines at high temperatures. NOx reacting with so many other compounds produces toxic gases such as ozone, and these are dangerous to the atmosphere. As with other pollutants, the regulation of NOx is a must in IC engines. Diesel engines suffer from more NOx due to more air availability and combustion occurring at leaner mixtures. The exhaust gas recirculation technique, though, reduces NOx effectively, but on the other side, the performance of the engine would be affected. Of the different technologies available, selective catalytic reduction is an effective method to reduce NOx and for meeting stringent emission regulations. In this method, ammonia (NH_3) is used to reduce the nitrogen oxides produced at high temperatures. When injected into the exhaust, ammonia undergoes reduction reactions with NOx and produces nontoxic components such as nitrogen (N_2) and H_2O. Other than ammonia, gases such as HC and CO are made to react with NOx. But in diesel engines, because of greater oxygen availability,

HC and CO can react with oxygen rather than reacting with NOx. Therefore, ammonia [11] is widely preferred as a reducing agent to react with nitrogen oxides. The following are the reactions when NH_3 reacts with NOx.

$$NH_3 + NO + \frac{1}{4}O_2 \rightarrow N_2 + 3/2\,H_2O$$
$$NH_3 + \frac{1}{2}NO + \frac{1}{2}NO_2 \rightarrow N_2 + 3/2\,H_2O$$
$$NH_3 + \frac{3}{4}NO_2 \rightarrow 7/8\,N_2 + 3/2\,H_2O$$

Investigations are under way to find an alternate to ammonia (NH_3) in deNOx systems. Different researchers have conducted surveys in such a way that the alternate reducing agent should be less costly, less corrosive, and should deoxidize the NOx, even at lower exhaust temperatures. Of the different substances, gaseous ammonia is the main contender, and it has a deNOx efficiency of up to 60%–80%. In another recent development, ammonium formate urea system containing 54% water was used as a reducing agent. A reduced freezing point and greater stability at high temperatures are the main advantages compared to standard urea mixture. In another development to increase the efficiency of the after-treatment devices, SCR is coupled with a diesel oxidation catalyst (DOC) and a diesel particulate filter (DPF) [12]. The reduction of NOx occurs at a faster rate when gases reacting with ammonia in SCR contain more quantities of NOx. This is the major reason behind the use of DOC in combination with SCR in the exhaust system. The first major use of DOC is the oxidation reactions of HC and CO occurring in DOC generate large amounts of exothermic heat, which are sufficient to burn the solid carbon particles and soot particles. The second advantage is oxidation reactions in DOC not only reduce HC, CO, and soot, but also oxidize NO formed at high temperatures in the engine cylinder to NO_2. The formed NO_2 when entered into the SCR can easily react with ammonia or gaseous ammonia compared to NO. Copper zeolite, iron zeolite, vanadium, and tungsten are used as SCR catalysts. Iron zeolite and copper zeolite were best for low-temperature performance, but they are more prone to HC and sulfur poisoning, thus minimizing their use. HC can accumulate and undergo combustion at higher temperatures, thus resulting in the thermal degradation of the catalyst. Even though vanadium is a cheaper catalyst, it cannot sustain high temperatures. Therefore, it is not suitable for SCR in combination with DPF as DPF requires high temperatures for its regeneration, as shown in Fig. 1.8. On the whole, selective catalytic reduction (SCR) is the most successful technique in reducing NOx today. However, a lot more scope for investigation is there in this method of reducing NOx emissions. One of

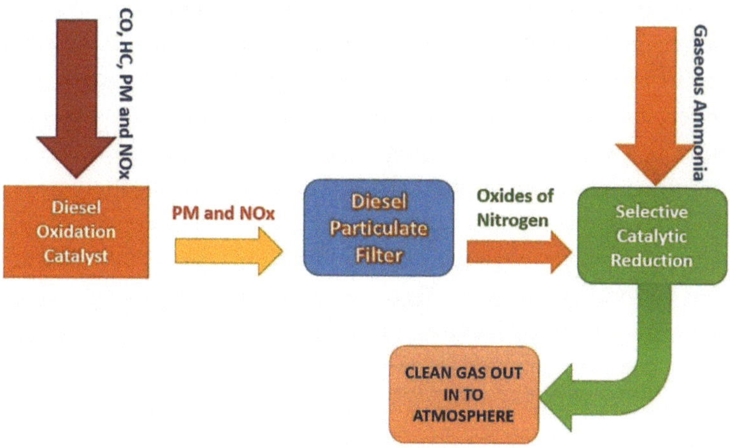

Fig. 1.8 Selective catalytic reduction in combination with DPF and DOC.

the main challenges is improving the low-temperature performance of SCR. On the other side, the behavior of various catalysts on HC and sulfur poisoning must be well addressed. NOx emissions from the engine cylinder will not decrease in the future as well since the modern day engines when compared to older engines generates more temperatures on account of better combustion. Therefore, a large scope is there in SCR technology regarding the poisoning of the catalyst and improving both low-temperature and high-temperature performance.

1.8.3 Other emission control techniques

Particulate matter (PM) is the other major toxic pollutant formed because of incomplete combustion in SI engines and due to the heterogeneous nature of the mixture in CI engines. The main source of PM is commercial vehicles, trucks, and locomotives, though a very small amount could also be emitted from passenger cars.

1.8.3.1 Diesel particulate filter

PM consists of carbon particles of different diameters with other foreign substances such as sulfur, hydrocarbons, and lubricating oil attached to it. The diameter of PM emitted from IC engines ranges from 10 to 150 nm. PM is basically categorized as PM2.5 and PM10 with 2.5 and 10 representing the diameter of particles in micrometers. Various investigations have revealed that smaller diameter particles have a more pronounced effect on human health than bigger diameter particles. The influence of PM on human health is far higher than any other pollutant. Smaller particles act as carcinogenic agents and can heavily damage the respiratory system. Therefore, in view of the effect of smaller particles on health and the environment, emission regulations for PM were introduced in the form of a number basis. DPF is the most widely used method for reducing soot or particulate matter released into the atmosphere. In this system, exhaust gases released from the engine are directed through DPF filter with passages consisting of various inlets and outlets. These passages were opened alternatively. The passages that were not opened consists of a plug required for the regeneration of soot. Each passage is separated by a silicon carbide ceramic honeycomb fiber or an alumina-coated wire mesh. Porous ceramic cordierite, silicon carbide, and aluminum titanate are used as filters because they have a low coefficient of thermal expansion. The exhaust gas entering the channel flows into the other adjacent channels via filters. Thus, the particulates and soot entering the DPF come into contact with the filters and get deposited on the honeycomb fiber structure or wire mesh, so clean exhaust gas is released into the atmosphere. The filtration efficiency of DPF filters is as high as 99% [13]. But the deposited particles on the ceramic structure must be periodically burnt, which otherwise results in exhaust backpressure. The burning of soot particles in the filter and bringing back the DPF to its original state are known as the regeneration of the filter [14]. However, the burning of soot particles in the DPF cannot occur on its own as higher temperatures in the range of 600°C are required. But the maximum exhaust temperatures, even for heavy-duty diesel engines, reaches only 400°C. Therefore, certain regeneration procedures must be adopted to clean the filter [15]. These procedures are further classified into active and passive regeneration systems.

Active regeneration system

In an active regeneration system, a burner and air compressor are used to bring the system back to its clean state. The exact period of regeneration can be determined by calculating the backpressure of the exhaust gases. When the backpressure of the exhaust gases reaches 10,000 Pa, regeneration will be initiated. Air circulated through the air compressor at high pressure is required for the combustion of soot, thereby converting soot into carbon dioxide. A drastic increase in temperatures occurs in the outlet of the particulate filter when soot undergoes combustion. This stage is the initiation of the regeneration process. Now, the burner is switched off to avoid the uncontrollable rise of temperatures, as very high temperatures may result in damaging the substrates [16].

Passive regeneration systems

Particulate oxidation can also be done by using catalyst technology [17]. Employing a catalyst enables soot oxidation, even at lower temperatures in the range of 200–300°C. Therefore, the cost for regeneration of the filter decreases significantly. Catalysts can be used as additives in the fuel or catalysts can be used as coatings (coating the ceramic honeycomb structure with a catalyst to accelerate the soot reaction with oxygen at lower temperatures in the range of 200–250°C).

Various catalysts such as cerium, copper, manganese, and iron are mixed with fuel. These catalysts convert into their corresponding oxides by reacting with oxygen in the combustion chamber. For example, iron by reacting with oxygen gets converted into iron dioxide. The iron dioxide leaving the exhaust of the engine enters the DPF, thus reacting with the carbon soot particles to form ferric oxide and carbon monoxide. Carbon monoxide, however, is again a toxic substance in the atmosphere. CO released during this process is converted into carbon dioxide by reaction with oxygen. The oxygen required for this reaction is sent in the form of compressed air. The problem of sending extra air does not occur in diesel engines as CI engines have abundant air availability. The reactions occurring during this process are presented below for the Fe catalyst. Similar reactions also occur using Mn, Cu, and Ce as catalysts.

$$2FeO_2 + Soot(C) \rightarrow Fe_2O_3 + CO$$
$$Fe_2O_3 + \tfrac{1}{2}O_2 \rightarrow 2FeO_2$$

Continuously regenerating trap

Soot oxidation can be done either by reacting carbon with oxygen or from the reaction of NO_2 with carbon soot particles [18]. However, NO_2 can oxidize soot with less trouble than oxygen. NO_2 is generated in more quantities, particularly in diesel engines, due to higher temperatures and extra air availability. On the other hand, NO_2 can also be generated in the monolith oxidation catalyst with a reaction of NO and oxygen. Therefore, the diesel oxidation catalyst is installed in the after-treatment system before the exhaust gases enter the DPF. NO unceasingly gets converted to NO_2 in the DOC and carbon soot is continuously oxidized by utilizing the formed NO_2 entering from the oxidation catalyst, as shown in Fig. 1.9. As the soot is continuously oxidized in this system, there is no problem with backpressure. On the other hand, the sulfur content in the fuel critically damages the conversion of NO to NO_2 in the DOC. A sulfur content of 40 PPM is desirable to avoid poisoning the catalyst in DOC. According to different investigations,

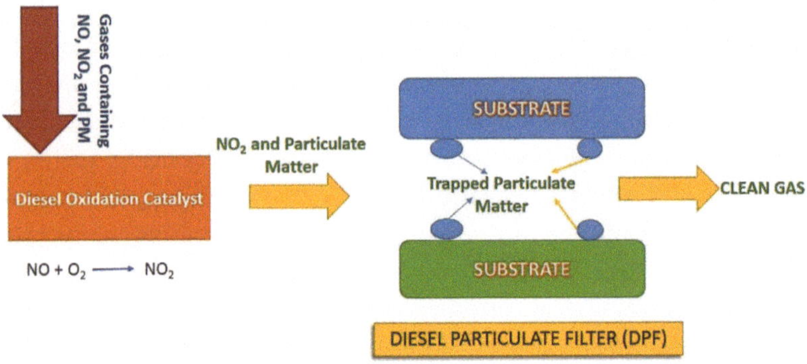

Fig. 1.9 Continuously regenerative trap system.

particulate matter is drastically reduced up to 90% when the sulfur content is less than 30 PPM in the fuel.

$$NO + \frac{1}{2}O2 \rightarrow NO_2 (\text{Reaction in Diesel Oxidation Catalyst})$$
$$2NO_2 + C \rightarrow CO_2 + 2NO$$
$$NO_2 + C \rightarrow CO + N$$

1.9 Conclusions

A brief view of different emissions in both SI and CI engines is presented in this chapter. Because of possible environmental and health effects, a detailed view of the requirement of stricter emission regulations is also illustrated in detail. The performance and combustion characteristics of the engine can be judged by different emissions formed. After-treatment devices have great potential in reducing different pollutants exhausted from the engine. But the cost increase because of the after-treatment device will also have to be considered by the manufacturer. Hence, the main aim of the futuristic research in this area is to efficiently reduce the emissions using exhaust treatment devices while keeping the cost of these systems low. A detailed summary of this chapter is presented as follows:

First, an increase in the living standards of people has drastically raised the number of vehicles, which has resulted in the combustion of vast quantities of petroleum-based fuels both in liquid and gaseous form. The combustion of petroleum-based fuels either in complete or partial form results in the formation of emissions such as hydrocarbons, carbon monoxide, particulate matter, nitrogen oxides, carbon dioxide, and other minor pollutants. The pollutants formed in IC engines can not only have an impact on human health but can also affect atmospheric and climatic conditions. Certain quantities of secondary pollutants are formed as well by the reaction of primary pollutants with atmospheric compounds. One of the harmful secondary pollutants is ozone (O_3). Ozone at higher levels of the atmosphere is advantageous, but when present at ground level can severely affect plant life and lead to skin disorders. The possible effects of the above-listed pollutants have an impact on both environment and human health. These include acid rains, damage to vegetation, respiratory disorders, and skin disorders while the inhalation of even minute carbon particles generated from IC engines can result in cancerous diseases. Regarding the perspective of the root cause of different emissions, NOx emissions are mainly formed due to the dissociation of diatomic nitrogen at high temperatures. Carbon monoxide emissions are formed due to a deficiency of oxygen. Hydrocarbon emissions are generated due to the absorption of a portion of the fuel by lubricating oil and the walls of the combustion chamber at higher temperatures. Particulate matter, on the other hand, is predominant in CI engines because of the heterogeneous nature of the mixture in CI engines. On the other side, emission control devices and engine parameter controls have huge potential to produce clean gases to the atmosphere. First, a three-way catalytic converter is used to reduce HC, CO, and NOx simultaneously. DPF is employed to filter the soot particles in the exhaust. But the absorbed soot particles must be removed periodically. Regarding this perspective, both active and passive regeneration systems employed in DPF were discussed extensively in this chapter. NOx, which is a major hurdle in CI engines, can be effectively removed by the use of the SCR technique. In SCR, ammonia is made to react with NOx formed during the combustion process.

In the next chapter, a detailed discussion regarding the formation of NOx is presented. Various NO generation pathways such as thermal NO, prompt NO, and NO generated due to in-built nitrogen in the air-fuel mixture are discussed in detail. Three mechanisms explaining the formation of NOx, that is, the Zeldovich mechanism based on the temperature of the reaction zone, the Fenimore mechanism, and the NNH mechanism are demonstrated. Of all the mechanisms in NOx formation, the Zeldovich mechanism is most familiar. To add on, the key concept explained in the next chapter is the factors affecting the formation of nitrogen oxides. Various factors such as ignition timing in SI engines, injection timing in CI engines, air-fuel ratio, ambient conditions, nitrogen content present in air and fuel, compression ratio, injection pressure, and EGR influence the formation of NOx.

References

[1] Kumar AN, Kishore PS, Raju KB, Kasianantham N, Bragadeshwaran A. Engine parameter optimization of palm oil biodiesel as alternate fuel in CI engine. Environ Sci Pollut Res 2019;26:6652–76. https://doi.org/10.1007/s11356-018-04084-z.

[2] Ashok B, Denis Ashok S, Ramesh KC. LPG diesel dual fuel engine – a critical review. Alex Eng J 2015;54:105–26. https://doi.org/10.1016/j.aej.2015.03.002.

[3] Nanthagopal K, Ashok B, Karuppa Raj RT. Influence of fuel injection pressures on *Calophyllum inophyllum* methyl ester fuelled direct injection diesel engine. Energ Conver Manage 2016;116:165–73. https://doi.org/10.1016/j.enconman.2016.03.002.

[4] Ni P, Wang X, Li H. A review on regulations, current status, effects and reduction strategies of emissions for marine diesel engines. Fuel 2020;279:118477. https://doi.org/10.1016/j.fuel.2020.118477.

[5] Leach F, Kalghatgi G, Stone R, Miles P. The scope for improving the efficiency and environmental impact of internal combustion engines. Transp Eng 2020;1:100005. https://doi.org/10.1016/j.treng.2020.100005.

[6] Nickischera A. Environmental impacts of internal combustion engines and electric battery vehicles. DUQuark 2020;4:21–31.

[7] Kim S, Xiao C, Platt I, Zafari Z, Bellanger M, Muennig P. Health and economic consequences of applying the United States' PM2.5 automobile emission standards to other nations: a case study of France and Italy. Public Health 2020;183:81–7. https://doi.org/10.1016/j.puhe.2020.04.024.

[8] Gambarotta A, Papetti V, Eggenschwiler PD. Analysis of the effects of catalytic converter on automotive engines performance through real-time simulation models. Front Mech Eng 2019;5. https://doi.org/10.3389/fmech.2019.00048.

[9] Sharma SK, Goyal P, Tyagi RK. Conversion efficiency of catalytic converter. Int J Ambient Energy 2016;37:507–12. https://doi.org/10.1080/01430750.2015.1020567.

[10] Kim J, Myung C-L, Lee K-H. Exhaust emissions and conversion efficiency of catalytic converter for an ethanol-fueled spark ignition engine. Biofuels Bioprod Biorefin 2019;13:1211–23. https://doi.org/10.1002/bbb.2007.

[11] Jiang Y, Gao W, Bao C, Yang Z, Lin R, Wang X. Comparative study of Ce-Nb-Ti oxide catalysts prepared by different methods for selective catalytic reduction of NO with NH_3. Mol Catal 2020;496:111161. https://doi.org/10.1016/j.mcat.2020.111161.

[12] Vignesh R, Ashok B. Critical interpretative review on current outlook and prospects of selective catalytic reduction system for De-NOx strategy in compression ignition engine. Fuel 2020;276:117996. https://doi.org/10.1016/j.fuel.2020.117996.

[13] Mokhri MA, Abdullah NR, Abdullah SA, Kasalong S, Mamat R. Soot filtration recent simulation analysis in diesel particulate filter (DPF). Procedia Eng 2012;41:1750–5. https://doi.org/10.1016/j.proeng.2012.07.378.

[14] Tang T, Zhang J, Cao D, Shuai S, Zhao Y. Experimental study on filtration and continuous regeneration of a particulate filter system for heavy-duty diesel engines. J Environ Sci 2014;26:2434–9. https://doi.org/10.1016/j.jes.2014.04.004.

[15] Rodríguez-Fernández J, Lapuerta M, Sánchez-Valdepeñas J. Regeneration of diesel particulate filters: effect of renewable fuels. Renew Energy 2017;104:30–9. https://doi.org/10.1016/j.renene.2016.11.059.

[16] Chen C, Yao A, Yao C, Qu G. Experimental study of the active and passive regeneration procedures of a diesel particulate filter in a diesel methanol dual fuel engine. Fuel 2020;264:116801. https://doi.org/10.1016/j.fuel.2019.116801.

[17] Bai S, Tang J, Wang G, Li G. Soot loading estimation model and passive regeneration characteristics of DPF system for heavy-duty engine. Appl Therm Eng 2016;100:1292–8. https://doi.org/10.1016/j.applthermaleng.2016.02.055.

[18] Zhang Y, Lou D, Tan P, Hu Z, Li H. Emission reduction characteristics of a catalyzed continuously regenerating trap after-treatment system and its durability performance. J Environ Sci 2019;84:166–73. https://doi.org/10.1016/j.jes.2019.05.001.

CHAPTER 2

NO_x formation chemical kinetics in IC engines

Avinash Alagumalai[a], Amin Jodat[b], Omid Mahian[c], and B. Ashok[d]
[a]Department of Mechanical Engineering, GMR Institute of Technology, Rajam, Andhra Pradesh, India
[b]Department of Mechanical Engineering, University of Bojnord, Bojnord, North Khorasan, Iran
[c]School of Chemical Engineering and Technology, Xi'an Jiaotong University, Xi'an, Shaanxi, China
[d]Engine Testing Laboratory, School of Mechanical Engineering, Vellore Institute of Technology, Vellore, Tamil Nadu, India

2.1 Introduction

Since the early 17th century, scientists have worked effortlessly to build an internal combustion (IC) engine. However, the successful IC engine was introduced by Lenoir in 1860. Following that, Nikolaus August Otto introduced the first four-stroke engine in 1878. In the same year, Sir Douglas Clerk introduced the two-stroke engine. Moving forward, several inventors worked to create efficient IC engines. The compression ignition (CI) engines are popular variants of IC engines because of their superior thermal efficiency, fuel economy, and durability [1]. However, one of the significant drawbacks of CI engines is the exhaust emissions [2]. In the previous chapter, the characteristics of exhaust emissions from IC engines were discussed. The formation of various emissions such as carbon monoxide (CO), hydrocarbon (HC), oxides of nitrogen (NO_x), soot, and sulfur dioxide (SO_2) from IC engines was presented. Recent advancements on the principal chemical kinetics that control the production of these emissions were explained. Also, the effects of engine design and operating parameters on these emissions were described. Additionally, the harmful effects of emissions on the environment and human health, such as global warming, ozone layer depletion, respiratory diseases, etc., were described in detail. In this chapter, the mechanism of NO_x formation, kinetics, and the factors affecting NO_x emissions in IC engines are discussed.

NO_x poses a serious threat to the environment and it plays a major role in global climate change. NO_x is primarily produced by fuel combustion in vehicles (more than 50%) and from various sources, as depicted in Fig. 2.1 [3–6]. Several studies revealed that NO_x content in several cities substantially increased over the specified limits [7–9]. A recent study in the United States

NOx Emission Control Technologies in Stationary and Automotive Internal Combustion Engines
https://doi.org/10.1016/B978-0-12-823955-1.00002-4

Fig. 2.1 Nitrogen oxide emission by source (fuel combustion in vehicle is the major contributor).

shows that brake-specific NO_x emissions (g/bhp-hr) from heavy-duty diesel vehicles during urban driving conditions produce two times more NO_x than suburban driving and nearly seven times more NO_x than highway driving. Furthermore, the study showed that the average NO_x emission released in urban driving is five times more than the certified limit [10]. Considering United Kingdom (UK) road transport, the recent report states that there is strong evidence of significant NO_x contribution by heavy-duty diesel vehicles. The average NO_2 concentration in the UK and other European countries exceeds $40\,\mu g m^{-3}$ [11]. The report by the International Council on Clean Transportation revealed that real world NO_x emissions from Euro 6 diesel cars are not meeting the set limits (Fig. 2.2) [12]. The report claimed than Euro 6 diesel cars emit NO_x emissions 6–7 times the limit. This shows the failure of Euro emission norms to achieve a set limit of NO_x emissions in real world driving conditions. The report also indicated the need to introduce a test that will likely cause manufacturers to change the way of calibrating after-treatment systems. Thus, real driving emissions should force performance improvements in moderate real-world NO_x emissions, even in the short term.

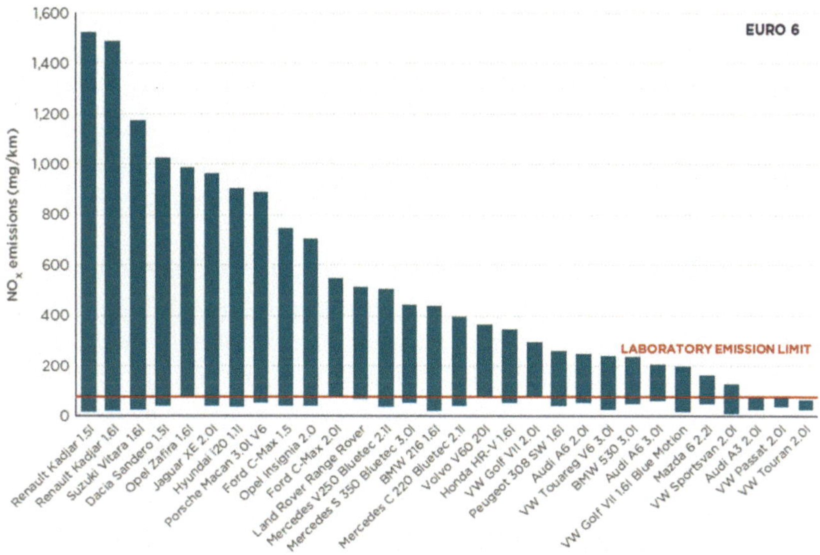

Fig. 2.2 Real world NO_x emissions of Euro 6 diesel cars (bottom value of the bar represents laboratory test result and top level represents on-road test result).

In engines, fuel combustion requires an oxidizer to initiate the combustion process. The widely used oxidizer for this purpose is atmospheric air, which contains more than 70% nitrogen. As a result, NO_x species are emitted through the engine exhaust. The factors that influence NO_x formation in IC engines are shown in Fig. 2.3 [13,14]. The primary factor is temperature. The crucial period for NO_x formation is when the burned gas

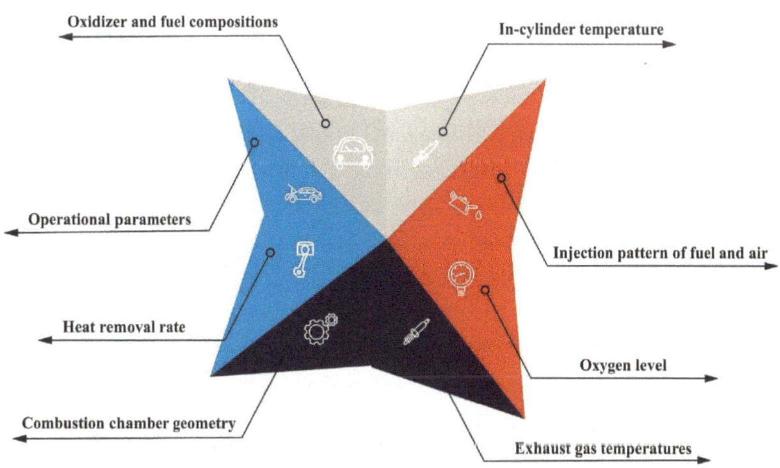

Fig. 2.3 Factors that influence NO_x formation in IC engines.

temperature is maximum. The maximum burned gas temperature is attained between the beginning of combustion and just after the occurrence of maximum cylinder pressure. Therefore, it is necessary to understand the combustion chemistry of NO_x formation to control its effects. Engine NO_x includes nitric oxide (NO) and nitrogen dioxide (NO_2) [15]. NO is the primary nitrogen oxide and constitutes up to 90% of total NO_x in IC engine emissions [15].

Different mechanisms were developed to understand the formation of NO. These mechanisms involve NO formation by the fixation of atmospheric nitrogen, the oxidation of nitrogen intermediates, or by the oxidation of fuel containing nitrogen [16]. In the case of low nitrogen-containing fuels (mainly gaseous fuels), NO formation occurs primarily by the fixation of atmospheric nitrogen. The nitrogen combustion reactions produce NO directly or form nitrogen intermediate products, which undergo subsequent reactions to produce NO [17].

In recent days, there has been extensive interest in exploiting biofuels such as biodiesel in CI engines [18]. Biodiesel is a biodegradable and nontoxic fuel. It is safe to store and transport biodiesel because of its high flash and fire points. It has chemically bound O_2, which can support combustion. The ignition ability of biodiesel is better than mineral diesel because of its higher cetane number. There is no sulfur or PAH content in biodiesel. The byproduct glycerol produced during biodiesel production has commercial value. On the other hand, biodiesel possesses high density and viscosity. Because of the higher density and viscousness, crude biodiesel (B100) can be used in CI engines. Instead, blended fuel (blending biodiesel with neat diesel) is suitable for practical applications. The most used biodiesel blend for acceptable engine performance and lower emissions is the B20 blend. Also, biodiesel possesses unfavorable cold flow properties compared to neat diesel. Also, the cost of biodiesel is higher than that of commercial diesel fuel. According to the American Society for Testing and Materials (ASTM), biodiesel is defined as "a fuel comprised of mono-alkyl esters of long-chain fatty acids derived from vegetable oils or animal fats, designated as B100" [19]. Research studies have reported that biodiesel usage positively impacts hydrocarbon, carbon monoxide, and particulate matter emissions. However, most studies reported that NO_x emissions increased with biodiesel-fueled engines [20,21]. Furthermore, it is evident from the available literature that the NO_x effect is difficult to quantify or accurately detect in case of low-concentration biodiesel blends such as B20 or below [20,22]. Therefore, it can be concluded that the NO_x effect is insignificant for low-concentration biodiesel blends. The biodiesel NO_x can be mitigated by engine modifications and fuel modifications. The engine modifications include late injection timing and higher exhaust gas recirculation. The fuel modification includes decreasing the aromatic content of the fuel and introducing cetane improvers. However, the mitigation of biodiesel NO_x by engine modification is more effective than fuel modifications.

The objective of this chapter is to provide brief and up-to-date information on nitrogen combustion chemistry. The main reactions involved in the formation of NO_x are explained and uncontrolled NO_x emission levels in SI and CI engines are presented. Furthermore, nitrogen oxide formation in biodiesel and the factors influencing biodiesel NO_x are explored.

2.2 Chemical kinetic model of NO formation

A detailed chemical kinetic model can be used to describe the macroscopic phenomena of the chemical reaction. Furthermore, the chemical kinetic mechanism is developed based on thermochemical and kinetic information. The thermochemical information includes elements such as measured values, semiempirical methods, and quantum chemistry data. On the other hand, kinetic information includes elements such as direct measurements, semiempirical methods, and information related to transition state theory derived from quantum chemistry. Once a detailed chemical kinetic mechanism with thermochemical and kinetic information is developed, the model is built by including transport parameters of the involved species. The developed model is then subjected to sensitivity and reaction path analysis to identify the uncertainties, boundary conditions, and reaction pathways of the reaction [23]. In the case of combustion, the nitrogen chemistry, along with submechanisms for hydrocarbon oxidation, reactive species oxidation, and interactions, has to be included [24].

The equilibrium composition and thermal NO formation mechanism can be studied using chemical kinetics simulation tools. The CHEMKIN II package is a widely used tool to perform equilibrium calculations and sensitivity analyses using EQUIL and PSR codes [25]. The CFD KIVA3V code is also used to study the origin of NO_x emissions in engines [25,26].

2.3 Thermodynamic properties

In thermochemistry, thermochemical tables are widely used to identify the thermodynamic properties of the chemical species [27]. The traditional thermochemical table is based on the sequential approach, which does not account for details about the selection of the particular value of the chemical species. On the other hand, Active Thermochemical Tables (ATcT) are significantly essential to derive reliable and accurate thermochemical values. The ATcT is developed based on the thermochemical network [28]. The shortcomings of the traditional thermochemical table and the advantages of ATcT are provided in Fig. 2.4 [27–29].

Ver. 1.110 of the ATcT was the early version released in 2012 [30]. In ver. 1.122 of the ATcT, the heat of formation of CH_n radicals was made consistent [31]. Later, the updated ver. 1.122b was released, which included details of the difference in available energy between HCN and HNC. In addition, this version provided details of the enthalpy of formation of HCN and HNC species and their positive and negative ions [32]. Ver. 1.22b was then expanded (ver. 1.22d) to provide details of the enthalpy of formation of three species (methylamine, dimethylamine, and trimethylamine) to obtain the bond dissociation energy of certain diatomic molecules [33].

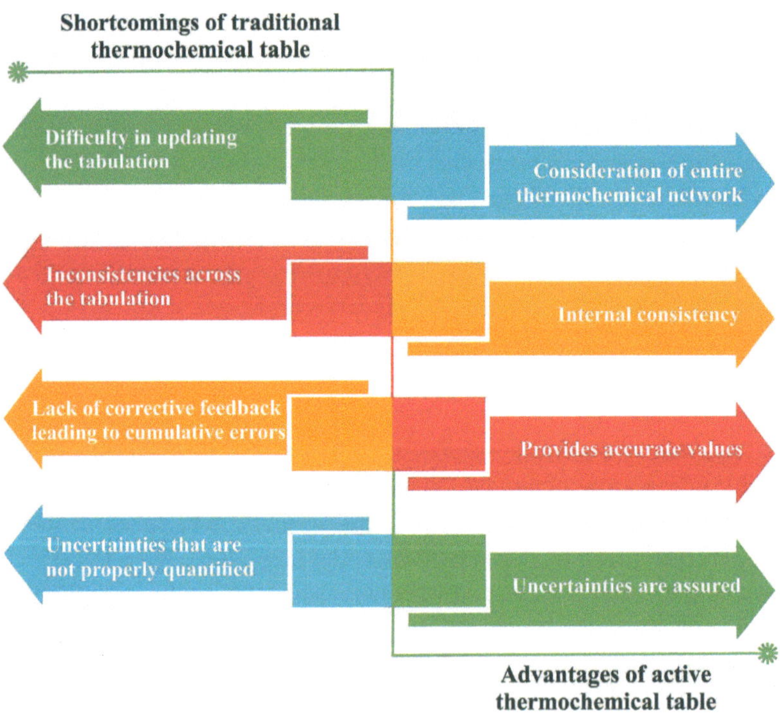

Fig. 2.4 Shortcomings of the traditional thermochemical table and the advantages of ATcT.

Following ver. 1.22d, ver. 1.22e was released to include methyl acetate and methyl formate species [34]. The updated ver. 1.22 g was then released in 2019 to include information on the appearance energy of CH_3^+ from CH_4 species [35]. One of the critical thermodynamic properties (enthalpy of formation) of some selected species related to NO formation is provided in Table 2.1 [35].

2.4 Reaction mechanism

The formation of NO is because of two significant reasons. One is the molecular nitrogen present in the oxidizer and the other is the presence of fuel-bound nitrogen. The formation of NO in IC engine combustion occurs from N_2 by three different mechanisms: the thermal mechanism of NO formation, the prompt mechanism of NO formation, and the fuel nitrogen mechanism of NO formation [36]. The formation of NO and NO_2 in the combustion process is provided in Table 2.2 [37]. It should be noted that thermal NO formation depends on O/H radicals, which are controlled by fuel oxidation chemistry. In the case of prompt NO formation, there is a direct relationship between fuel oxidation and nitrogen chemistry. Almost all fuel radicals obey this relationship and the most predominant one is the CH radical [17,24]. The fuel NO formation is based on the oxidation of nitrogen-containing compounds present in the fuel. The formation of NO through the three mechanisms is discussed in detail in the following.

2.5 NO_x formation in IC engines

NO is the predominant NO_x produced inside the engine cylinder due to the oxidation of atmospheric (molecular) nitrogen. Atmospheric nitrogen occurs as a stable diatomic molecule at low temperatures. Thus, small quantities of nitrogen are detected. However, some diatomic nitrogen (N_2) breaks down to monatomic nitrogen (N), which is reactive at high temperatures in the engine's combustion chamber. Diesel fuel contains more nitrogen than gasoline. However, the current levels are not significant. Furthermore, fuel may contain NH_3, NC, and HCN, but their contribution to NO emissions is negligible. It is worth noting that the formation of NO_x is highly temperature-dependent. In the temperature range of 2500–3000 K, which can occur in an engine, a significant amount of nitrogen oxide is produced. Other gases that are stable at low temperatures but become reactive, including oxygen and water vapor, lead to the formation of

Table 2.1 Enthalpies of formation of some selected species related to NO formation (ver. 1.22g available at ATcT.anl.gov).

Species name	$\Delta_f H^\circ$ (0 K)	$\Delta_f H^\circ$ (298.15 K)	Uncertainty	Units	Relative molecular mass
Ammonia	−38.565	−45.557	± 0.030	kJ/mol	17.03056 ± 0.00022
Amidogen	188.92	186.03	± 0.11	kJ/mol	16.02262 ± 0.00016
Imidogen	358.75	358.79	± 0.17	kJ/mol	15.014680 ± 0.000099
Nitrogen atom	470.579	472.442	± 0.024	kJ/mol	14.006740 ± 0.000070
Diazenyl	252.17	249.25	± 0.47	kJ/mol	29.02142 ± 0.00016
Nitric oxide	90.619	91.123	± 0.065	kJ/mol	30.00614 ± 0.00031
Nitrogen dioxide	36.859	34.052	± 0.065	kJ/mol	46.00554 ± 0.00060
Nitroxyl	79.40	74.14	± 0.19	kJ/mol	62.00494 ± 0.00090
Nitrous oxide	85.997	82.567	± 0.097	kJ/mol	44.01288 ± 0.00033
Hydrogen cyanide	129.661	129.276	± 0.090	kJ/mol	27.02538 ± 0.00081
Hydrogen isocyanide	191.98	192.36	± 0.37	kJ/mol	27.02538 ± 0.00081
Nitrilomethyl	436.73	440.01	± 0.15	kJ/mol	26.01744 ± 0.00080
Methanetetraylbisamidogen	450.42	450.83	± 0.64	kJ/mol	40.02418 ± 0.00081
Nitrosyl hydride	109.93	106.96	± 0.11	kJ/mol	31.01408 ± 0.00032
Hydroxyimidogen	218.00	215.12	± 0.89	kJ/mol	31.01408 ± 0.00032
Nitrous acid	−73.005	−78.662	± 0.079	kJ/mol	47.01348 ± 0.00061
Nitric acid	−124.50	−134.21	± 0.18	kJ/mol	63.01288 ± 0.00091
Methane	−66.557	−74.526	± 0.055	kJ/mol	16.04246 ± 0.00085
Methyl	149.872	146.458	± 0.060	kJ/mol	15.03452 ± 0.00083
Methylidyne	592.825	596.159	± 0.099	kJ/mol	13.01864 ± 0.00080
Carbon atom	711.399	716.884	± 0.047	kJ/mol	12.01070 ± 0.00080

Table 2.2 Formation of NO and NO_2 in combustion.

Nitrogen oxide	Area of formation	Mainly dependent on
Thermal NO	1. Flame 2. After burner (all kinds of fuel)	1. Concentration of O-atoms from O_2-dissociation 2. Residence time ($T>1300°C$)
Prompt NO	Flame (all kinds of fuel)	1. Concentration of O-atoms from combustion 2. O_2-concentration
Fuel NO	Flame (coal, heavy oil)	1. O_2-concentration 2. Resistance time
NO_2	Flame	Quenching of combustion reaction
	1. Smoke ducts 2. Chimneys	1. O_2 and NO concentration 2. Residence time ($T<650°C$)
	3. Free atmosphere	1. O_2 concentration 2. Light intensity 3. Resistance time 4. Air pollution

NO_x at high temperatures. The higher the temperature of the combustion reaction, the greater the dissociation of N_2 with N and the more NO_x that will be formed. There is very little NO_x produced at a low temperature.

NO is produced in both flame front and post flame gases. It can be said that NO production in post flame gases dominates the flame front region because the flame reaction region is thin and the resistance time in this region is short. Moreover, this trend can be explained by the rate of pressure rise inside the engine cylinder. Thus, the burned gases are produced early. Considerations of chemical equilibrium imply that NO_2/NO ratios should be negligible for burned gases at normal flame temperatures. Although experimental results suggest that this is valid for spark ignition engines, up to 30% of the total exhaust oxides of nitrogen emissions can be NO_2 in diesel fuel. NO produced in the flame zone can be transformed rapidly to NO_2 [14].

N_2O emissions from the automotive industry occur at low temperatures (less than 950°C) during combustion and are influenced by the type of fuel, the engine operating parameters, excess air levels, fuel-bound oxygen, and catalytic operation. For instance, a promising technology such as exhaust gas recirculation (EGR) is used to improve the regulation of NO_x emissions by reducing the temperature of combustion, which can lead to high N_2O emissions [13,14].

2.6 Thermal NO formation

The primary source of thermal NO formation in the IC engine is the oxidation of molecular nitrogen. The available techniques to control thermal NO formation involve the reduction of burned gas temperature or available oxygen. Exhaust gas treatment is also a possible method to reduce NO emissions. However, the combustion process modifications such as temperature reduction and oxygen availability reduction can control NO emissions economically. The combustion process modification to reduce temperature or available oxygen can be achieved by different techniques such as retardation of injection timing, water injection, and exhaust gas recirculation. All the above-mentioned techniques will help reduce the burned gas temperature or minimize the contact between nitrogen and oxygen. These modifications create a fuel-rich zone in which NO can be reduced to N_2. However, the modifications in the combustion process to reduce NO emissions require considerable experimental work. Therefore, it can be said that the thermal NO formation rate is influenced by several factors, among which the flame temperature and molecular oxygen concentration are predominant. The other factors influencing the thermal NO formation rate are the fuel density, the ignition lag period, and the equivalence ratio [38]. The significance of different factors on thermal NO formation is illustrated in Fig. 2.5 [13,38,39].

Considering the effect of flame temperature on thermal NO formation, combustion in the near-stoichiometric zone greatly influences the NO production rate, as expressed in Eq. (2.1) [13].

$$[NO_X] \propto \exp\left(\frac{E_a}{R_u T_f}\right) \tag{2.1}$$

where E_a is the activation energy in J/mol, R_u is the universal gas constant in J/molK, and T_f is the adiabatic flame temperature in K. The flame temperature can be determined by calculating the in-cylinder temperature.

In CI engines, the availability of oxygen at each load remains the same. However, the quantity of fuel admitted varies with respect to load. Therefore, changes in equivalence ratio with respect to load influence oxygen availability. It should be noted that the relationship between equivalence ratio and NO formation is not linear. This is because NO formation increases with equivalence ratio and attains the maximum at stoichiometric conditions. Even at a higher flame temperature, the reduction in oxygen availability can control NO formation. Furthermore, considering the effect

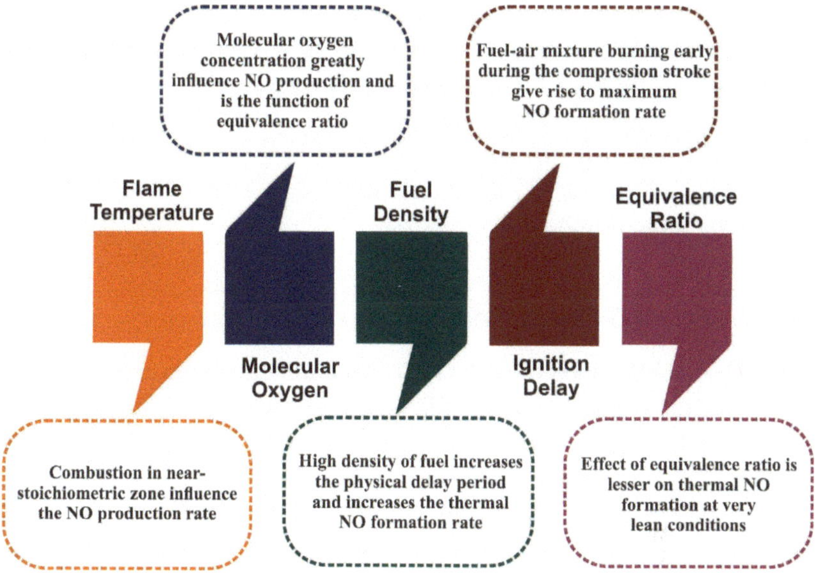

Fig. 2.5 Significance of different factors on thermal NO formation rate [13,38,39].

of molecular oxygen concentration on thermal NO formation, the rate constants of the reaction significantly influence the thermal NO formation rate [40–42]. However, the determination of the rate constants and oxygen concentration is tedious for different alternative fuels. Therefore, the availability of molecular oxygen can be determined by assuming that the molecular oxygen concentration varies with respect to equivalence ratio and the fuel-air mixture is homogenous [13]. Pertaining to fuel density, the thermal NO concentration increases linearly with an increase in the density of the fuel under given engine operating conditions [39]. This trend is because the fuel atomization is very dependent on the density of the fuel. This results in a larger fuel droplet diameter. If the fuel density is high, then the atomization worsens. Moreover, the fuel with high density increases the physical delay period (the period between the start of fuel injection and the attainment of chemical reactions) and in-cylinder temperature, which increases the thermal NO formation rate.

Generally, significant NO formation occurs between the start of ignition and the occurrence of maximum in-cylinder pressure. Therefore, the fuel-air mixture burning early during the compression stroke gives rise to the maximum NO formation rate [13]. For this reason, when the ignition delay period is shortened, the compression duration of the burning mixture after the ignition lag phase is more. Considering the factor equivalence ratio, it can be said that there is a strong relationship between the equivalence ratio and the thermal NO production rate [13]. However, there is little effect of the equivalence ratio on NO production at lean conditions. This is due to the prevalence of low flame temperature at very lean conditions. Therefore, in diesel engines, the NO production is negligible at no load condition and significantly high at high load condition.

2.7 Prompt NO formation

In the case of fuel-rich regions, the NO formation rate can exceed that of the thermal NO mechanism. This rapidly formed NO is referred to as "prompt NO." Thus, prompt NO formation is influenced mainly by the fuel chemistry [43,44]. Furthermore, the prompt NO production rate is sensitive to the free radical concentration. Generally, free radicals are reactive independent species that consist of one or more unpaired electrons [45,46]. Therefore, the free radical produced during the burning of the air-fuel mixture decides the reaction rate as well as prompt NO formation. Practically, the prompt NO levels can range from a few ppm to greater than 100 ppm [47]. Some studies indicate that prompt NO forms because of O and OH at super equilibrium concentrations. However, most of the studies confirmed that the reaction between hydrocarbon radicals and molecular nitrogen forms amines/cyano compounds and upon further reaction, prompt NO is produced [48,49]. Among different hydrocarbons, C_2H_2 produces twice as much as prompt NO [50]. Compared to the parent hydrocarbon, prompt NO formation is significantly influenced by carbon atoms. In addition, the equivalence ratio plays a very vital role in prompt NO formation. It is reported that prompt NO increases up to a certain equivalence ratio, attains a peak, and then decreases afterward due to oxygen deficiency.

2.8 NO production from fuel nitrogen

The primary source of NO production in IC engines is atmospheric nitrogen. Besides, the fuel-bound nitrogen remains the second major source of NO production. For example, the solid fuel (coal and its derivatives) has up to 2 wt% of fuel-bound nitrogen. In the case of heavy oils, the fuel nitrogen is present in the form of heterocycles [51]. Pyridine is the major heterocyclic compound found in heavy oils [52]. Different measurement methods are available to determine NO production from fuel nitrogen; the

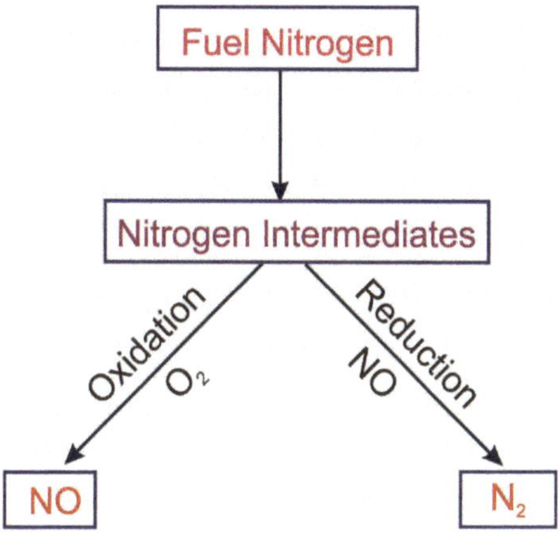

Fig. 2.6 Reaction pathway of fuel nitrogen in the formation of NO and N_2.

concentration of NO production from fuel nitrogen is dependent on the reaction conditions and the level of fuel nitrogen [53]. A simple reaction pathway representing the formation of NO from fuel nitrogen is illustrated in Fig. 2.6 [54,55].

The available literature suggests that gas phase oxidation and the heterogeneous oxidation of fuel nitrogen produces NO [37,56]. In gas phase oxidation of fuel nitrogen, HCN and NH_3 are the primary compounds. The oxidation of HCN is expressed in reactions (R1), (R2).

$$HCN + O \leftrightarrow NCO + H \tag{R1}$$

$$HCN + O \leftrightarrow NH + CO \tag{R2}$$

The subsequent reaction of NCO and NH with a hydrogen atom produces atomic nitrogen, which upon further reactions (R3), (R4) produces NO and N_2.

$$N + OH \leftrightarrow NO + H \tag{R3}$$

$$N + NO \leftrightarrow N_2 + O \tag{R4}$$

Furthermore, the oxidation of NH_3 in the presence of oxygen and hydrogen atoms can be expressed as (R5), (R6)

$$NH_3 + O \leftrightarrow NH_2 + OH \tag{R5}$$

$$NH_3 + H \leftrightarrow NH_2 + H_2 \tag{R6}$$

The subsequent reaction of NH_2 produces NO through a reaction with HNO, as expressed in (R7) (R10).

$$NH_2 + O \leftrightarrow HNO + H \quad \text{(R7)}$$
$$HNO + M \leftrightarrow NO + H + M \quad \text{(R8)}$$
$$HNO + OH \leftrightarrow NO + H_2O \quad \text{(R9)}$$
$$HNO + NH_2 \leftrightarrow NO + NH_3 \quad \text{(R10)}$$

2.9 Mechanisms for the formation of NO

In the literature, different mechanisms for NO formation, such as the Zeldovich mechanism, the nitrous oxide mechanism, the Fenimore mechanism, and the NNH mechanism, were proposed to explain the IC engine combustion process. The Zeldovich mechanism of NO formation is based on the temperature of the reaction zone [57,58]. The nitrous oxide mechanism portrays the formation of NO from intermediate species [59]. The NO formation in rich flames (smallest contribution) is expressed in the Fenimore mechanism [60], and the formation of NO for fuels with a large carbon-to-hydrogen ratio is depicted in the NNH mechanism [61,62].

2.9.1 Zeldovich mechanism

In 1946, Russian scientist Zeldovich proposed a pair of reactions (R11), (R12) to explain the formation of NO at higher temperatures. These high-temperature reactions govern the thermal NO formation from molecular nitrogen [57,58]. The third reaction (R13) was later added to the mechanism, which accounts for the influence of OH radicals during combustion [63].

$$O + N_2 \leftrightarrow NO + N \quad \text{(R11)}$$
$$N + O_2 \leftrightarrow NO + O \quad \text{(R12)}$$
$$N + OH \leftrightarrow NO + H \quad \text{(R13)}$$

In the case of lean premixed combustion, the rate of formation of NO (neglecting the backward reaction rate) can be expressed as given in Eq. (2.2).

$$d\frac{[NO]}{dt} = k_{f1}[N_2][O] + k_{f2}[N][O_2] + k_{f3}[N][OH] \quad (2.2)$$

where k_{f1}, k_{f2}, and k_{f3} are the rate constants of the reaction and are given as:

$$k_{f1} = 2 \times 10^{14}\, exp^{\left(-\frac{315}{R_u T}\right)}$$

$$k_{f2} = 6.4 \times 10^9 \, exp^{\left(-\frac{26}{R_u T}\right)}$$

$$k_{f3} = 3.8 \times 10^{13}$$

where T is the temperature in K, R_u is the universal gas constant, and the activation energy has unit kJ/gmol.

The evaluation of rate coefficients of reactions (R11)–(R13) can be done by both direct and indirect measurement methods. Moreover, it can be seen from reactions (R11)–(R13) that the presence of O and OH radicals can give details about fuel oxidization reactions. However, compared to the fuel oxidation reaction, thermal NO formation is slow. In this regard, the thermal NO formation mechanism should be separated from the fuel oxidation mechanism.

2.9.2 Nitrous oxide mechanism

In the nitrous oxide mechanism, NO is formed via the intermediate species N_2O that can be produced by fossil fuel combustion (R14) and (R15) and natural gas combustion, as given by reaction (R16) [59].

$$NCO + NO \rightarrow N_2O + CO \quad (R14)$$

$$NH + NO \rightarrow N_2O + H \quad (R15)$$

$$O + N_2 + M \rightarrow N_2O + M \quad (R16)$$

The reaction (R16) is an important form that occurs when the mixture is fuel-lean at low temperature and high pressure. Furthermore, the lifetime study of N_2O reveals that the average lifetime of N_2O is less than 10 ms for temperatures greater than 1500 K [59]. This reveals that the N_2O emission is significant in low-temperature combustion systems and not significant in combustors and other combustion systems.

The major path of the N_2O mechanism is illustrated in Fig. 2.7.

2.9.3 Fenimore mechanism

The formation of NO in the flame zone is the thrust area of study because the concentration of free radicals in the flame zone is significantly higher than in the post flame region. Fenimore first proposed the formation of NO in the flame zone in 1971 [60]. His study disclosed that NO formation could be captured in the flame zone, other than the mechanism proposed by Zeldovich. The Fenimore study proved that the reaction of CH radicals

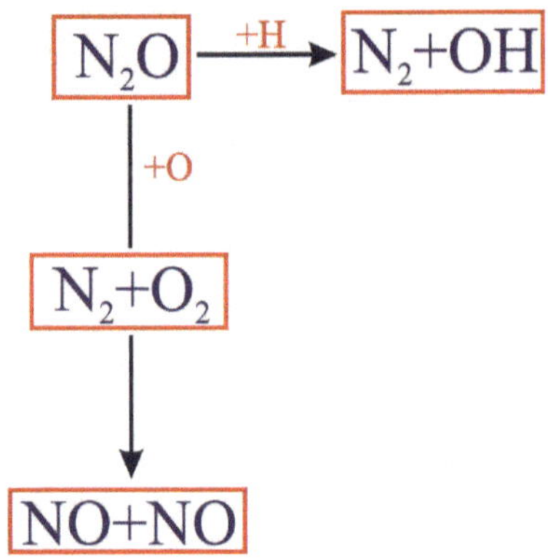

Fig. 2.7 Steps in N_2O removal.

with N_2 forms NO. However, his research is limited to hydrocarbon flames (R17) and (R18), and no trace of NO formation is evident with nonhydrocarbon flames.

$$CH + N_2 \rightarrow HCN + N \tag{R17}$$

$$C_2 + N_2 \rightarrow 2 \times CN \tag{R18}$$

The products of reactions (R17), (R18) (HCN, N, and CN) form NO when they react with free radicals. Moreover, comparing reaction (R17) with (R18), (R17) is the significant contributor in methane air combustion because of the presence of species CH.

2.9.4 NNH mechanism

Although different mechanisms were proposed to capture the formation of NO, additional reaction steps to form NO from N_2 are possible by recombining N_2 with O or H atoms followed by N_2 intermediate oxidation. This concept was proposed by Bozzelli and Dean in 1995 [62]. The formation of NO by the NNH mechanism is expressed in R19.

$$NNH + O \rightarrow NH + NO \tag{R19}$$

In reaction (R19), the formation of NO is significantly influenced by three parameters. The first parameter is the NNH's lifetime, the second one is the heat of formation of NNH, and the last one is the rate constant of the reactants (NNH + O). Furthermore, it should be noted that there is an inverse relationship between the lifetime of NNH and the rate constant and a direct association between NO formation and the rate constant [64]. The formation of the dominant source of NO by the NNH mechanism is affected by the experimental conditions. From the literature, it is evident that the NNH mechanism is most significant at low temperatures or shorter resistance time [65,66]. However, the formation of NO via the NNH mechanism under low pressure is also evident [67].

2.10 Uncontrolled NO_x emission levels in IC engines

The operating pressure and temperature are the two key factors determining the rates of formation of NO_x in IC engines. Besides, the Zeldovich mechanism is the predominant mechanism by which NO_x emissions are produced because most conventional fuels used in IC engines have little or no nitrogen [68]. Hence, the formation of fuel NO_x is minimal in IC engines. The uncontrolled NO_x emission levels in IC engines are provided in Table 2.3 [68]. It can be seen that rich-burn SI engines have an average NO_x emissions range from 13.1 to 14 g/hp-hr. On the other hand, the average NO_x emissions for lean-burn SI engines range from 7.9 to 16.5 g/hp-hr. In the case of diesel engines, the average NO_x emissions range from 11.2 to 12 g/hp-hr, whereas for dual fuel engines, the average NO_x emissions range from 4.9 to 10 g/hp-hr.

2.11 Factors influencing NO_x emissions from IC engines

NO_x emissions from IC engines are influenced by three major factors: engine design and operating parameters, fuel effects, and ambient conditions. The influence of each factor is discussed in the following sections. Also, the significance of each factor is illustrated in Fig. 2.8.

2.11.1 Engine design and operating parameters

Engine design and operating parameters significantly influence the NO_x formation rate. These parameters include the air-to-fuel ratio (A/F) and charging method, ignition timing, combustion chamber and valve design, engine combustion cycle, and operating load and speed.

2.11.2 Air-to-fuel ratio (A/F) and charging method

It is well known that the NO_x emission rate is dependent on combustion temperature and the emission increases with an increase in combustion temperature. The maximum combustion temperature occurs when the A/F is just above the stoichiometric [69]. This trend discloses that NO_x emission

Table 2.3 Uncontrolled NO_x emissions for IC engines.

Engine size, hp.	No. of engines	Highest NO_x emissions (g/hp-hr)	Lowest NO_x emissions (g/hp-hr)	Average NO_x emissions (g/hp-hr)
Rich-burn SI engines				
0–200	8	15.8	19.1	13.1
201–400	13	23.5	9.1	16.4
401–1000	31	22.4	10.4	16.3
1001–2000	19	25	13	16.3
2001–4000	10	18	13	15.0
4001 +	2	14	14	14.0
Lean-burn SI engines				
0–400	7	17.5	3	7.9
401–1000	17	27	15.5	18.6
1001–2000	43	27	14	17.8
2001–4000	30	27	10	17.2
4001 +	25	17.5	10	16.5
Diesel engines				
0–200	12	17.1	10	11.2
201–400	8	19	7.6	11.8
401–1000	22	19	9	13.0
1001–2000	14	19	8.5	11.4
2001–4000	6	14	9.3	11.4
4001 +	6	12	12	12.0
Dual fuel engines				
700–1200	5	13	9.3	10.0
1201–2000	3	13	6.2	10.7
2001–4000	5	13	5	8.4
4001 +	4	5	4.5	4.9

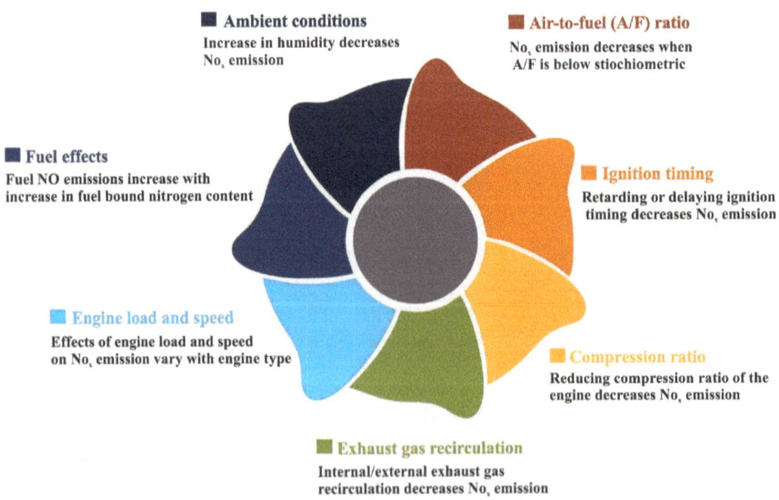

Fig. 2.8 Significance of factors influencing NO_x formation from IC engines.

decreases when A/F is below stoichiometric and increases when A/F is just above stoichiometric. The reason for the decrease in NO_x emissions when A/F is below stoichiometric is the lack of excess oxygen. On the contrary, the increase in NO_x emissions when A/F is just above stoichiometric is due to the availability of excess oxygen. However, the NO_x emissions steadily decrease when A/F increases beyond the stoichiometric condition [69].

The charging method influences the A/F, which affects the NO_x emission rate in IC engines. In turbocharged fuel-injected engines, A/F can be precisely controlled, whereas in naturally aspirated engines, the carbureted control of A/F is not precise and can result in misfiring.

2.11.3 Ignition timing

Combustion is initiated by external spark assistance in SI engines and by autoignition in CI engines. In this regard, advancing or retarding the ignition timing will significantly affect the combustion process. Advancing the ignition timing could result in early combustion, which would increase in-cylinder pressure, temperature, and NO_x levels. On the contrary, retarding the ignition timing will lower the in-cylinder pressure and temperature due to an increase in combustion chamber volume and help to lower the NO_x levels. Although retarding the ignition timing lowers NO_x emissions, it negatively affects engine efficiency, fuel consumption, and soot levels [68].

2.11.4 Combustion chamber and valve design

The design of the engine combustion chamber as well as the inlet and exhaust valves plays a crucial role in exhaust emissions. However, the effects are difficult to quantify because the variation in engine design variables can influence other favorable effects. The NO_x levels can be lowered by varying combustion chamber geometry and valve design to improve the air motion. The improved air motion enhances air-fuel mixing inside the engine cylinder and helps to improve fuel atomization. Furthermore, lower NO_x levels can be achieved by reducing the effective compression ratio of the engine. However, lowering the compression ratio will affect the cycle efficiency [70].

2.11.5 Engine combustion cycle

Based on the engine cycle, IC engines can be classified as two-stroke and four-stroke engines. Compared to four stroke engines, the NO_x variation in two-stroke engines is interesting because of the series of events occurring during the charging process. In carbureted two-stroke engines, the scavenge air purges the exhaust gas. During this process, some part of the unreacted fuel can also escape the exhaust. If the exhaust gas purging is not complete, then some exhaust gases will be retained inside the engine cylinder. This effect is most commonly referred to as internal exhaust gas recirculation (EGR). The internal EGR aids in lowering the cylinder temperature by absorbing combustion energy, thereby lowering NO_x levels [68].

2.11.6 Engine load and speed

Variations in engine load and speed affect the NO_x formation rate, and such variations are engine specific. In SI engines, the total NO_x emissions increase with an increase in power output, and the brake-specific NO_x emissions decrease with an increase in power output. Furthermore, it should be noted that NO_x emissions from an SI engine decrease with the increasing load when the speed decreases with decreasing load. Likewise, brake-specific NO_x emissions from the SI engine decrease with decreasing load at a constant speed [71]. Considering the CI engine, the brake-specific NO_x emissions decrease with increasing load at a constant speed. This trend is because of the changes in the A/F ratio. However, a reverse trend can be noted in turbocharged diesel engines. The brake-specific NO_x emission level increases with increasing load for turbocharged diesel engines. Overall, it should be noted that the effects of engine load and speed on NO_x emissions vary with the engine type [68].

2.12 Effects of alternative fuel (biodiesel)

As discussed in Section 2.7, thermal NO is more predominant in IC engines than the fuel NO. However, the fuel NO emissions can increase with an increase in fuel-bound nitrogen content. Generally, crude oil consists of organic nitrogen or fuel-bound nitrogen, and the use of residual crude oil in engines can increase fuel NO emissions. Furthermore, gaseous fuels such as coal gas and waste stream gases will increase the fuel NO emissions. The fuel NO emissions from coal gas or waste stream gas are higher than that of natural gas. A wide range of alternative fuels is available for IC engines. The alternative fuels used in engines should have a low combustion temperature to reduce NO_x emissions. Pertaining to CI engines, biodiesel is considered one of the potential surrogates to mineral diesel. The use of biodiesel in engines reduces almost all emissions except NO_x, as reported in previous

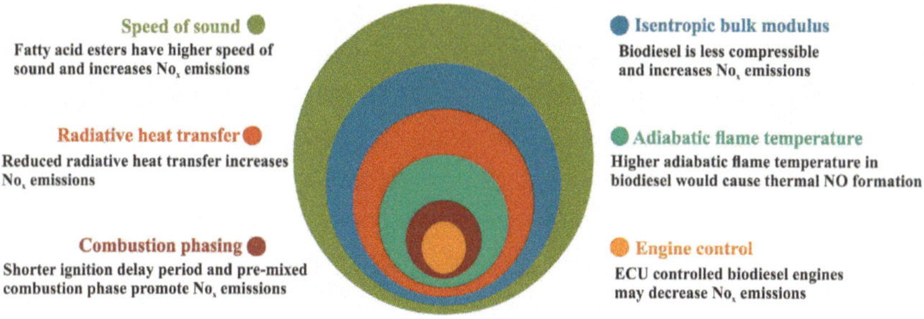

Fig. 2.9 Significance of factors influencing biodiesel NO_x.

studies. Biodiesel NO_x is not based on a single factor, whereas several factors such as engine design, duty cycles, fuel injection strategy, emission reduction strategy, etc., influence biodiesel NO_x. Several theories are proposed in the literature to explain specific causes. All the theories explain the difference in in-cylinder temperature between biodiesel and pure diesel [72–74]. This discloses that thermal NO dominates the total NO_x formation in CI engines fueled with biodiesel. The significance of each factor on biodiesel NO_x is depicted in Fig. 2.9.

2.12.1 Speed of sound

The measurement of ultrasonic velocity for biodiesel fuels can provide useful information regarding the physiochemical behavior of the fuel. An ultrasound cell can be used to measure the ultrasonic velocity of biodiesel fuels [75]. In general, the speed of sound of longer-chain paraffin is higher than the shorter-chain paraffin. Comparing the carbon chain lengths of biodiesel and mineral diesel, biodiesel consists of carbon chain lengths of 17–19. Thus, the speed of sound of biodiesel is expected to be higher than mineral diesel (No. 2 diesel), which has the carbon chain length of 13 16 [76]. In case of No. 1 diesel, the carbon chain length is lower compared to No. 2 diesel. This reveals that all fatty acid esters have a higher speed of sound than mineral diesel and the higher speed of sound of biodiesel is the major reason for the advancement of injection timing and the increase in NO_x emissions.

2.12.2 Isentropic bulk modulus

The bulk modulus of compressibility is one of the critical properties affecting the hydraulic behavior of the fuel injection process. The study of bulk modulus of compressibility provides information about dilatation [77]. The bulk modulus of compressibility can be measured by observing dilatations in fuels under pressure or determined from the speed of the sound, as there is a direct relationship between the bulk modulus of compressibility and speed of sound of fluid. From the literature, it is evident that larger hydrocarbon molecules are less compressible than the smaller hydrocarbon molecules [78]. As a result, biodiesel exhibits lower compressibility than conventional diesel. The less-compressible nature of biodiesel paves the way for early injection, which increases the in-cylinder pressure, temperature, and NO_x.

2.12.3 Radiative heat transfer

During the diesel engine combustion process, the presence of soot within the combustion chamber helps to reduce the flame temperature because of the radiative heat transfer instigated by soot particles. Using biodiesel, however, greatly decreases soot emissions by oxidizing the soot particles using fuel-bound oxygen. Furthermore, the length of the fatty acid chain has a significant impact on particle size. Biodiesel with a longer chain length produces more particles compared to biodiesel with a shorter chain length. Therefore, the soot reduction effect of biodiesel leads to an increase in the overall flame temperature, thus increasing thermal NO formation [73]. However, some studies have indicated that there is a simultaneous increase in soot and NO_x emissions [79]. In this regard, this theory needs further investigation to bridge the research gap.

2.12.4 Adiabatic flame temperature

Adiabatic flame temperature is the temperature when heat released by the chemical reaction heats the combustion products [80]. Studies have reported that adiabatic flame temperature is higher for biodiesel with high unsaturated fatty acids and double bonds [74,81]. Based on this hypothesis, it can be stated that biodiesel exhibits higher adiabatic flame temperature than mineral diesel because of the presence of a high degree of unsaturation [81,82]. Moreover, the higher adiabatic flame temperature in biodiesel would cause the thermal NO formation rate to increase. However, further experimental evidence is needed to understand that higher adiabatic flame temperatures in biodiesel induce thermal NO formation.

2.12.5 Combustion phasing

To understand the effects of combustion phasing on biodiesel NO_x, the three main phases of combustion—the ignition delay period, the premixed combustion phase, and the diffusion combustion phase—must be acknowledged. The ignition delay period is defined as the internal time between the start of fuel injection and the start of combustion. This period can be related to the cetane number of the fuel. The higher the cetane number of the fuel, the shorter the ignition delay period. Because biodiesel exhibits a higher cetane number than mineral diesel fuel, its ignition delay period is shorter [19]. The shorter ignition delay period increases peak cylinder pressure and temperature, thus increasing NO_x emissions. However, the shorter ignition delay period is not the sole factor responsible for the increase in biodiesel NO_x.

In the premixed combustion phase, the maximum pressure and temperature attained inside the engine cylinder significantly promote NO_x emissions. In the diffusion burning phase, NO species are formed in the region of burned gases [13]. However, the NO species formed in the diffusion burning phase is minimal compared to the premixed combustion phase due to the lower in-cylinder pressure and temperature prevailing during the diffusion burning period.

2.12.6 Engine control strategy

In modern IC engines, the electronic control unit (ECU) is attached to control the series of actuators and provide optimum engine performance. The ECU controls the A/F ratio, engine idle speed, valve timing, injection timing, and other important events. In ECU-controlled diesel engines, better cycle efficiency, greater fuel economy, and lower emissions can be realized when diesel is used as fuel. Instead of diesel, biodiesel in ECU-controlled diesel engines can decrease engine performance and increase emissions. This is because the ECU designed for diesel engines controls the actuators by reading values already fed in the lookup tables. In this perspective, programming the ECU by designing lookup tables for biodiesel is desirable.

2.13 Ambient conditions

Variations in atmospheric pressure and temperature have a significant effect on NO_x emissions. This is because the density of air changes when ambient conditions change. The density change affects the A/F ratio and thus influences NO_x formation. It is reported that up to 25% of NO_x emissions can be influenced by the atmospheric temperature changes and up to 40% by atmospheric pressure changes [68,69]. Furthermore, humidity has a crucial role in decreasing NO_x emissions. This is because the high moisture content in air helps to reduce the peak cylinder temperature and lower NO_x levels. It is

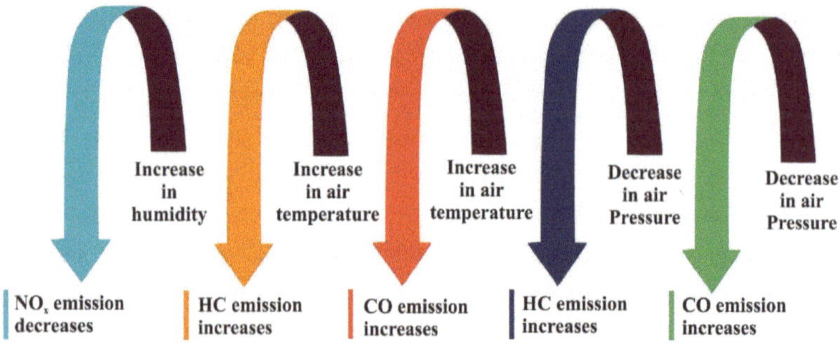

Fig. 2.10 General effects of ambient conditions on IC engine emissions.

reported that humidity has the potential to decrease NO_x emissions by 25% [68]. Moreover, it should be noted that the effect of ambient conditions on emissions varies from engine to engine. However, the general effects on emissions depicted in Fig. 2.10 can be observed when the engine operates close to stoichiometric conditions.

2.14 Concluding remarks

NO_x formation mechanisms have been discussed in some detail in this chapter, and useful summaries of this area are provided by several literature references. The three process mechanisms of NO formation—thermal NO, prompt NO, and fuel NO—are generally believed to be important. In this chapter, we tried to present the NO formation mechanisms such as the Zeldovich mechanism, nitrous oxide, Fenimore, and NNH mechanisms that can be applied to IC engines. The survey shows that the Zeldovich mechanism of NO formation strongly relies on the temperature, which remains a significant and crucial factor in the control of NO formation. Therefore, the modification of the present engine combustion process is needed to reduce the peak temperature to decrease NO_x emissions. Although methods such as exhaust gas recirculation and micro/nanoemulsions minimize NO_x emissions, they lead to a high fuel penalty and reduced thermal efficiency. As a result, strategies such as limiting oxygen and molecular nitrogen concentration near the peak flame temperature and reducing the resistance time of the peak flame temperature can be preferred to minimize thermal NO formation. Unlike thermal NO, prompt NO formation cannot be substantially reduced simply by reducing the peak flame temperature. To control prompt NO formation, it is indispensable to bring some modifications in the engine design to reduce the substoichiometric region in the flame.

Concerning the control of fuel-bound nitrogen emissions, renewable biofuels such as biodiesel are strongly recommended. However, the thermal NO and prompt NO formation are dominant in biodiesel-powered engines.

The theories ofbiodiesel's impact on NO_x emission disclose that biodiesel NO_x depends on various factors, including engine design and operating parameters. One should note that these theories are not mutually exclusive. Instead, they can be viewed as influential variables that partly explain the biodiesel NO_x. This is owing to the fact that combustion processes are complex and variable in nature. Therefore, a single theory cannot completely explain biodiesel's impact on NO_x emissions. In addition to engine design and operating parameters, fuel modifications can mitigate biodiesel NO_x; however, such benefits will be limited with modern engines. Although after-treatment systems can be used to meet the requirements, there is little real world explanation available on exhaust after-treatment systems. Even with US 2010 heavy duty emission standards, the after-treatment systems were in use for a short time. This shows that long-term study is needed. The impurities present in modern alternative fuels can significantly impact the performance of after-treatment systems. However, their real world situation is not known. Moreover, pre- and posttreatment methods to reduce NO_x could hinder business development. They potentially prevent a company from growing or investing in modern/more productive manufacturing processes. As a result, new jobs are not created and current jobs can even be put at risk.

Overall, mechanisms described in the present work will help study the formation ofNO in IC engines. However, the extension of the mechanisms to modern fuels such as biofuels is needed to explain the nitrogen chemistry thoroughly. Furthermore, the exploitation of oxygenated biofuels has a significant impact on particulate matter (PM) emissions. In IC engines, the study of NO_x and PM emissions is the thrust area because of the trade-off nature between these two emissions. It is perceived that exploiting oxygenated alternative fuels can reduce both NO_x and particulate matter emissions simultaneously. Besides, different in-cylinder and after-treatment techniques are reported in the literature to address the NO_x and PM trade-off in IC engines, which will be dealt with in the following chapter.

References

[1] Folkson R. Alternative fuels and advanced vehicle technologies for improved environmental performance: towards zero carbon transportation. Woodhead Publishing; 2014. https://doi.org/10.1533/9780857097422.

[2] Sindhu R, Amba Prasad Rao G, Madhu Murthy K. Effective reduction of NOx emissions from diesel engine using split injections. Alexandria Eng J 2018; 57:1379–92. https://doi.org/10.1016/j.aej.2017.06.009.
[3] Klingstedt F, Arve K, Eränen K, Murzin DY. Toward improved catalytic low-temperature NOx removal in diesel-powered vehicles. Acc Chem Res 2006;39:273–82. https://doi.org/10.1021/ar050185k.
[4] Roy S, Baiker A. NOx storage-reduction catalysis: from mechanism and materials properties to storage-reduction performance. Chem Rev 2009;109:4054–91. https://doi.org/10.1021/cr800496f.
[5] Boningari T, Smirniotis PG. Impact of nitrogen oxides on the environment and human health: Mn-based materials for the NOx abatement. Curr Opin Chem Eng 2016;13:133–41. https://doi.org/10.1016/j.coche.2016.09.004.
[6] Alberto Bernabeo R. Health and environmental impacts of Nox: an ultra-low level of Nox (oxides of nitrogen) achievable with a new technology. Glob J Eng Sci 2019;2 (3):1–6. https://doi.org/10.33552/gjes.2019.02.000540.
[7] Dell RM, Moseley PT, Rand DAJ. Chapter 3: Unconventional fuels. In: Dell RM, Moseley PT, DAJ R, editors. Towards sustainable road transport. Boston: Academic Press; 2014. p. 86–108. https://doi.org/10.1016/B978-0-12-404616-0.00003-7.
[8] Simmons WA, Seakins PW. Estimations of primary nitrogen dioxide exhaust emissions from chemiluminescence NOx measurements in a UK road tunnel. Sci Total Environ 2012;438:248–59. https://doi.org/10.1016/j.scitotenv.2012.08.050.
[9] Wang X, Khlystov A, Ho K-F, Campbell D, Chow JC, Kohl SD, et al. Real-world vehicle emissions characterization for the Shing Mun Tunnel in Hong Kong and Fort McHenry tunnel in the United States. Res Rep Health Eff Inst 2019;199:5–52.
[10] Badshah H, Posada F, Muncrief R. Current state of NOx emissions from in-use heavy-duty diesel vehicles in the United States; 2019.
[11] AIR U. Air pollution in the UK, 2016. Department for Environment, Food, and Rural Affairs (DEFRA); 2017. 2018.
[12] ICCT. NOX emissions from heavy-duty and light-duty diesel vehicles in the EU: comparison of real-world performance and current type-approval requirements; 2016.
[13] Heywood. IC engine fundamentals. New York: McGraw-Hill; 1989.
[14] Subramanian KAGBM. Alternative transportation fuels: utilisation in combustion engines; 2013.
[15] Miller B. Chapter 5: Nitrogen oxides formation and control. In: Miller B, editor. Fossil fuel emissions control technologies. Butterworth-Heinemann; 2015. p. 243–80. https://doi.org/10.1016/B978-0-12-801566-7.00005-1.
[16] Miller JA, Bowman CT. Erratum: mechanism and modeling of nitrogen chemistry in combustion. Prog Energy Combust Sci 1990;16. https://doi.org/10.1016/0360-1285(90)90043-3.
[17] Glarborg P. Hidden interactions-trace species governing combustion and emissions. Proc Combust Inst 2007;31:77. https://doi.org/10.1016/j.proci.2006.08.119.
[18] Du W, Kamal R, Zhao ZK. Biodiesel. In: Compr Biotechnol. 3rd ed. Pergamon: Elsevier; 2019. https://doi.org/10.1016/B978-0-444-64046-8.00151-8.
[19] Hoekman SK, Robbins C. Review of the effects of biodiesel on NOx emissions. Fuel Process Technol 2012;96:237–49. https://doi.org/10.1016/j.fuproc.2011.12.036.
[20] Yanowitz J, McCormick RL. Effect of biodiesel blends on North American heavy-duty diesel engine emissions. Eur J Lipid Sci Technol 2009;111:763–72. https://doi.org/10.1002/ejlt.200800245.
[21] Sun J, Caton JA, Jacobs TJ. Oxides of nitrogen emissions from biodiesel-fuelled diesel engines. Prog Energy Combust Sci 2010;36:677–95. https://doi.org/10.1016/j.pecs.2010.02.004.

[22] McCormick RL. The impact of biodiesel on pollutant emissions and public health. Inhal Toxicol 2007;19:1033–9. https://doi.org/10.1080/08958370701533509.

[23] Senkan SM. Detailed chemical kinetic modeling: chemical reaction engineering of the future. Adv Chem Eng 1992;51:1595–614. https://doi.org/10.1016/S0065-2377(08)60123-4.

[24] Glarborg P, Miller JA, Ruscic B, Klippenstein SJ. Modeling nitrogen chemistry in combustion. Prog Energy Combust Sci 2018;67:31–68. https://doi.org/10.1016/j.pecs.2018.01.002.

[25] Yang J, Golovitchev VI, Redón Lurbe P, López Sánchez JJ. Chemical kinetic study of nitrogen oxides formation trends in biodiesel combustion. Int J Chem Eng 2012;2012. https://doi.org/10.1155/2012/898742, 898742.

[26] Ni P, Wang X, Wang Z, Mao G, Wei S. Numerical modeling of soot and NOx emissions of a diesel/methanol dual fuel engine. Proc Int Conf Comput Distrib Control Intell Environ Monit CDCIEM 2011;2011. https://doi.org/10.1109/CDCIEM.2011.164.

[27] Ruscic B, Pinzon RE, Morton ML, Von Laszevski G, Bittner SJ, Nijsure SG, et al. Introduction to active thermochemical tables: several "key" enthalpies of formation revisited. J Phys Chem A 2004;108:9979–97. https://doi.org/10.1021/jp047912y.

[28] Ruscic B, Pinzon RE, Von Laszewski G, Kodeboyina D, Burcat A, Leahy D, et al. Active thermochemical tables: thermochemistry for the 21st century. J Phys Conf Ser 2005;16:561–70. https://doi.org/10.1088/1742-6596/16/1/078.

[29] Ruscic B, Feller D, Peterson KA. Active thermochemical tables: dissociation energies of several homonuclear first-row diatomics and related thermochemical values. Theor Chem Acc 2014;133:1415. https://doi.org/10.1007/s00214-013-1415-z.

[30] Ruscic B. Updated active thermochemical tables (ATcT) values based on ver. 1.110 of the thermochemical network. Argonne Natl Lab; 2012. http://atct.anl.gov/.

[31] Ruscic B. Active thermochemical tables: sequential bond dissociation enthalpies of methane, ethane, and methanol and the related thermochemistry. J Phys Chem A 2015;119:7810–37. https://doi.org/10.1021/acs.jpca.5b01346.

[32] Nguyen TL, Baraban JH, Ruscic B, Stanton JF. On the HCN - HNC energy difference. J Phys Chem A 2015;119:10929–34. https://doi.org/10.1021/acs.jpca.5b08406.

[33] Ruscic B, Bross DH. Active Thermochemical Tables (ATcT) values based on ver. 1.122d of the Thermochemical Network. Argonne Natl Lab; 2018. http://atct.anl.gov/.

[34] Porterfield JP, Bross DH, Ruscic B, Thorpe JH, Nguyen TL, Baraban JH, et al. Thermal decomposition of potential ester biofuels. Part I: methyl acetate and methyl butanoate. J Phys Chem A 2017;121:4658–77. https://doi.org/10.1021/acs.jpca.7b02639.

[35] Ruscic B, Bross DH. Active thermochemical tables (ATcT) values based on ver. 1.122g of the thermochemical network. Argonne Natl Lab; 2019. http://atct.anl.gov/.

[36] Bowman CT. Gas-phase reaction mechanisms for nitrogen oxide formation and removal in combustion. In: Pollutants from combustion. Dordrecht: Springer; 2000. https://doi.org/10.1007/978-94-011-4249-6_7.

[37] Anon. Formation and control of nitrogen oxides. Catal Today 1988;2:369–79. https://doi.org/10.1016/0920-5861(88)80002-6.

[38] Saravanan S, Nagarajan G, Anand S, Sampath S. Correlation for thermal NOx formation in compression ignition (CI) engine fuelled with diesel and biodiesel. Energy 2012;42:401–10. https://doi.org/10.1016/j.energy.2012.03.028.

[39] Gopinath A, Puhan S, Nagarajan G. Numerical modeling of oxides of nitrogen based on density of biodiesel fuels. Int J Energy Environ 2010;1:313–20.

[40] Blauvens J, Smets B, Peters J. 16th symposium (international) on combustion. Pittsburgh, PA: The Combustion Institute; 1977.

[41] Flower WL, Hanson RK, Kruger CH. 15th symposium (international) on combustion. Pittsburgh, PA: The Combustion Institute; 1975. p. 823.

[42] Hanson RK, Salimian S. Survey of rate constants in the N/H/O system. In: Combustion chemistry. New York, NY: Springer; 1984. https://doi.org/10.1007/978-1-4684-0186-8_6.
[43] Balaji G, Cheralathan M. Experimental reduction of NOx and HC emissions in a CI engine fuelled with methyl ester of neem oil using p-phenylenediamine antioxidant. J Sci Ind Res (India) 2014;73:177–80.
[44] Balaji G, Cheralathan M. Simultaneous reduction of NOx and HC emissions in a CI engine fueled with methyl ester of neem oil using ethylenediamine as antioxidant additive. Energy Sources Part A Recover Util Environ Eff 2015;37:2684–91. https://doi.org/10.1080/15567036.2012.749314.
[45] Lobo V, Patil A, Phatak A, Chandra N. Free radicals, antioxidants and functional foods: impact on human health. Pharmacogn Rev 2010;4:118–26. https://doi.org/10.4103/0973-7847.70902.
[46] Pham-Huy LA, He H, Pham-Huy C. Free radicals, antioxidants in disease and health. Int J Biomed Sci 2008;4:89–96.
[47] Miller JA, Bowman CT. Mechanism and modeling of nitrogen chemistry in combustion. Prog Energy Combust Sci 1989;4:287–338. https://doi.org/10.1016/0360-1285(89)90017-8.
[48] Hayhurst AN, Vince IM. Production of "prompt" nitric oxide and decomposition of hydrocarbons in flames. Nature 1977;266:521–2. https://doi.org/10.1038/266524a0.
[49] Matsui Y, Nomaguchi T. Spectroscopic study of prompt nitrogen oxide formation mechanism in hydrocarbon-air flames. Combust Flame 1978;32:205–14. https://doi.org/10.1016/0010-2180(78)90094-9.
[50] Hayhurst AN, Vince IM. The origin and nature of "prompt" nitric oxide in flames. Combust Flame 1983;50:41–57. https://doi.org/10.1016/0010-2180(83)90047-0.
[51] Prado GHC, Rao Y, De Klerk A. Nitrogen removal from oil: a review. Energy Fuels 2017;31:14–36. https://doi.org/10.1021/acs.energyfuels.6b02779.
[52] Santos RG, Loh W, Bannwart AC, Trevisan OV. An overview of heavy oil properties and its recovery and transportation methods. Brazilian J Chem Eng 2014;31:571–90. https://doi.org/10.1590/0104-6632.20140313s00001853.
[53] Fenimore CP. Reactions of fuel-nitrogen in rich flame gases. Combust Flame 1976;26:249–56. https://doi.org/10.1016/0010-2180(76)90075-4.
[54] Liu X, Luo Z, Yu C, Jin B, Tu H. Release mechanism of fuel-N into NOx and N_2O precursors during pyrolysis of rice straw. Energies 2018;11:520. https://doi.org/10.3390/en11030520.
[55] Houser TJ, McCarville ME, Biftu T. Kinetics of the thermal decomposition of pyridine in a flow system. Int J Chem Kinet 1980;12:555–68. https://doi.org/10.1002/kin.550120806.
[56] Haynes BS. Reactions of ammonia and nitric oxide in the burnt gases of fuel-rich hydrocarbon-air flames. Combust Flame 1977;28:81–9. https://doi.org/10.1016/0010-2180(77)90010-4.
[57] Zeldovich YB. The oxidation of nitrogen in combustion Explosions. Acta Physicochim USSR 1946;21:577–628.
[58] Zeldovich YB, Barenblatt GI, Librovich VB, Makhviladze GM. Mathematical theory of combustion and explosions; 1985.
[59] Bowman CT. Control of combustion-generated nitrogen oxide emissions: technology driven by regulation. Symp Combust 1992;24:859–78. https://doi.org/10.1016/S0082-0784(06)80104-9.
[60] Fenimore CP. Formation of nitric oxide in premixed hydrocarbon flames. Symp Combust 1971;13:373–80. https://doi.org/10.1016/S0082-0784(71)80040-1.
[61] Rutar T, Lee JCY, Dagaut P, Malte PC, Byrne A. NOx formation pathways in lean-premixed-prevapourized combustion of fuels with carbon-to-hydrogen ratio between

0.25 and 0.88. Proc Inst Mech Eng Part A J Power Energy 2007;221:387–98. https://doi.org/10.1243/09576509JPE288.

[62] Bozzelli JW, Dean AM. O + NNH: a possible new route for NOx formation in flames. Int J Chem Kinet 1995;27:1097–109.

[63] Lavoie GA, Heywood JB, Keck JC. Experimental and theoretical study of nitric oxide formation in internal combustion engines. Combust Sci Technol 1969;1:313–26.

[64] Klippenstein SJ, Harding LB, Glarborg P, Miller JA. The role of NNH in NO formation and control. Combust Flame 2011;158:774–89. https://doi.org/10.1016/j.combustflame.2010.12.013.

[65] Konnov AA. On the relative importance of different routes forming NO in hydrogen flames. Combust Flame 2003;134:421–4. https://doi.org/10.1016/S0010-2180(03)00135-4.

[66] Frassoldati A, Faravelli T, Ranzi E. A wide range modeling study of NOx formation and nitrogen chemistry in hydrogen combustion. Int J Hydrogen Energy 2006;31:2310–28. https://doi.org/10.1016/j.ijhydene.2006.02.014.

[67] Harrington JE, Smith GP, Berg PA, Noble AR, Jeffries JB, Crosley DR. Evidence for a new no production mechanism in flames. Symp Combust 1996;26:2133–8. https://doi.org/10.1016/S0082-0784(96)80038-5.

[68] Snyder RB. Alternative control techniques document - NOx emissions from stationary reciprocating internal combustion engines; 1993.

[69] U.S. EPA. Stationary internal combustion engines. Standards support and environmental impact statement, Volume I: Proposed standards of performance; 1979.

[70] El-Kassaby M, Nemit-Allah MA. Studying the effect of compression ratio on an engine fueled with waste oil produced biodiesel/diesel fuel. Alexandria Eng J 2013; 52:1–11. https://doi.org/10.1016/j.aej.2012.11.007.

[71] Helmich MJ, Schleigh MA. C-B reciprocating cleanburn update. Cooper-Bessemer reciprocating products division of cooper industrie. In: Sixth Annu. Reciprocating Mach. Conf; 1991. p. 23.

[72] Mueller CJ, Boehman AL, Martin GC. An experimental investigation of the origin of increased NOx emissions when fueling a heavy-duty compression-ignition engine with soy biodiesel. SAE Int J Fuels Lubr 2009;2:789–816. https://doi.org/10.4271/2009-01-1792.

[73] Cheng AS, Upatnieks A, Mueller CJ. Investigation of the impact of biodiesel fuelling on NOx emissions using an optical direct injection diesel engine. Int J Engine Res 2006;7:297–318. https://doi.org/10.1243/14680874JER05005.

[74] Ban-Weiss GA, Chen JY, Buchholz BA, Dibble RW. A numerical investigation into the anomalous slight NOx increase when burning biodiesel; a new (old) theory. Fuel Process Technol 2007;88:659–67. https://doi.org/10.1016/j.fuproc.2007.01.007.

[75] Lopes AFG, Talavera-Prieto MDC, Ferreira AGM, Santos JB, Santos MJ, Portugal ATG. Speed of sound in pure fatty acid methyl esters and biodiesel fuels. Fuel 2014;116:242–54. https://doi.org/10.1016/j.fuel.2013.07.044.

[76] Gouw TH, Vlugter JC. Physical properties of fatty acid methyl esters IV Ultrasonic sound velocity. J Am Oil Chem Soc 1964;41:142–5. https://doi.org/10.1007/BF02898122.

[77] Lapuerta M, Agudelo JR, Prorok M, Boehman AL. Bulk modulus of compressibility of diesel/biodiesel/HVO blends. Energy Fuels 2012;26:1336–43. https://doi.org/10.1021/ef201608g.

[78] Tat ME, Van Gerpen JH, Soylu S, Canakci M, Monyem A, Wormley S. Speed of sound and isentropic bulk modulus of biodiesel at 21°C from atmospheric pressure to 35 MPa. J Am Oil Chem Soc 2000;77:285–9. https://doi.org/10.1007/s11746-000-0047-z.

[79] Schönborn A, Ladommatos N, Williams J, Allan R, Rogerson J. The influence of molecular structure of fatty acid monoalkyl esters on diesel combustion. Combust Flame 2009;156:1396–412. https://doi.org/10.1016/j.combustflame.2009.03.011.

[80] Zheng L. Oxy-fuel combustion for power generation and carbon dioxide (CO_2) capture. Woodhead Publishing; 2011. https://doi.org/10.1533/9780857090980.

[81] Knothe G, Sharp CA, Ryan TW. Exhaust emissions of biodiesel, petrodiesel, neat methyl esters, and alkanes in a new technology engine. Energy Fuels 2006;20:403–8. https://doi.org/10.1021/ef0502711.

[82] Nabi MN, Akhter MS, Shahadat MMZ. Improvement of engine emissions with conventional diesel fuel and diesel-biodiesel blends. Bioresour Technol 2006; 97:372–8. https://doi.org/10.1016/j.biortech.2005.03.013.

CHAPTER 3

NO_x and PM trade-off in IC engines

Ashwin Jacob[a], B. Ashok[a], R. Vignesh[a], Saravanan Balusamy[b], and Avinash Alagumalai[c]
[a]Engine Testing Laboratory, School of Mechanical Engineering, Vellore Institute of Technology, Vellore, Tamil Nadu, India
[b]Department of Mechanical and Aerospace Engineering, Indian Institute of Technology Hyderabad, Hyderabad, India
[c]Department of Mechanical Engineering, GMR Institute of Technology, Rajam, Andhra Pradesh, India

3.1 Introduction

For the past decade, global air quality has been degrading at an alarming level due to the emissions originating from the transportation sector as well as extensive industrialization. In a survey by the US Environmental Protection Agency (EPA), it was estimated that gaseous pollutants and particulate pollutants from transportation vehicles were responsible for at least 80% of greenhouse gases and smog, and 11% of this was methane based. Apart from these gases, emissions such as nitrogen oxides (NO_x) and particulate matter (PM) from internal combustion (IC) engines are responsible for climate change while also affecting human health in concentrated forms [1]. NO_x and PM emissions from IC engines are considered to be among the most dangerous pollutants, as they can cause photochemical smog and acid rain as well as pose a major threat to the ozone layer that protects the Earth from harmful ultraviolet radiation. Also, excessive amounts of NO_x and PM emissions can cause cardiovascular and respiratory ailments to humans and other animals. The nature of combustion in IC engines is of utmost importance, as it intricately affects emissions and performance. There are ways to optimize the pollutants emitted to adapt to a greener and cleaner regime of emission regulations [2]. The formation of NO_x, its kinetics, and its formation are discussed in depth in the previous chapter. However, the formation of PM and its relationship with NO_x requires a deeper understanding. In a spark ignition (SI) engine, the combustion process involves a premixed charge intake that is ignited by a spark plug after sufficient compression. Once the air-fuel mixture is ignited, a flame front is produced that consumes the charge ahead until it reaches the walls, where it is extinguished. Hence, the initial reactant gas mixture gets converted into a high-temperature and low-density gas mixture on combustion. The flame compresses the unburnt

NOx Emission Control Technologies in Stationary and Automotive Internal Combustion Engines
https://doi.org/10.1016/B978-0-12-823955-1.00003-6

gases ahead of it into an ever-decreasing volume, resulting in nonlinear combustion. This nonlinear combustion together with compression action is responsible for the higher temperatures produced toward the end. Such high temperatures are favorable for NO_x formation in a spark-ignition engine [3]. Similarly, in compression ignition (CI) engines, liquid fuel is injected into a hot compressed-air chamber. The fuel is then atomized and vaporized, leading to combustion. The combustion mode is nonpremixed and occurs at a wide range of equivalence ratios. Initially, due to lower temperatures for auto-ignition, the diesel fuel evaporates much more quickly than combustion takes place because of the high temperature and energy needed to form intermediate species. This results in uncontrolled combustion and is thus responsible for the high heat release rate, localized high-temperature zones, and NO_x formation. Then comes mixing controlled combustion, where most of the heat is released. During this phase, if the combustion is retarded, the low temperature in-cylinder conditions facilitate PM formation. The PM, which mainly consists of graphite-such as carbon rings, tends to have poor affinity for oxidation. It is evident that there is an inversely opposing relationship between NO_x and PM emissions, as conditions that increase one emission tend to decrease the other. In order to control these emissions, a trade-off should be finalized so that at least one emission can be effectively controlled [4]. Hence, this chapter intends to provide insight into how NO_x and PM can be controlled simultaneously and also as a trade-off with certain precombustion and postcombustion techniques in IC engines. In recent years, various studies have been done to control these emissions in IC engines. One of the most prominent ways to improve the trade-off relationship is by incorporating oxygenated alternative fuels at various compositions with conventional fuels to improve the combustion characteristics. This can be assisted by the homogenization of fuel additives to further enhance the composition of the fuel. Furthermore, engine input parameters such as injection pressure, injection timing, injection mass, and exhaust gas recirculation (EGR) rate can be calibrated effectively to achieve the desired emissions without sacrificing other performance parameters such as thermal efficiency and fuel consumption [5]. Of late, injection strategies such as split and multiple injections for the pilot, main, and post phases are being widely explored to control NO_x and PM emissions during combustion. In addition, facilitating low-temperature combustion using advanced combustion modes such as homogeneous charge compression ignition (HCCI), reactivity controlled compression ignition (RCCI), and premixed charge compression ignition (PCCI) modes is widely being explored as an effective method to inhibit NO_x formation during combustion. Similar to the precombustion emission control techniques, the NO_x and PM emissions that escape the combustion chamber can effectively be controlled using after-treatment devices such as a selective catalytic reactor (SCR) and a regenerative diesel particulate filter (DPF). The simultaneous reduction of NO_x and PM emissions can be facilitated by incorporating a combination of pre-/postcombustion emission control techniques at various operating conditions in IC engines [6]. This chapter intends to discuss the trade-off relationship between NO_x and PM emissions and their variations with respect to the type of fuel and engine input parameters used at various operating conditions. Furthermore, studies that have incorporated these techniques to improve the NO_x-PM trade-off during and after the combustion process to control these emissions are tabulated and discussed. Also, the limitations and challenges in the simultaneous reduction of NO_x and PM emissions are briefly discussed to portray some possibilities for future scope.

3.2 Legislative norms aimed at controlling vehicular emissions

In the last couple of decades, organizations such as the International Union of Air Pollution Prevention and Environmental Protection Associations (IUAPPA) and the World Health Organization (WHO) have voiced their concerns over the alarming rise of harmful gases in the atmosphere, predominantly due to vehicular emissions, and their adverse effects on human health and the climate. In a report from WHO in 2017, nearly 4.2 million deaths are estimated due to the presence of particulate matter smaller than 2.5 μm ($PM_{2.5}$) in the atmosphere [7]. As per the air quality guidelines (AQG) developed by WHO, the $PM_{2.5}$ concentration should not exceed 10 μg/m^3 as the annual mean to stabilize the death rate and the climatic changes incurred due to this pollutant. To achieve this target, legislative policies were implemented to improve air quality. Of the multiple long-term policies enforced, emission policies such as the Bharat Stage (BS) and European emission norms play major roles in curbing vehicular emissions. BS norms are emission standards established by the Indian government to monitor the exhaust output from IC engines. The standards were implemented by the Central Pollution Control Board (CPCB) in reference to European regulations. These stringent norms were progressively enforced in the course of the last decade with varying control levels of harmful gases and pollutants such as NO_x, HC, CO, and PM with respect to the fuel used to power the vehicle. From inception, three levels of BS norms—BS 3, BS 4, and BS 6—have been enacted in lieu of the Euro 4, Euro 5, and Euro 6d, norms, respectively. As the levels progress, the stringency of the norms becomes evident based on the allowable levels of gaseous and particulate pollutants. For instance, Table 3.1 highlights the allowable levels of BS 4 and BS 6 emission standards for gasoline-fueled spark ignition (SI) engines and diesel-fueled compression ignition (CI) engines. As described, the levels of BS 4 emissions are more lenient compared to BS 6 and similar to Euro 4 and Euro 6d norms. Hence, it is important to understand the concepts that assist in reducing NO_x, PM, and other emissions in IC engines to meet these stringent norms and subsequently improve the air quality. The following sections will delve into the control strategies involved in reducing NO_x and PM emissions, as they contribute to at least 70% of all emissions.

3.3 NO_x reduction techniques in IC engines

Generally, about 56% of NO_x emissions are produced by IC engine vehicles, whereas the rest can be attributed to power-generating plants or chemical factories. During combustion, about 95% of the NO_x formed is nitric oxide (NO) and only 5% is nitrogen dioxide (NO_2). An in-depth explanation of

Table 3.1 Current legislative emissions of BS 6/Euro 6d and their differences with the BS 4/Euro 4.

Emission norm	Fuel type	Vehicular pollutants	Allowable levels (mg/km)
Bharat Stage 6 and Euro 6d	Diesel	Nitrogen Oxides (NO_x)	80
		Particulate Matter (PM)	4.5
		Carbon Monoxide (CO)	50
		Hydrocarbons and NO_x Combined	170
	Gasoline	Nitrogen Oxides (NO_x)	60
		Particulate Matter (PM)	4.5
		Carbon Monoxide (CO)	100
Bharat Stage 4 and Euro 4	Diesel	Nitrogen Oxides (NO_x)	250
		Particulate Matter (PM)	25
		Carbon Monoxide (CO)	50
		Hydrocarbons and NO_x Combined	300
	Gasoline	Nitrogen Oxides (NO_x)	80
		Particulate Matter (PM)	–
		Carbon Monoxide (CO)	100

NO_x emission formation during combustion and its kinetics is provided in Chapter 2. However, both these compositions are dangerous to the atmosphere and human health. Therefore, controlling any form of NO_x emission is crucial. In this context, NO_x reduction techniques in IC engines can be divided into two groups. The first group involves precombustion processes such as the calibration of engine input parameters and the utilization of alternative fuels and oxygenated additives. The second group involves postcombustion processes such as the employment of after-treatment devices and other techniques. Table 3.2 highlights some of the techniques used in reducing NO_x emissions along with their effectivity in terms of the percentage of NO_x reduced and the fuel consumed during the process. This section briefly discusses the reduction of NO_x emissions using different techniques in IC engines.

3.3.1 Role of precombustion engine parameters and oxygenated fuels on NO_x control

As discussed in the earlier chapter, the formation of NO_x emissions is primarily due to high-temperature combustion leading to the enormous formation of NO and NO_2. The formation of NO_x emissions can be restricted by effective calibration and strategic utilization of fuel injection parameters in CI engines as well as other parameters such as ignition timing and equivalence ratio in SI engines before combustion takes place. Moreover, of these parameters, certain common engine operating parameters and design factors such as injection pressure, injection timing, split injections, and combustion chamber design contribute to controlling NO_x emissions effectively. Furthermore, the use of oxygenated alternative fuels and emulsion techniques can also help reduce NO_x emissions in IC engines [8]. In an SI engine, the effective calibration of ignition timing is

Table 3.2 Techniques to reduce NO_x emissions in IC engines with their effectivity.

NO_x control technique	Change in fuel consumption	NO_x emissions reduction
EGR	4%–5%	46%–54%
Indirect injection	3%–4%	45%–50%
Injection retardation	2%–4%	20%–35%
Inlet temperature reduction	2%–5%	12%–16%
Pre-chamber designs	9%–15%	45%–50%
Water injection	0%–3%	38%

crucial, as it controls spark activation with respect to the piston position and crankshaft angle at the end of the compression stroke to combust the injected fuel. Upon retarding the ignition timing, the gas residence within the combustion chamber is reduced, which in turn lowers the peak in-cylinder temperature and pressure. Ideally, up to 50% of NO_x emissions can be controlled with respect to the piston position, favorably at the top dead center. However, this usually comes at a cost of compromising the power generated as well as reduced engine efficiency. Similarly, the equivalence ratio in the SI engine plays an important role in influencing engine emissions, as it quantifies the air-fuel mixture ratios to be either lean or rich. Generally, rich mixture combustion leads to a significant reduction of NO_x emissions with increased emissions of HC, CO, and PM. A lean mixture generally reduces all emissions nominally because most engines do not operate at the ideal stoichiometric condition. However, significant improvements in prechamber designs might enable traditional IC engines to operate in lean-burn conditions and reduce NO_x emissions without compromising any other engine output characteristics. The reduction of NO_x emissions in CI engines is more complicated than SI engines due to the inability to control the air-fuel mixture ratio accurately and the diffusive combustion of liquid fuels, which generates high combustion temperatures [9]. In a CI engine, calibrating split injection strategies with respect to injection mass, injection timing, and injection pressure optimally with oxygenated operating alternative fuel can drastically reduce NO_x emissions developed during combustion. An injection mass of 0%–75% has the ability to form a homogenous mixture, thereby facilitating better mixing with the oxygen and consequently expediting low-temperature complete combustion, which reduces NO_x, HC, CO, and PM emissions in CI engines. Similarly, advancing the injection timing duration between the split injections results in the conversion of the heterogeneous mixture to a homogenous mixture due to low-temperature combustion. Although advancing the injection timing reduces NO_x emissions, the performance output characteristics of the engine are drastically affected. Ideally, an injection pressure of 300–350 bar tends to deliver optimal performance, and emission characteristics such as advancing the pressure further might result in higher peak pressure, which impacts NO_x emissions negatively. Utilizing alternative fuels such as oxygenated biodiesel at 20% (v/v) with conventional diesel fuel in combination with advanced injection timing in a split injection strategy can induce low-temperature combustion and consequently reduce NO_x emissions of up to 100 ppm. Table 3.3 highlights some studies that have incorporated

Table 3.3 Notable studies that have utilized oxygenated fuels to control NO_x emissions in CI engines.

Feedstock	Blend composition	Test conditions	NO_x emissions compared to diesel	Reference
Camelina oil	B0, B20, B30, B100	Varied engine speed (1000–4000 rpm)	Increases with blend percentage and engine speed	[8]
Soapnut oil	B0, B10, B20, B30, B40	Varied engine load (0%–100%)	Decreases with increase in blend percentage	[9]
Microalgae oil	B0, B20	Varied engine speed (1200–2700 rpm)	Decreases while operating on B20 blend	[10]
Soybean	B0, B20	Varied engine load (0%–100%)	Increases with blend percentage and engine load	[11]
Swine Lard	B0, B25, B50, B75	Varied load (50, 100, 150 Nm) and speed (1500, 3000 rpm)	Decreases with increase in speed	[12]
Rapeseed oil	B0, B20	Varied engine speed (1400, 2400 rpm), at full load	Decreases at low speed, increases at high speed	[13]
Waste oil	B0, B50,B100	Varied engine load (0, 3, 6, 9 KW)	Increases with blend percentage and engine load	[14]
Pongamia oil	B0, B10, B20, B30	Constant speed (1500 rpm), varied load (0, 25, 50, 75, 100%)	Increases with blend percentage and engine load	[15]
Waste cooking oil	B0, B20, B50	Varied speed (1200–2400 rpm)	Increases with blend percentage and engine speed	[16]
Rice bran	B0, B20, B100	Constant speed (1500 rpm)	Decreases with blend percentage	[17]

NO_x emission control techniques in CI engines [10–17]. The trend of the studies shows that employing oxygenated biofuels in combination with pre-combustion NO_x emission control techniques has proven effective in abating these emissions. In addition, facilitating low-temperature combustion using advanced combustion modes such as homogeneous charge compression ignition (HCCI), reactivity controlled compression ignition (RCCI), and premixed charge compression ignition (PCCI) modes in tandem with water injection or emulsion is widely being explored as an effective method to inhibit NO_x formation during combustion. An in-depth explanation of this subject is provided in Chapter 14. The next section discusses the NO_x control techniques postcombustion in IC engines.

3.3.2 Postcombustion NO_x emission control techniques in IC engines

NO_x emissions emanating from the combustion chamber through the exhaust manifold contain a concentrated species of NO_x compound and are extremely harmful to human health and the environment. Hence, it is imperative that these gases be regulated and controlled. The commonly used NO_x control techniques postcombustion are exhaust gas recirculation (EGR), selective catalytic reduction (SCR), and catalytic converters. EGR works by recirculating the exhaust gases back into the intake manifold after removing the excess oxygen in the residual exhaust, thereby lowering the in-cylinder temperature and consequently creating an anti-NO_x environment. However, EGR tends to increase PM and smoke emissions due to incomplete combustion. This conflict of compromise is also known as the NO_x-PM trade-off, as when one is suppressed, the other pollutant is not controllable. EGR can be optimized and calibrated for the effective control of NO_x emissions when this system is combined with precombustion NO_x control parameters. Most of the studies that have used EGR and oxygenated fuels such as biodiesel or alcohols have shown a reduction of about 50%–75% for NO_x emissions with a sharp increase of PM and smoke emission of about 6%. SCR is an advanced after-treatment device that is usually installed between the exhaust manifold and the end of the tailpipe [18]. It works by injecting a liquid reducing agent such as urea onto the exhaust gases through a catalyst and reduces them to harmless neutral gases, as shown in Fig. 3.1. For instance, when NO_x emissions are emitted, the reducing liquid agent such as urea is injected through a metal oxide catalyst such as

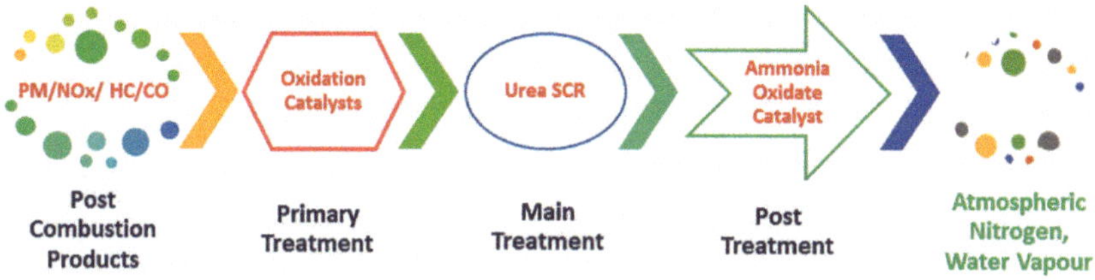

Fig. 3.1 General working principle of a urea SCR for reducing NO_x emissions.

platinum oxide or aluminum oxide onto the gases, reducing them into free nitrogen and water vapor, as shown in Eq. (3.1).

$$NO + NO_2 + 2NH_3\, N_2 + H_2O \tag{3.1}$$

$$NO + O_2 + HC\, N_2 + H_2O + CO_2 \tag{3.2}$$

The conventional SCR uses urea as its reducing agent. However, when gaseous reductants such as methane are used, they are called HC-SCR and they work with a similar principle as shown in Eq. (3.2). The effectiveness of controlling NO_x emissions depends on the type of catalyst and reducing agent used against the volume of gases given out. Typically, SCR systems are capable of controlling 70%–90% of the NO_x emissions. A catalytic converter is another postcombustion after-treatment device to control gaseous emissions with a similar working principle as an SCR system. However, a catalytic converter works by using a catalyst such as platinum or palladium to initiate a redox (reduction and oxidation) reaction to convert the harmful gases into neutral gases. Catalytic converters are of two types, that is, a two-way catalytic converter and a three-way catalytic converter. A two-way catalytic converter works by combining oxygen with harmful emissions such as CO and HC into CO_2 and water vapor. However, a three-way catalytic converter, in addition to reducing gases such as CO, also reduces NO_x emissions into free nitrogen and water vapor while the engine is operating at stoichiometric or rich mixture conditions. In CI engines, a diesel oxidation catalyst (DOC) is used that is mostly comprised of palladium, aluminum oxide, and platinum, which also oxidizes PM, HC, and CO with oxygen to give out CO_2 and water vapor. Although NO_x emissions are a major pollutant, certain particulate contaminants such as PM, smoke, and soot are secondary pollutants that are harmful as well. The next section briefs the different particulate contaminants from IC engines along with their nature and morphology.

3.4 Differences in PM emissions based on their nature and size

The emissions from any IC engine contain a mixture of both gaseous and particulate pollutants. The particulate pollutants vary in size based on their nature; they can be classified as soot, smoke, and PM. Particulate matter comprises the particle pollutant that is the smallest and inhalable; they are small enough to not be visible at a scale of 2.5 μm ($PM_{2.5}$). Soot and smoke emissions are a collection of bigger, dense particulate pollutants that are visible to the human eye, as they are larger than 10 μm. Smoke emissions are an unwanted byproduct of combustion that takes place in the presence of oxygen. Usually, high-temperature combustion in the cylinder facilitates complete combustion and lowers smoke emission formation, as it is formed due to the incomplete combustion of the injected fuel and low-temperature combustion. However, these conditions favor the formation of NO_x emissions, hence the trade-off effect arises. For the most part, smoke emissions mostly comprise ash and particles of condensed aerosols. Moreover, combustion that occurs with a minimum supply of oxygen tends to produce harmful unregulated emissions along with the particulates such as aldehydes and ketones due to the partial oxidation of carbon and nitrogen-based compounds. The particle sizes in smoke emissions might vary from 2.5 to 250 nm depending on the type of combustion that produces them. On the other hand, soot emissions are primarily dependent on the fuel composition undergoing combustion within the engine, as concentrations of naphthenes, benzene, and alkenes in the fuel tend to produce more soot. Also, the order of soot-producing tendencies varies drastically on the flame topography for aliphatic fuels. Similar to smoke emissions, soot is produced due to the incomplete combustion of fuels from precursory acetylene at the corners of the flame. The average size of agglomerated soot nanoparticles ranges between 6 and 30 nm in diameter, which is smaller compared to smoke particles. Nanoadditives or metal oxide additives can be added to the fuel to reduce or inhibit soot formation. PM emissions are considered the most dangerous aerosols as they are small and can be easily inhaled without notice. These particle pollutants comprise a mixture of soot and smoke in smaller quantities as compared to higher quantities of particles below 2.5 μm. PM composition has both solid and liquid fractions upon condensing during the dilution process. Most are made of carbonaceous solid hydrocarbons and polynuclear aromatic hydrocarbons (PAH). PM is primarily produced by low-temperature combustion and is also influenced by the fuel

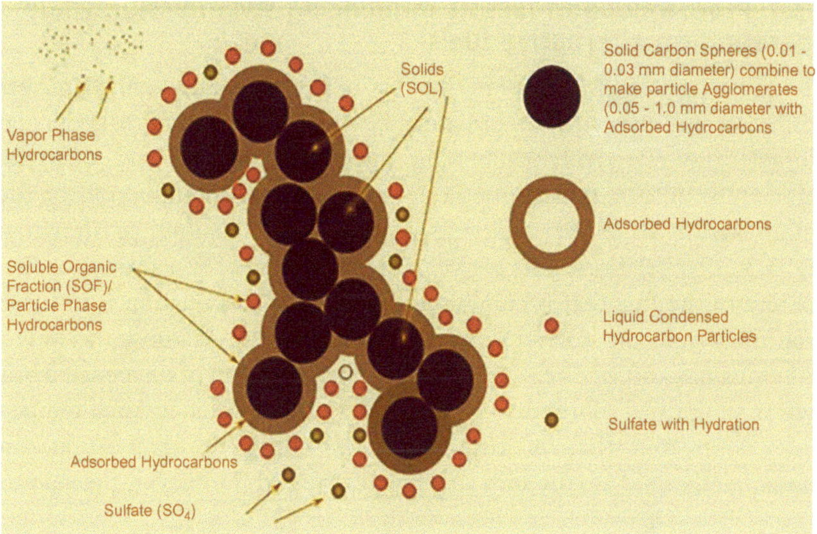

Fig. 3.2 Modes of PM emission formed after combustion.

composition and residual oxygen content in the cylinder. CI engines produce particulate emissions such as smoke and PM more than in SI due to the aliphatic nature of the fuel combusted. Fig. 3.2 illustrates the PM modes formed during combustion such as nucleation and agglomeration. Also, the contrasting effects of PM formation and NO_x emissions have been explored in recent years to arrest both harmful emissions without any trade-off between them. As discussed in the earlier section, techniques similar to NO_x control are available for PM during pre-/postcombustion, which are discussed in the following section.

3.5 PM control techniques in IC engines

PM emissions released from IC engine tailpipes are extremely dangerous and need to be effectively controlled to prevent devastating effects on air quality and human health. In this context, certain control pre-/postcombustion techniques are employed similar to the techniques followed for NO_x control in IC engines. This section discusses the various techniques followed for controlling PM before and after the combustion process.

3.5.1 Precombustion factors influencing PM emission while operating on alternative fuels

The formation of PM emissions can be inhibited by employing precombustion techniques such as utilizing oxygenated additives in alternative biofuels and effectively calibrating injection strategies and operating parameters. Also, auxiliary methodologies such as air induction or fuel emulsification have the ability to control PM emissions, as shown in Fig. 3.3. A correlation exists between the formation of PM and the oxygen content of the fuel that is combusted. Generally, this relation is inversely proportional, as the inherent oxygen content in fuels is known to reduce PM emissions and vice versa. Hence, the incorporation of oxygenated biofuels is an optimal solution to reduce PM because a higher flame temperature is created that in turn facilitates complete combustion and consequently oxidizes the soot and negates the PM. However, the amount

Fig. 3.3 Different PM control techniques during pre-/postcombustion.

of PM that can be controlled depends upon the physicochemical properties such as viscosity and density [19]. Apart from the oxygen content in the fuel, factors such as the carbon chain length, unsaturation degree, and aromatic concentration of the fuel constituents affect PM formation. To boost the effectivity of the fuel utilized in PM control, oxygenated additives such as dimethyl ether, diethylene glycol, dimethyl carbonate, etc., are homogenized with the fuel to enhance combustion characteristics. This consequently improves the cetane number and reduces the ignition temperature, which creates particulates. Ultimately, the reformulated fuel will be comprised of a low carbon-to-hydrogen ratio and be rich in oxygen content, which is adequate to reduce PM formation upon combustion. To get optimal emission characteristics, it is necessary to calibrate the engine operating parameters such as injection pressure, injection timing, load, and engine speed. Injection pressure is one of the important parameters in controlling PM emissions, as higher pressure leads to finer atomization and air-fuel mixture homogeneity of the injected fuel, which results in finer combustion characteristics leading to significant PM reduction. However, a lower injection pressure results in inferior fuel atomization, which results in partial low-temperature combustion that favors PM formation. Similarly, injection timing is an important factor that is responsible for PM control, as retarding injection timing facilitates early injection. This consequently reduces the combustion temperature and pressure, as the start of combustion occurs earlier and fuels burns after they go beyond the top dead center. On the contrary, with advancing injection timing, the adiabatic flame temperature is at its maximum; also, the time for soot oxidation is reduced in the expansion stroke [20]. Hence, superior mixture formation for homogenous combustion is facilitated, which consequently reduces primary and secondary PM emissions due to complete combustion. Also, while operating at full load and speed, the oxygen content in the biofuel influences the soot more than the viscosity; hence, the PM formation is reduced with the increase in oxygen content. However, at lower loads, the fuel viscosity overpowers the oxygen factor, which consequently increases PM emissions. An auxiliary air induction technique at the end of injection also alters the nature of turbulence mixing within the cylinder and ensures high-temperature combustion, which reduces PM emissions. Similarly, emulsion techniques with water facilitate microexplosion events that increase the premixed combustion duration and prolong ignition delay, thereby facilitating more time for the air-fuel mixture and eventually providing a favorable environment for PM reduction. Another technique to control PM is by formulating a strategy for split or multiple injections and calibrating the associated injection parameters to achieve the desired engine output characteristics. Post injection timing and a fuel mass of at least 10% between 10 and 50° crank angle (CA) is more important than the pilot and main injections to sufficiently increase the exhaust gas temperature, which in turn assists in the regeneration phase in diesel particulate filters (DPF) in the postcombustion phase. The techniques of controlling PM after the combustion process are discussed in the next section.

3.5.2 Influence of postcombustion PM emission control techniques in IC engines

It is important to control the PM emissions channeling through the exhaust manifold after the combustion process, as these harmful pollutants deteriorate the atmosphere. In addition to the precombustion techniques, after-treatment devices such as DPF and its regeneration strategies are primarily employed to control PM emissions. Apart from the primary techniques, secondary methods such as regeneration-assisted fuel additives have also been explored in recent years. DPF is an after-treatment device that consists of a complex net of honeycomb-designed ceramic wall flow filters along with a coated oxidation catalyst placed at the start of the unit, as shown in Fig. 3.4. The particulate matter that is trapped in the porous walls is eliminated by a process called active regeneration, where the temperature of the exhaust gases flowing through the DPF is increased using a controlled signal from the engine electronic control unit (ECU), which incinerates the soot particles by oxidizing them. This rise in exhaust gas temperature of about 500°C

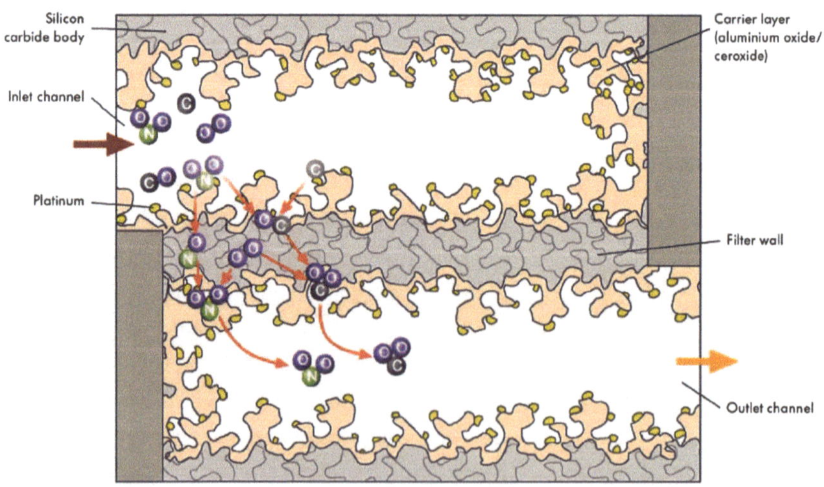

Fig. 3.4 Structure of DPF for trapping PM emissions.

is facilitated by postinjection fuel mass in most cases toward the end of the expansion stroke to activate regeneration without damaging the walls of the DPF. Another method to activate regeneration is by throttling the engine enough to let more intake air inside and allowing more fuel to be injected as well as increasing the exhaust gas temperature to clear the walls of the DPF. Similarly, an auxiliary diesel burner equipped with a differential pressure sensor maintains the back-pressure value of the filter and signals the ECU when the filter is loaded with soot. The burner is activated and incinerates the soot at 650°C when the pressure sensor indicates 75 mmHg. Alternatively, an electric regeneration system can be used without any complex connections to the ECU. The heating element is powered by the engine alternator, which turns on periodically when the back-pressure value crosses its limit and initiates the regeneration cycle. Apart from the complex and expensive active regeneration DPF strategies, passive regenerative DPF strategies use catalysts to inhibit the oxidation temperatures of soot, which is equivalent to regular exhaust gas temperature, thereby eliminating complex extra hardware additions. These catalysts can be coated within the passage of the DPF channel or by oxidizing metal oxide additives such as cerium oxide, zinc oxide, etc., to the fuel to lower the soot oxidation temperatures [21]. Another passive DPF strategy is to put oxidation catalysts at the entrance of the ceramic particulate filter, which works continuously without any mass accumulation within the filter. This technique is also known as a continuous regeneration trap (CRT). It can be modified to facilitate the simultaneous reduction of NO_x-PM with the help of a combination of after-treatment devices such as SCR and catalytic converters as well as calibrating the injection and operating parameters optimally. From the above sections, it is clear that the control of NO_x-PM is based on a concept of trade-off. The following section discusses in depth the trade-off relationship between NO_x and PM emissions.

3.6 Trade-off relationship between NO_x and PM emissions in IC engines

As discussed in the previous sections, it is evident that NO_x and PM emissions have an opposing relationship while functioning at the same operating conditions. Due to this inverse relationship, a trade-off between the two emissions is imperative to effectively control at least one of the two emissions. This is because the conditions favorable to reduce NO_x emissions tend to boost the production of PM emissions and vice versa. For instance, combustion conditions that facilitate high in-cylinder temperature and pressure enable thermal oxidation between the atmospheric nitrogen from the intake air and the oxygen content in the fuel to produce nitrogen oxides (NO and NO_2). However, at the same higher in-cylinder temperature and pressure, the emissions decrease drastically due to complete combustion of the fuel and the attainment of the soot oxidation temperature. This trade-off relationship depends upon the type of fuel used, the status of the engine input parameters, and other operating conditions. This section intends to discuss the trade-off relationship using various techniques during the combustion process.

3.6.1 Improving NO_x-PM trade-off in IC engines

Generally, the trade-off between NO_x and PM emissions comes at a cost of excessive fuel consumption and reduced power delivered by the engine. Hence, the effective trade-off between NO_x and PM emissions also requires achieving optimal output performance characteristics. The trade-off relationship can be achieved with techniques and approaches discussed in the sections above for NO_x and PM emissions. These approaches would actively control either NO_x emissions or PM emissions individually. However, by combining certain emission control approaches and utilizing them strategically, a reasonable level of NO_x and PM emissions can be controlled, even though only one of the two can be dominantly controlled [22]. For the most part, the trade-off between NO_x and PM is facilitated using injection strategies and EGR at optimal levels by calibrating the injection parameters and EGR rate. Typically, NO_x and PM emissions are highly influenced by the air-fuel mixture and the gas temperature within the combustion chamber. Split injection or multiple injection strategies are the most popular techniques to control NO_x and PM emissions, as they improve the mixture homogeneity within the cylinder. Similarly, EGR rates can be optimally calibrated to regulate the oxygen concentration in the combustion mixture and dilute the intake air. Also, the in-cylinder temperature can be reduced using recirculated carbon dioxide and water vapor, which are the products of combustion, back into the engine to lower the combustion temperature. This recirculation also delays the oxidation temperature of nitrogen by reducing the peak firing temperature and consequently reducing the formation of NO_x components. However, the use of this technique aggravates the production of PM emissions [23]. To put it in perspective, a study done using these two techniques is briefly discussed to understand the NO_x-PM trade-off relationship. In this study by Sindhu et al., a two-stage split injection strategy was proposed to improve the emission and performance characteristics by comparing and contrasting conventional emission control techniques. Generally, retarding injection timing leads to a decrement in NO_x emissions at the cost of extensive fuel consumption as well as higher smoke emissions and exhaust gas temperatures. Similarly, higher EGR rates usually tend to reduce NO_x emissions but deteriorate the combustion quality due to the incomplete combustion of the injected fuel. In this study, the injection strategy followed involves two split injections and a single injection at 16 °bTDC where the first pulse injects 25% of the fuel and the second pulse injects 75% of the fuel. The dwell time for the study was maintained at 8° of CA with a standard pilot and main injection fuel mass. On operation, the results show that NO_x emissions were reduced significantly for split injection mode but had a marginal reduction with single injection. However, PM emissions increased for split injection when the first injection was 25% and 75% for the second whereas the contrary was observed when 75% was injected first and 25% fuel was injected in the second pulse. This is because the premixed combustion phase is at its highest for single injection followed by 75%–25% injection and 25%–75%, consequently increasing the in-cylinder temperature. This can also be explained by the higher gas resident time for larger fuel quantities in the combustion chamber. At these conditions, NO_x emissions increase and PM emissions reduce, whereas upon reversing these conditions, PM emissions increase and NO_x emissions decrease [24]. Hence, this relationship requires a trade-off to effectively control at least one of these emissions. The next section discusses the role of oxygenated fuels as a candidate to facilitate the NO_x-PM trade-off.

3.6.2 Role of oxygenated additives and alternative fuels in NO_x-PM trade-off

Apart from the calibration of engine input parameters to control engine emissions, the factor that is directly proportional to the emissions produced is the type of fuel being combusted. Because different fuel compositions dissociate into multiple fractions due to their respective chemical composition, the nature of the fuel being burnt is important to gauge the emission outputs. The conventional liquid fuels and alternative biofuel compositions can be classified under aromatic or aliphatic groups. Unlike conventional fuels such as gasoline and diesel, biofuels tend to contain inherent oxygen content, which makes them an ideal candidate to control emissions as they facilitate complete combustion. These fuels are blended with conventional fuels to upgrade their physicochemical properties and molecular structure, which provides optimal engine output characteristics on operation. Furthermore, certain fuel additives such as antioxidants, nanoparticles, and combustion enhancers further boost the efficiency of these biofuels in obtaining optimal emission characteristics [25]. In order to understand the role of these reformulated fuels in controlling NO_x and PM emissions and their trade-off relationship in IC engines, certain studies are described. Upon the addition of oxygenated lower carbon chained alcohols in small volumes in diesel fuel, the PM emissions from CI engines were reduced due to the superior combustion characteristics delivered due to the longer ignition delay and lower cetane number of these fuels. Moreover, upon adding incremental volumes of these oxygen fractions, the PM emissions further decreased without increasing the NO_x emissions. Upon incorporating EGR with lower carbon chained alcohols, a significant reduction in NO_x emissions is observed as compared to the reductions achieved without EGR. However, a slight increase in PM emissions is notable, especially at higher EGR rates. However, larger volumes of oxygenated alcohol additives reduced the PM emissions due to the facilitation of the postflame oxidation of the PM emitted, which is boosted by improved diffusion combustion during the late expansion stroke. The trade-off relationship between NO_x and PM emissions is evident upon using this combination of fuel mixtures. More importantly, the combination of these additives assisted by effective calibration of the engine input parameters has the ability to control NO_x and PM emissions simultaneously [26]. In another study, a combination of emulsified alcohol-biodiesel-diesel blends coupled with EGR proved an effective mixture for controlling NO_x and PM emissions in CI engines. This emulsion combination reduced the NO_x emissions at a range of loads and speeds due to the higher latent heat and lower heating value of the mixture. This produces a quenching effect within the combustion chamber, consequently reducing the cylinder temperature and the NO_x emissions. Furthermore, the resident time of the high temperature gas within the cylinder is reduced drastically due to the water-alcohol emulsion mixture. Similarly, PM emissions are effectively controlled upon using the emulsion mixture combination as compared to pure diesel operation. This can be explained by the inherent oxygen content of the aromatic fuel mixture, which completely oxidizes itself due to complete combustion. On the other hand, the longer delay caused by this fuel mixture would result in fuel accumulation in the combustion chamber and result in mass combustion, which increases NO_x emissions [27]. Hence, the trade-off relationship between NO_x and PM emissions is evident to some extent mainly in higher loads. However, recent studies have explored the possibilities of reducing NO_x and PM emissions simultaneously in IC engines. The next section will discuss the techniques used to reduce NO_x and PM emissions simultaneously along with their limitations and challenges.

3.7 Simultaneous reduction of NO_x and PM emissions

In recent years, extensive research has been carried out to reduce NO_x and PM emissions simultaneously. The emission control strategy followed depends upon the type of fuel used and the pre-/postcombustion factors involved in the operation. In this context, this section briefly discusses the techniques followed for controlling NO_x and PM emissions simultaneously.

3.7.1 Combined influence of alternative fuels and NO_x-PM control techniques

In order to facilitate the simultaneous reduction of NO_x and PM emissions in IC engines, the individual techniques discussed under NO_x and PM emission control approaches in the previous sections can be combined and optimized for various operating conditions. The simultaneous reduction of NO_x and PM emissions is the facilitation of a suitable environment during the pre-/postcombustion stages that hampers the formation of these harmful pollutants without any significant losses in power delivered or excessive fuel consumption. For instance, a combination of biofuel-oxygenated additive conventional fuel can be used to power an IC engine by calibrating engine input parameters such as injection mass, timing, angle, and EGR rate for split or multiple injection strategies and equipping after-treatment devices such as SCR and DPF to simultaneously reduce NO_x and PM emissions [28]. In recent years, multiple researchers have explored using these emission control techniques in combination and assessed their capabilities in controlling NO_x and PM emissions, as shown in Table 3.4. Furthermore, because the primary source of the origin of these emissions is during the combustion period, modes such as RCCI, PCCI, and HCCI are incorporated with the combination of control techniques. For instance, in a recent study, the impact of the split injection strategy along with the PCCI combustion mode and EGR were explored to assess their ability to control NO_x and PM emissions simultaneously. Engine input parameters such as EGR rate, pilot, and main injection timing were calibrated while operating in PCCI combustion mode. It was found that advancing the start of main injection timing resulted in higher NO_x emissions and lower PM emissions. This can be attributed to the longer duration of fuel mixing, which improved the combustion

Table 3.4 Simultaneous reduction of NO_x and PM emissions in IC engines.

Fuel blend composition	EGR Rate	NO_x emissions	PM emissions	Other emissions	Reference
Gasoline/diesel blends	35%–50%	↓	↓	–	[29]
Isobutanol/n-pentanol/diesel blends	0%–30%	↓≤41.7%	↓≤90.7%	↓ CO, ↑ HC	[30]
n-heptane/gasoline/n-butanol blends	Up to 42%	↓	–	↑ CO, ↑ HC	[31]
Butanol/diesel-B30 blend	30%–50%	↓	↓ 0.19 ppm	↑ CO, ↑ HC	[32]
Gasoline/diesel blends	–	↓≤5 ppm	↓ 2.5 mg/m^3	–	[33]
Diesel/biodiesel/ethanol blend	0%–40%	↓ below 25 ppm	↓	↑ CO, ↑ HC	[34]
Diesel/biodiesel (B20, B30)	0% and 25%	↓	↓	–	[35]

efficiency and consequently increased the NO_x emissions slightly. For a specific start of main injection timing, advancing the start of pilot injection timing results in a simultaneous reduction of NO_x and PM emissions. However, advancing it beyond 36 °bTDC led to an increase in NO_x emissions due to the longer residence time of the air-fuel mixture, which consequently lowers PM emissions even further. Also, introducing the EGR rate in smaller amounts up to 15% resulted in the simultaneous reduction of NO_x and PM emissions to an extent. The NO_x emissions were reduced primarily due to the lowered in-cylinder pressure caused by the increase in EGR rate and further retarded the combustion phasing due to the lowered chemical kinetics of the air-fuel mixture. This phenomenon further increases the combustion duration and provides sufficient time to oxidize the soot development that consequently reduced PM emissions [36]. Hence, at certain conditions, the simultaneous reduction of NO_x and PM emissions is possible. In another study, the use of large amounts of oxygenated additives such as dimethoxy methane with diesel fuel at advanced injection timing and EGR rate can simultaneously reduce NO_x and PM emissions due to the oxygen-rich nature of the additive, which increases the tolerance of the EGR and prevents a trade-off relationship [37]. Similarly, the use of higher carbon chained alcohols in large amounts as oxygenated additives with conventional diesel has a significant effect on controlling NO_x and PM emissions. These mixtures facilitate leaner operation, as these alcohols tend to produce a cooling effect within the cylinder and negate the effects of the longer ignition delay, which tends to increase the in-cylinder temperature produced by the same mixture and consequently reduces PM emissions. Hence, to equalize the negative effects of the oxygenated mixture, EGR is introduced in lower levels, which reduces NO_x emissions. Therefore, at optimal levels of blend composition and EGR rates, the simultaneous reduction of NO_x and PM emissions is possible [38]. Although there are emerging techniques to control both NO_x and PM emissions at the same time, various limitations and challenges are inevitable. The next section highlights some of the limitations and challenges in controlling NO_x and PM emissions simultaneously in IC engines.

3.7.2 Limitations and challenges in simultaneous control of NO_x-PM emissions

The simultaneous reduction of NO_x and PM emissions in IC engines remains a major challenge due to the complexity in understanding the nature of their formation. NO_x formation can be primarily attributed to the thermal oxidation of atmospheric nitrogen in the presence of oxygen in the intake air or the oxygenated fuel while under combustion. Also, high-temperature combustion of the oxygenated or conventional fuel fractions tends to produce NO_x emissions. Approaches such as reducing the compression ratio, ambient gas temperature, and injection timing as well as increasing the charge dilution amount can control NO_x emissions apart from low-temperature combustion modes. However, these low-temperature environments tend to produce more PM emissions and affect thermal efficiency. Hence, this trade-off relationship can be challenging to rectify. From the discussions in the previous sections, it is evident that the possibility of achieving the simultaneous reduction of NO_x and PM emissions is possible without the trade-off relationship. However, through effective calibration of the engine inputs and employing suitable alternative fuels or additives and after-treatment devices, simultaneous reduction is possible in one sitting. Hence, at only one specific range of engine operating conditions, the NO_x and PM emissions will be controlled simultaneously. It is impossible to predict the NO_x and PM emission control efficiency in transient or dynamic conditions. The strategic utilization of DPF and SCR might be effective in controlling both emissions postcombustion process with the penalty of thermal efficiency and excessive fuel consumption. Extensive research on the catalysts for SCR and flow arrangement within the DPF might reduce these emissions to some extent and assist in active regeneration. Even though the possibility of completely negating these emissions altogether is minimal, it is imperative that a novel technique be established to overcome this challenge.

3.8 Conclusion

This chapter discussed how NO_x and PM can be controlled simultaneously and also as a trade-off with certain precombustion and postcombustion techniques in IC engines. Although pollutants from the combustion processes are inevitable, the techniques discussed in this chapter should ideally keep these pollutants within the tolerable limit of the legislative laws. For the most part, the key to control the NO_x and PM emissions is to understand the nature of combustion further and to dedicate substantial research to explore the combustion modes further. A CI engine burns in a homogeneous mode, which produces NO_x and PM at the same time. As the combustion process is highly nonlinear, our efforts to curb pollutants are skewed. From this discussion, it can be inferred that a combination of low EGR with water-alcohol emulsion techniques assisted by split injection strategies and advanced combustion modes is adequate to project an improved NO_x-PM tradeoff relationship with slight penalties in PM emissions. This trade-off relationship can be further improved by using oxygenated biofuels as a secondary fuel along with the primary conventional fuels. The effective reduction of NO_x and PM emissions was enhanced when after-treatment devices such as SCR and DPF with active/passive regeneration techniques were coupled with injection strategies and EGR while operating on oxygenated biofuels. The simultaneous reduction of NO_x and PM emissions was feasible at a set range of operating conditions when pre-/postcombustion emission control techniques were used in combination. However, in real-time operations, the prediction of NO_x and PM emission formation is complex and clearly requires a novel methodology or model to identify and control these emissions. Novel combustion modes such as HCCI, PCCI, and RCCI will have a greater chance of mitigating the pollutants in the future when they are effectively combined with other emission control techniques. The facilitation of these combustion modes would need some altering of design parameters to effectively control NO_x and PM emissions. The next chapter will delve deep into the influence of engine design parameters in controlling NO_x emissions.

References

[1] Nanthagopal K, Raj RT, Ashok B, Elango T, Saravanan SV. Influence of exhaust gas recirculation on combustion and emission characteristics of diesel engine fuelled with 100% waste cooking oil methyl ester. Waste Biomass Valoriz 2019;10(7):2001–14.

[2] Bragadeshwaran A, Kasianantham N, Balusamy S, Muniappan S, Reddy DM, Subhash RV, Pravin NA, Subbarao R. Mitigation of NOx and smoke emissions in a diesel engine using novel emulsified lemon peel oil biofuel. Environ Sci Pollut Res 2018;25(25):25098–114.

[3] Costagliola MA, De Simio L, Iannaccone S, Prati MV. Combustion efficiency and engine out emissions of a SI engine fueled with alcohol/gasoline blends. Appl Energy 2013;111:1162–71.

[4] Ashok B, Nanthagopal K, Vignesh DS. Calophyllum inophyllum methyl ester biodiesel blend as an alternate fuel for diesel engine applications. Alex Eng J 2018;57(3):1239–47.

[5] Jacob A, Ashok B. An interdisciplinary review on calibration strategies of engine management system for diverse alternative fuels in IC engine applications. Fuel 2020;278:118236.

[6] Karthickeyan V, Thiyagarajan S, Geo VE, Ashok B, Nanthagopal K, Chyuan OH, Vignesh R. Simultaneous reduction of NOx and smoke emissions with low viscous biofuel in low heat rejection engine using selective catalytic reduction technique. Fuel 2019;255:115854.

[7] Khodke A, Watabe A, Mehdi N. Implementation of accelerated policy-driven sustainability transitions: case of Bharat stage 4 to 6 leapfrogs in India. Sustainability 2021;13 (8):4339.

[8] Özçelik AE, Aydoğan H, Acaroğlu M. Determining the performance, emission and combustion properties of camelina biodiesel blends. Energ Conver Manage 2015;96:47–57.
[9] Misra R, Murthy M. Performance, emission and combustion evaluation of soapnut oilediesel blends in a compression ignition engine. Fuel 2011;90:2514–8.
[10] Uludamar E. Effect of hydroxy and hydrogen gas addition on diesel engine fuelled with microalgae biodiesel. Int J Hydrogen Energy 2018;43:18028–36.
[11] Shaafi T, Velraj R. Influence of alumina nanoparticles, ethanol and isopropanol blend as additive with dieselesoybean biodiesel blend fuel: combustion, engine performance and emissions. Renew Energy 2015;80:655–63.
[12] Mikulski M, Duda K, Wierzbicki S. Performance and emissions of a CRDI diesel engine fuelled with swine lard methyl estersediesel mixture. Fuel 2016;164:206–19.
[13] Aldhaidhawi M, Chiriac R, Bădescu V, Descombes G, Podevin P. Investigation on the mixture formation, combustion characteristics and performance of a Diesel engine fueled with Diesel, Biodiesel B20 and hydrogen addition. Int J Hydrog Energy 2017;42:16793–807.
[14] Akar MA, Kekilli E, Bas O, Yildizhan S, Serin H, Ozcanli M. Hydrogen enriched waste oil biodiesel usage in compression ignition engine. Int J Hydrogen Energy 2018;43:18046–52.
[15] Kumar S, Dinesha P, Bran I. Influence of nanoparticles on the performance and emission characteristics of a biodiesel fuelled engine: an experimental analysis. Energy 2017;140:98–105.
[16] Aydın S, Sayın C. Impact of thermal barrier coating application on the combustion, performance and emissions of a diesel engine fueled with waste cooking oil biodiesediesel blends. Fuel 2014;136:334–40.
[17] MohamedMusthafa M, Sivapirakasam S, Udayakumar M. Comparative studies on fly ash coated low heat rejection diesel engine on performance and emission characteristics fueled by rice bran and pongamia methyl ester and their blend with diesel. Energy 2011;36:2343–51.
[18] Khair M, Lemaire J, Fischer S. Integration of exhaust gas recirculation, selective catalytic reduction, diesel particulate filters, and fuel-borne catalyst for NO x/PM reduction. SAE Trans 2000;1:1607–13.
[19] Mohankumar S, Senthilkumar P. Particulate matter formation and its control methodologies for diesel engine: a comprehensive review. Renew Sustain Energy Rev 2017;80:1227–38.
[20] Park SH, Yoon SH, Lee CS. Effects of multiple-injection strategies on overall spray behavior, combustion, and emissions reduction characteristics of biodiesel fuel. Appl Energy 2011;88(1):88–98.
[21] Davies C, Thompson K, Cooper A, Golunski S, Taylor SH, Macias MB, Doustdar O, Tsolakis A. Simultaneous removal of NOx and soot particulate from diesel exhaust by in-situ catalytic generation and utilisation of N2O. Appl Catal Environ 2018;239:10–5.
[22] Sukjit E, Herreros JM, Dearn KD, Tsolakis A, Theinnoi K. Effect of hydrogen on butanol–biodiesel blends in compression ignition engines. Int J Hydrogen Energy 2013;38(3):1624–35.
[23] Jung S, Ishida M, Yamamoto S, Ueki H, Sakaguchi D. Enhancement of NOx-PM trade-off in a diesel engine adopting bio-ethanol and EGR. Int J Automot Technol 2010;11(5):611–6.
[24] Sindhu R, Rao GA, Murthy KM. Effective reduction of NOx emissions from diesel engine using split injections. Alex Eng J 2018;57(3):1379–92.
[25] Bertola A, Boulouchos K. Oxygenated fuels for particulate emissions reduction in heavy-duty DI-diesel engines with common-rail fuel injection. SAE Trans 2000;1:2705–15.

[26] Chaichan MT. Improvement of NOx-PM trade-off in CIE though blends of ethanol or methanol and EGR. Diesel Engine 2015;2(12):121–8.
[27] Chang YC, Lee WJ, Wu TS, Wu CY, Chen SJ. Use of water containing acetone–butanol–ethanol for NOx-PM (nitrogen oxide-particulate matter) trade-off in the diesel engine fueled with biodiesel. Energy 2014;64:678–87.
[28] Nakatani K, Hirota S, Takeshima S, Itoh K, Tanaka T, Dohmae K. Simultaneous PM and NOx reduction system for diesel engines. SAE Trans 2002;1:362–9.
[29] Han D, Ickes AM, Bohac SV, Huang Z, Assanis DN. Premixed low temperature combustion of blends of diesel and gasoline in a high speed compression ignition engine. Proc Combust Inst 2011;33(2):3039–46.
[30] Rajesh Kumar B, Saravanan S. Effects of iso-butanol/diesel and n-pentanol/diesel blends on performance and emissions of a di diesel engine under premixed LTC (low temperature combustion) mode. Fuel 2016;170:49–59.
[31] Mao B, Liu H, Zheng Z, Yao M. Influence of fuel properties on multicylinder PPC operation over a wide range of EGR and operating conditions. Fuel 2017;215:352–62.
[32] Han J, Wang S, Somers B. Effects of different injection strategies and EGR on partially premixed combustion. SAE Tech Pap 2018;2018:1–12.
[33] Torregrosa AJ, Broatch A, Novella R, Gomez-Soriano J, Mónico LF. Impact of gasoline and diesel blends on combustion noise and pollutant emissions in premixed charge compression ignition engines. Energy 2017;137:58–68.
[34] Fang Q, Fang J, Zhuang J, Huang Z. Effects of ethanol-diesel-biodiesel blends on combustion and emissions in premixed low temperature combustion. Appl Therm Eng 2013;54(2):541–8.
[35] Huang H, Liu Q, Yang R, Zhu T, Zhao R, Wang Y. Investigation on the effects of pilot injection on low temperature combustion in high-speed diesel engine fueled with n-butanol-diesel blends. Energ Conver Manage 2015;106:748–58.
[36] Jain A, Singh AP, Agarwal AK. Effect of split fuel injection and EGR on NOx and PM emission reduction in a low temperature combustion (LTC) mode diesel engine. Energy 2017;122:249–64.
[37] Huang ZH, Ren Y, Jiang DM, Liu LX, Zeng K, Liu B, Wang XB. Combustion and emission characteristics of a compression ignition engine fuelled with diesel–dimethoxy methane blends. Energ Convers Manage 2006;47(11–12):1402–15.
[38] Kumar BR, Saravanan S, Rana D, Anish V, Nagendran A. Effect of a sustainable biofuel–n-octanol–on the combustion, performance and emissions of a DI diesel engine under naturally aspirated and exhaust gas recirculation (EGR) modes. Energ Conver Manage 2016;118:275–86.

CHAPTER 4

Effect of engine design parameters in NO_x reduction

R. Sakthivel[a], S. Sidharth[b], P. Ganesh Kumar[c], T. Mohanraj[a], A. Tamilvanan[d], and B. Ashok[e]

[a]Department of Mechanical Engineering, Amrita School of Engineering, Amrita Vishwa Vidyapeetham, Coimbatore, Tamil Nadu, India

[b]Robert Bosch Engineering and Business Solutions Pvt Ltd, Coimbatore, India

[c]Alstom Transport India Ltd, Chennai, India

[d]Department of Mechanical Engineering, Kongu Engineering College, Erode, Tamil Nadu, India

[e]Engine Testing Laboratory, School of Mechanical Engineering, Vellore Institute of Technology, Vellore, Tamil Nadu, India

4.1 Introduction

The trade-off between nitrogen oxides (NO_x) and particulate matter (PM) discussed in the previous chapter shows a clear view of how various fuel and design-based techniques affect pollutant formation. The difficulty in maintaining the trade-off paves the way to search for novel techniques to reduce NO_x without sacrificing the PM augmentation. A major part of NO_x is mainly produced due to high in-cylinder temperature, forcing the cleavage of the nitrogen triple bond. The in-cylinder and after-treatment are two important technologies for emission reduction in engine exhaust. Among these technologies, in-cylinder technology provides two methods for reducing pollution: (i) fuel modification and (ii) engine modification. The former deals with the modification of the fuel composition whereas the latter deals with changing the design aspects of the engine parameters. Fuel modification focuses on either finding a fuel to replace fossil fuels entirely or combining a part of another fuel with fossil fuels. The viscosity, calorific value, oxidation stability, and extractability of other fuels put forth an array of problems, including engine modification, which is why blending is considered a suitable alternative [1]. Higher alcohols such as n-pentanol and n-butanol with their high cetane number, energy density, and lower hygroscopic nature assist in blending and facilitate minimizing the drawbacks of biofuels [2]. It also results in better combustion leading to higher in-cylinder temperatures vis-à-vis higher NO_x emissions. Fuel modification faces certain issues such as aging during long storage periods. On the other

NOx Emission Control Technologies in Stationary and Automotive Internal Combustion Engines
https://doi.org/10.1016/B978-0-12-823955-1.00004-8

hand, engine design modifications have a broad array of advantages to provide reduced emissions over a wide variety of fuels.

The primary idea in the engine design modification technique is to control the basic aspects of combustion. This includes optimizing the compression ratio, thermal barrier coating, charge induction, valve design, injector design, manifold design, and the power cycle. The dynamics of charge inside the cylinder are transient, primarily depending on chamber geometry and design. The modern design aspect in the combustion chamber is mainly focused on inducing turbulence to the charge by enhancing the swirl-squish interaction during compression. In addition to that, proper modification of the chamber would have significant effects on phases of heat release owing to the variation in bulk airflow and turbulence in the mixture. This control over heat release phases leads to an effective reduction in NO_x emissions as well as noise. Most current engines are designed with a hemispherical combustion chamber that gives adequate performance when fueled with diesel, whereas it may not prove to be good for alternate fuel tractions. The reentrant-type combustion chamber showed predominant variations in turbulent kinetic energy and swirl rate as compared to that of baseline operation [3]. It is noteworthy that modifying the combustion chamber design positively affects the emissions without affecting the performance in RCCI engines. Meanwhile, heat loss during engine operation leads to performance drops and hikes in certain emissions, irrespective of the fuel used. This can be countered by using thermal barriers inside the cylinders. This barrier is meant to hold the heat inside the chamber and make additional energy available during combustion, thereby positively affecting the performance and emission of the engine. The thermal conductivity of the primary layer plays a vital role because it is responsible for the reduction of heat loss inside the cylinder. Using titanium oxide to coat the engine has proven to reduce NO_x emissions with a slight increase in specific fuel consumption [4].

In addition, a strong correlation exists between the degree of air-fuel mixing and combustion. A lean air-fuel mixture will initiate a higher in-cylinder temperature, thereby increasing the NO_x emissions but lowering the CO emissions [5]. A rich air-fuel mixture will have lower NO_x along with unburnt carbon and CO emissions on the higher side due to incomplete combustion. Therefore, the optimization of the air-fuel ratio is necessary to balance the emissions and reduce them simultaneously. Along with this, phenomena such as atomization of fuel, droplet breakup, collision, momentum exchange with the air, and interaction between the wall and fuel droplet play vital roles in combustion while also influencing NO_x emissions. Also, the NO_x emissions vary directly concerning the compression ratio in all working conditions owing to the augmented in-cylinder temperature. Furthermore, the use of the Miller cycle to decrease NO_x emissions is a viable option. The Miller cycle focuses on longer expansion strokes and shorter compression strokes, thereby inflicting an ignition delay that causes a lower in-cylinder temperature. The use of the Miller cycle can also be extended to petrol engines. It can be used to decrease the NO_x emissions in petrol engines due to the decreased pressure and temperature at the end of the compression stroke [6]. The use of the Miller cycle increases NO_x emissions when paired with a turbocharger. This can be countered by using an ethanol blend, which has been shown to decrease NO_x emissions [7].

Alternatively, using variable exhaust valve timing has been shown to decrease NO_x emissions. The variation in valve overlapping to reduce the in-cylinder temperature can also be considered for reducing NO_x emissions during engine operation. The use of valve overlapping and the control of injection timing significantly affect combustion, which in turn affects NO_x emissions. The optimization of these design parameters has a positive effect on reducing harmful pollutants such as NO_x and smoke. Forced induction of the intake charge employing a turbocharger can be effective in enhancing the performance of the engine and reducing emissions. Turbocharging generally increases the volumetric efficiency of the engine and provides more oxygen molecules for better combustion of the fuel droplets.

Apart from these design parameters, the injection system of the engine provides a significant contribution toward emissions because it influences the charge mixing to a greater extent. The narrow spray cone angle is favorable for better performance, whereas the NO_x emission seems to be augmented at this condition [8]. While using biofuel blends, there is a chance for wall impingement effects that cause increased NO_x during operation. This issue can be addressed by implementing a variable nozzle configuration, which can alter the spray characteristics such as air entrainment, Sauter mean diameter, and penetration distance. It is well known that the NO_x magnitude is very much related to spray penetration distance. So, modifying the nozzle holes can neutralize this problem at the source level. Along with this, controlling the fluid flow through the intake and exhaust manifold is vital for emission reduction to meet the requirements of international standards. The manifold shape has a significant effect on the flow pattern of the charge inside the cylinder, which in turn affects the volumetric efficiency of the engine [9]. The intake flow can be modified by providing blades that venture along the intake manifold to induce turbulence. The pressure variations on the intake manifold influence the NO_x magnitude, as high-pressure systems generate more heat that in turn augments the thermal NO_x. Low-temperature combustion (LTC) is yet another strategy for reducing NO_x emissions by decreasing the flame temperature.

This chapter shines light on the effect of various engine design parameters on NO_x formation. The novelty of the chapter lies in analyzing the pros and cons of the most important design considerations of engine components with respect to NO_x emissions in detail, which is not discussed in previous literature works. The flow of the chapter is framed in such a way that it is similar to the flow of charge in the engine cycle, that is, from intake to exhaust. The intake system design modification on NO_x is taken as the first topic of interest followed by injection system modification. The flow of the chapter progresses to combustion chamber design, where the geometry and piston profile play predominate roles in combustion characteristics. Along with that, the compression ratio, barrier coating, and valve timing are also covered to give better clarity in the view of NO_x reduction. Finally, modern combustion strategies to achieve low-temperature combustion for reducing NO_x are discussed.

4.2 Role of engine design parameters on NO_x emission

From the above discussion, it is quite clear that the design aspects of the engine provide consistent results in terms of emission reduction. The design aspects discussed in this chapter are illustrated in Fig. 4.1. Primarily, the flow of charge can be altered by modifying component design along the intake path and combustion chamber till exhaust manifold, which in turn affects the combustion process and varies the NO_x level. The modifications in the intake manifold can induce swirl and squish to the flow, thereby enhancing the flow at the initial stages. On the other hand, employing forced induction devices such as a turbocharger augments the volumetric efficiency of the engine and leads to better combustion. Next to that, the injection system plays a vital role in achieving better combustion because the crucial properties such as spray penetration, atomization, and spray tip length depend upon it. The proper injector configuration has the potential to reduce NO_x by 10%–20% at the source level. Once the fuel enters the cylinder, the combustion chamber and the allied design aspects inside the cylinder

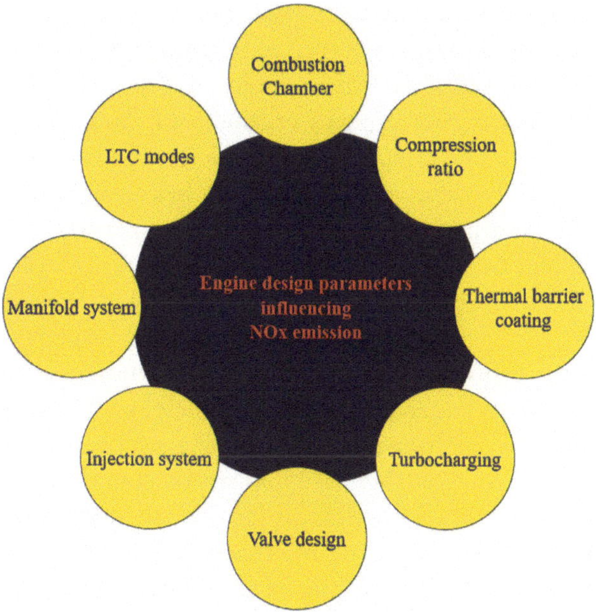

Fig. 4.1 Engine design parameters influencing NO_x.

such as the compression ratio and valve timing take control over flow properties. In view of this, the subsequent sections deal with the effect of the in-cylinder designs on NO_x emissions. The change in combustion chamber design plays a pivotal role in increasing the cylinder temperature through enhancing the turbulence of the mixture. Altering the chamber for a traditional fueled engine could lead to the enhancement of NO_x up to 10%, whereas for biofueled engines, this augmentation may extend up to 15%–20%. A separate section is dedicated to give a view to barrier coating and its effects on NO_x formation. Apart from design modifications, the concept of low-temperature combustion is also added as one of the topics in this chapter to provide extensive subject coverage to the readers.

4.3 Effect of intake system design on NO_x emissions

Generally, an increase in the turbulence of charge within the combustion chamber has a significant impact on engine attributes. Before generating turbulence within the chamber, it is advisable to induce turbulence during the intake charge. The intake charge motion is induced during the suction stroke and later altered during the compression stroke. The generation of swirl motion at the intake is one of the best ways to control turbulence and influence NO_x emissions. The intensity of the swirl at the end of the compression stroke mainly depends on the swirl motion formed during the suction process. The directed ports in the intake manifold tangentially direct the charge into the cylinder, whereas helical or spiral ports have the tendency to cause preswirl owing to the flow over curved geometry during the intake stroke. The induction of swirl motion can cause improved combustion due to the enhanced mixing of air-fuel, which in turn augments the NO_x level. The design parameters such as the inlet and outlet angles of the manifold play major roles in exhaust emission characteristics. It is evident that increasing the outlet angle and inner diameter of the intake manifold negatively affects the performance by decreasing the combustion temperature. This also reduces the thermal NO_x emission and increases the CO magnitude [9]. Meanwhile providing a groove of the different pitches in the intake manifold is one of the most effective methods in controlling the swirl intensity in the view of reducing emissions. Increasing the pitch of the groove decreases the NO_x to a certain limit, above which it increases due to an increase in temperature [10].

Apart from the manifold design, certain forced induction systems can be designed to control NO_x emissions. Turbochargers are one such forced induction device employed in all types of modern engines. Owing to the high mass flow rate in turbocharged engines, the engine power increases along with the decrease in specific emissions as compared to that of naturally aspirated engines. Turbocharging along with the combination of aftercooling and retarded injection plays a vital role in reducing NO_x emission. The increase in air density and high temperature at the end of compression leads to a shorter ignition delay period, which in turn reduces the quantity of fuel accumulated in premixed combustion. The reduction in the temperature of combustion in the premixed phase leads to lower NO_x emissions. At the same time, employing a retarded injection time in a turbocharged engine curbs NO_x emissions without compromising the fuel efficiency. Also, the intercooling of the boosted intake charge increases the density further and lowers the intake temperature, which in turn reduces NO_x emissions.

In the case of a traditional turbocharger, a nonlinear relationship between the boost pressure ratio (BPR) with respect to the mass flow rate of air is observed. The BPR shows the least values at a lower mass flow rate of air charge whereas it surges exponentially with respect to flow rate. Meanwhile, at low engine speeds, the energy of the exhaust will be low enough to produce a small magnitude of BPR, which poses complex problems such as higher smoke emissions due to reduced combustion characteristics. This poor combustion results in a lower NO_x emission level. As the engine speed accelerates, an additional portion of the total exhaust gas is made to bypass the turbine directly to the atmosphere to decrease the turbine power. These are "fixed geometry turbochargers" (FGT) and their use results in a negotiation between low- and high-speed NO_x characteristics. Variable geometry turbochargers (VGT) were developed to overcome the limitations of the FGT. The VGT turbine has mobile vanes that can vary the turbine flow area or the angle at which the exhaust gas arrives or departs the turbine rotor. A reduction in the turbine flow area augments the upstream exhaust gas pressure and leads to a rise in the speed of the turbocharger as well as advanced boost pressure. The increased boost pressure reduces the overall equivalence ratio of the cylinder, which decreases the particulate emission with augmentation in the NO_x level. It should also be noted that the use of advanced start of injection timings is required to maintain the same NO_x magnitude.

4.4 Effect of injection system design on NO_x emissions

The forced induction of charge with proper aftercooling techniques has proved to be a better way to reduce NO_x. Although the forced induction system increases the amount of oxygen by compressing the intake charge, it is important to have a suitable fuel injection system to achieve better combustion. The optimum design of the injection system supports the intake system for reducing NO_x. In order to reduce NO_x emissions, multiple injections are utilized, including pilot injection, postinjection, and after injection. Modern advancements in the ability to adjust the fuel pressure, flexible injection timing, and multiple injection system have led to the optimization of NO_x emissions without affecting performance. The influence of injection system design in NO_x limitation is illustrated in Fig. 4.2.

The major phenomena such as the atomization of fuel, combustion, and the formation of pollutants in a diesel engine can be significantly altered by the design of the 000 fuel injector nozzle. Better mixing of air-fuel inside the combustion chamber and enhanced combustion are achieved when the size of the fuel injector hole is modified. The oversized fuel injector and multiple fuel injectors attributes to the improved mixing results in faster combustion along with increased temperature which leads to augmented NO_x emission

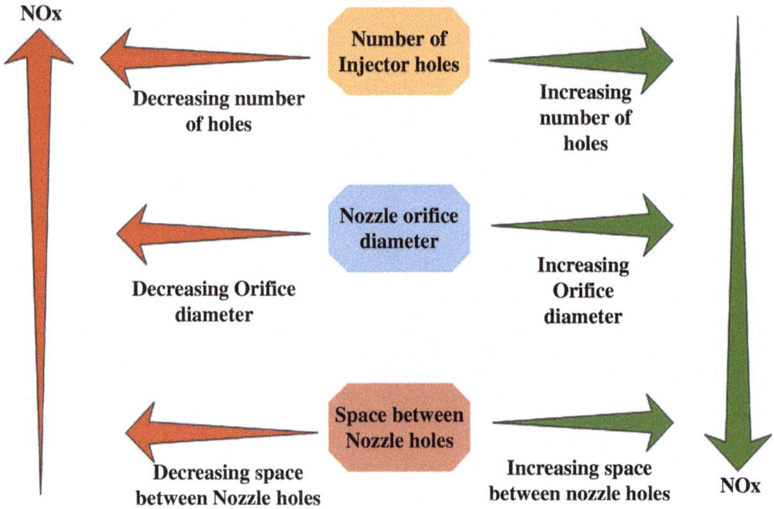

Fig. 4.2 Influence of injection system design on NO_x emission.

[11]. Apart from this, wall impingement increases NO_x emission levels due to the high temperature. Meanwhile, by altering the injector hole number, NO_x emissions can be reduced because the wall impingement is decreased during the process. On the spark plug end, a twin spark plug configuration showed improved performance and fuel economy as compared to a single spark plug configuration. Meanwhile, BTE enhancement and a reduction of cyclic variations can be found with the use of a dual surface discharge electrode spark plug. The above configuration also increases the rate of heat release and pressure along with a reduction in the combustion duration. This in turn increase NO_x emissions and reduce HC and CO emissions under lean clause. It is worth noting that the NO_x level during engine operation is indirectly proportional to the injector hole number due to the deprived combustion characteristics. Thus, the reduced nozzle hole diameter and increased number of holes have a greater influence on the reduction of NO_x emissions. At the same time, care has to be taken in performance attributes. This leads to a situation to optimize the nozzle hole diameter and number of holes in order to balance the NO_x level and the performance of the engine [12]. Next to that, increased injector pressure along with a reduction in nozzle hole diameter leads to increased NO_x emissions. This is because better atomization of fuel at high pressure with good spray penetration due to a small hole diameter leads to the complete combustion of fuel. Also, the premix burn is another factor that can lead to

NO_x formation. In multiple injection systems, the premix burn is reduced at a significant rate, which helps the system reduce NO_x emissions [13].

Apart from nozzle hole modifications, nozzle geometry and position have a significant effect on NO_x emissions. For instance, a conical-type nozzle produces lower NO_x emissions compared to a hydroground nozzle due to the reduction of diffusion combustion. In addition, the high rate of fuel injection in a hydroground nozzle leads to a rich air-fuel mixture and lower base flame speeds [14]. On comparing round-edged and sharp-edged nozzles, the difference in the coefficient of discharge leads to a lower injection pressure for a round-edged nozzle, which further results in higher NO_x emissions. Along with this, the surge in the rate of the injection profile of the round-edged nozzle is yet another reason for the augmentation of the NO_x level. In addition to that, the geometry of the nozzle has a direct effect on the spray angle, which can influence NO_x emissions by a significant margin. At high loads, the influence of the spray angle on NO_x emissions is small compared to lower loads. At lower loading conditions, combustion is dominated by the premix burn whereas at higher loads, there is an extended duration of injection that allows the fuel to be injected a little later during the cycle of diffusion. The diffusion phase is the region where the majority of the particulate matter is formed. With the lower spray angle, the wall impingement on the piston bowl wall is high compared to the engine operation with a high spray angle. From the above discussion, it is clear that the higher extent of wall impingement leads to higher NO_x emissions. It is also evident that a lesser spray angle produces high NO_x emissions during the combustion cycle [13]. It should also be noted that increasing the number of holes in a single injection system leads to a decrease in the penetration length and spray angle, which results in higher NO_x emissions. This increase in NO_x emissions can be attributed to the reduced liquid fuel moment as the number of holes increases [15]. Additionally, the reduction in nozzle orifice diameter directly increases NO_x emissions and decreases soot emissions. This hike in NO_x emissions is because the higher mixing rate of fuel and air leads to reduced combustion duration [16]. Apart from the nozzle hole diameter, the position of the nozzle hole also plays a predominant role in reducing NO_x emissions.

The phenomenon of flame extinction as a result of the reduced air-fuel ratio beyond the minimum or threshold point at which the flame can be sustained by geometry is termed a lean blowout. Increased spacing between nozzles at a lower equivalence ratio has a high chance of blowout. Therefore, NO_x emissions have been observed to decrease when the space between nozzle holes is increased; this varies inversely when the spacing is decreased. This is because the flame front from the nozzle at large spacing will not interact in a proper way, leading to a reduced cylinder temperature. While moving from near lean blowout to stoichiometric, there is a significant change in the NO_x emission index [17]. Considering the multiple injection system, triple injection provides greater flexibility in controlling NO_x emissions at lesser loads compared to the single and double injection systems. Thus, the type of nozzle hole geometry, the number of holes, the hole diameter, the spray angle, and the hole spacing have considerable influence on NO_x emissions.

4.5 Design of combustion chamber

The turbulence created by the intake and injection system helps the fuel mix thoroughly with air to a certain extent. The oxidation of fuel takes place in the combustion chamber whose design plays a vital role in further inducing the swirl and squish motion. The combustion chamber design provided for a certain engine is based on the fuel and operating conditions on which the engine is supposed to work. The need for a modification in chamber design arises when any of the fuel or working conditions is varied. Due to the higher viscosity and oxygen content, biofuels tend to produce a higher amount of emissions. So, the combustion chamber design optimization reduces the magnitude of certain emissions at the exhaust. The major aspect of the combustion geometry is the piston bowl geometry. Further, it is important to get an optimum design based on the requirement. Another aspect of the combustion chamber is to change the piston profile. Generally, the use of full skirted pistons is to increase the durability and wear resistance. These kinds of pistons are used in diesel engines due to their ability to handle more loads during operation. Due to their full profiles, these pistons can increase performance, that is, the combustion profile of the engine, which in turn augments the NO_x magnitude. Slipped skirt pistons are used in reciprocating engines to increase the ability to perform optimally at higher engine speeds. The major aspect associated with the piston geometry is the tumble motion. This secondary flow enhances combustion inside the chamber and improves efficiency. The tumble motion also has a swirl associated with it, leading to the increase in the in-cylinder temperature that elevates NO_x. Curved piston profiles have decreased angular momentum during the engine operation. Flat pistons maintain the tumble until the start of the compression stroke, which enhances the fuel mixing rate during injection.

Numerical results also showed how a mix of both top surface and pit surface geometries can be used to enhance performance [18].

4.6 Effects of chamber geometry on NO_x emission

The various combustion chamber geometries are given in Fig. 4.3A–G. The toroidal combustion chamber is one of the most efficient chambers owing to its double-curved design. This curved design initiates a double swirl, causing better mixing of the fuel and thereby improving combustion. Along with this, the squish motion ensures better movement of the charge inside the chamber. The use of biofuels in the toroidal chamber has shown better combustion and lower smoke and CO emissions, but increased NO_x emissions. This may be attributed to the better swirl motion of the charge, which leads to augmentation in combustion temperature. The toroidal chamber with modified injection parameters is recommended for optimized use to reduce NO_x emissions. The lower ignition delay in the toroidal chamber due to the swish motion is also a reason for augmentation in in-cylinder temperature and pressure during combustion, thereby increasing NO_x. On the other

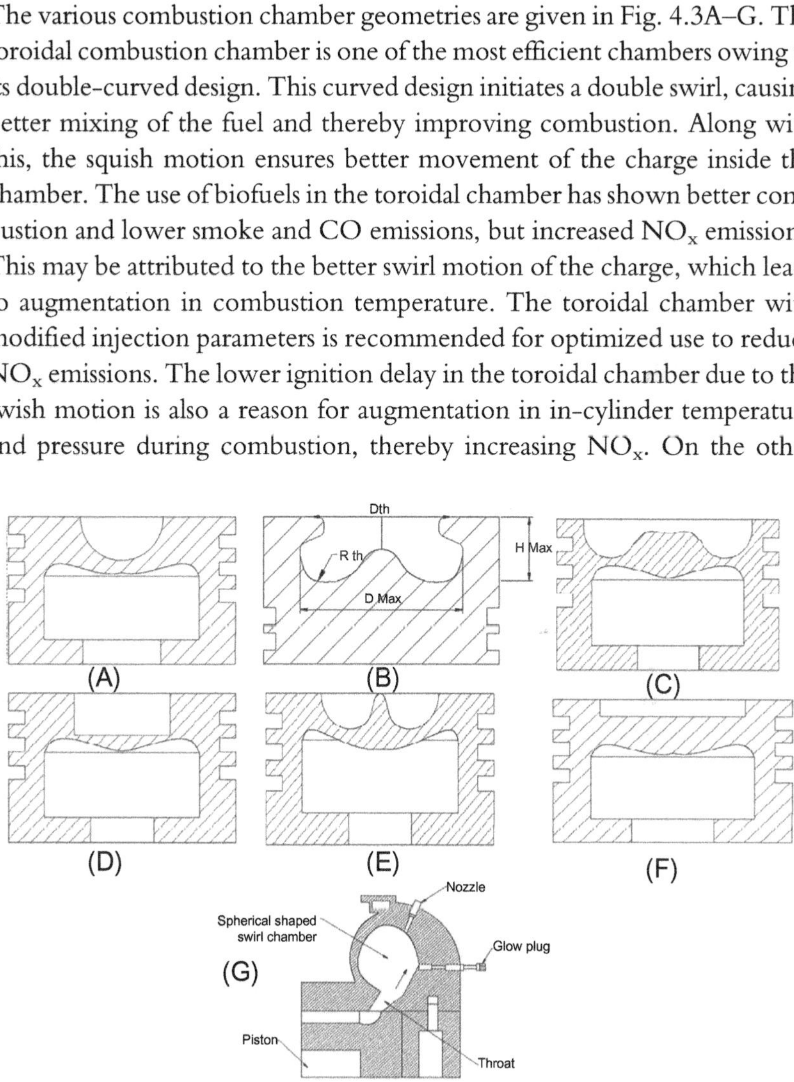

Fig. 4.3 Schematic diagram of (A) a hemispherical combustion chamber, (B) a reentrant combustion chamber, (C) a shallow depth combustion chamber, (D) a cylindrical combustion chamber, (E) a toroidal combustion chamber, (F) a square combustion chamber, and (G) a swirl combustion chamber.

hand, the swirl chamber is another type of combustion chamber that focuses on the mixing of fuel and air. During the operation of the swirl chamber, the upstroke of the piston causes the fuel to enter the chamber [19]. The turbulent flow along with the guidance walls causes the air to move in all directions inside the chamber, thereby thoroughly mixing the air and fuel. The heat loss is more excessive and direct compared to other combustion chambers. This heat loss reduces the in-cylinder temperature and consequently, the magnitude of NO_x gets reduced at the tailpipe. The use of these chambers is mostly where conditions are adverse and fuel economy isn't the priority. The reentrant combustion chamber is another type of chamber geometry used widely for renewable fuels. These chambers are designed in such a manner that the lip of the combustion chamber protrudes beyond the bowl walls to enhance the mixing of the charge. This type of chamber enhances combustion, leading to increased NO_x emissions as compared to a traditional hemispherical chamber. The reentrant chamber has a high mixing rate, which allows for retarded injection that allows us to reduce the NO_x emissions. This chamber also assures proper combustion and enhances soot oxidation, thereby reducing soot emissions. Meanwhile, the shallow depth combustion chambers are used mainly in large engines operating at low speeds. This is due to the low cavity depth, which also makes the swish negligible. At low speeds, the geometry gives good efficiency with reduced emission characteristics. To be more specific, CO and UBHC emissions are less for swirl-chambered engines at lower rpms than hemispherical chambers due to the better combustion and lower ignition delays [20]. This also increases the NO_x emission geometry at lower speed operation for shallow depth chambers. But as the engine speed increases, the NO_x emissions and heat release rate decrease. Table 4.1 summarizes some of the salient results of employing various CCs for NO_x emissions.

Next to that, the hemispherical combustion chamber is one of the most commonly used geometries for IC engines. The chamber provides good and efficient combustion due to minimal heat loss to the piston head. It should be noted that the compression ratio should be lower to achieve minimal heat loss [27]. This minimal heat loss leads to the minimization of NO_x and an increase in smoke at the exhaust. The alteration of the depth-to-diameter ratio in a hemispherical chamber can give the desired swish motion, whereas it is not as efficient as that of swish chamber geometry. The viscous nature of certain fuels causes poor atomization of fuel, thereby retarding combustion phenomena as well as increasing CO along with decreased NO_x at the tailpipe. The optimum chamber design uses the turbulence from the piston and

Table 4.1 Effect of CC geometries on NO_x emissions.

Geometry	Parameter studied	Salient findings	Reference
Spherical, reentrant, and toroidal	Tested 20%–100% Jatropha biodiesel blends.	Toroidal CC gives better efficiency for 20% blend but the NO_x emission was observed to be high.	[21]
Hemispherical, toroidal, reentrant, and shallow depth	Fixed compression ratio for all CCs.	Toroidal reentrant CC recorded best thermal efficiency (33%) with increased NO_x (784 ppm).	[22]
Reentrant bowl	Fixed compression ratio, squish clearance, and injection rate.	Reentrant bowl creates better swirl so that it produces high BTE with low emissions such as HC, CO, and NO_x.	[3]
Swirl chamber with reentrant	Swirl ratio is varied between 0.2 and 0.32.	The best air motion in CC has been observed for 0.8 swirl ratio. At the same condition, lower NO emission is observed.	[23]
Reentrant and toroidal CC	Swirl number is varied from 1 to 3.5.	Reentrant bowl without projection in center produced better swirl motion with reduced emissions.	[24]
Stepped bowl	Injection timing, spray angle, and fraction of fuel per injection.	Stepped bowl chamber recorded low CO and soot with increased NO_x due to the enhanced swirl rate.	[25]
Omega CC	Three types were discussed: central projection, shallow W, and pataloid type.	Pataloid type displayed least reverse squish motion. The performance is best in the central projection type.	[26]

cylinder arrangement and induces the swirl. Meanwhile, the square combustion chamber has a narrow squish region that forces a strong swish of the fuel charge, particularly in the corner regions. The intake stroke allows a great number of swirl forces to enhance the turbulence of the charge. But this decreases the momentum of the air inside the chamber, thereby reducing the combustion efficiency and leading to NO_x reduction to a minimum extent. The cylindrical combustion chamber has a typical cylindrical cavity in which the swish is produced by altering the depth. The cylindrical combustion chamber takes the shape of a truncated cone with a base angle of 90 degrees. The cylindrical combustion chamber generally has a higher combustion efficiency and promises better efficiency than a hemispherical chamber. It generally has lower NO_x emissions accompanied by higher UBHC and soot emission compared to the toroidal combustion chamber. On the whole, the selection of a suitable chamber design is up to the requirements toward the performance or emission point of view.

4.7 Effects of chamber design parameters on NO_x emissions

Combustion chamber design also focuses on the efficient utilization of fuel and air to ensure proper combustion. The cylinder is comprised of the piston bowl, the clearance between the piston and cylinder head, the valve recess, the top land-crevice, and the head gasket clearance on which the air is contained. Among the components of the engine cylinder, the clearance volume is distributed predominately in the piston bowl (55%) and piston-cylinder head (30%). Meanwhile, less than 15% of the air contained in the remaining portion is utilized poorly during the combustion phase. Also, the air pockets entrapped in the gap between the piston crown and head at the top dead center are utilized poorly. The complete utilization of the air volume refers to employing a lower clearance volume to increase the volumetric efficiency of the engine. This may improve the combustion, therefore decreasing HC and soot emissions. On the other hand, this increases NO_x emissions because of the high in-cylinder temperatures. The crevice volume stores more air due to its decreased temperature. Therefore, when reduced, this will increase the air interacting with the fuel during combustion. This leads to an augmented combustion temperature, resulting in higher efficiencies and higher NO_x emissions. Therefore, air utilization requires a balance between improving combustion efficiency and decreasing emissions. The augmented heat transfer from the hot gases at the vicinity of the wall results in the deprivation of NO_x emissions. On the other hand, the low surface-to-volume ratio of the compact hemispherical chamber reduces engine knocking.

4.8 Effect of compression ratio on NO_x emissions

Even though the combustion chamber geometry influences the combustion efficiency, the extent to which the intake charge is compressed before combustion plays a crucial role in the augmentation of cylinder temperature. The increase in compression ratio directly affects certain other parameters such as the rate of combustion, friction, and heat transfer, which have significant contributions to the emissions. This influence of compression ratio has its significance across the entire range of engine operation, irrespective of load and speed. The compression ratio should not be varied randomly and its value is limited by the fuel quality and knocking. The following section gives details on the effect of the compression ratio on NO_x emissions.

4.9 Role of compression ratio in NO_x mitigation for CI engines

The compression ratio modification can come in handy in a situation such as cold starting to reduce tailpipe emissions. Under cold starting conditions, it is advisable to employ high compression ratios. An increase in the compression ratio of the CI engine shortens the ignition delay period and the rapid combustion of accumulated fuel increases the temperature inside the combustion chamber, which favors thermal NO_x with reduced particulate matter. On the other hand, this shorter ignition delay period reduces the overmixing of the charge, thereby resulting in declined HC along with augmented NO_x magnitude at the exhaust. Contradictorily, at low compression range, the ignition delay period is too long, due to which the fuel combustion may start at the beginning of the power stroke, leading to a reduction in in-cylinder pressure and temperature. This long delay period leads to overmixing of the charge and a lower combustion temperature while also reducing NO_x emissions at the exhaust. The longer delay period due to the lower compression ratio during the warm-up period causes a high unburned fuel emission called white smoke. The engine operated at a higher compression ratio is susceptible to soot formation. Meanwhile, it should be noted that a high in-cylinder temperature favors soot oxidation. Every engine design should consider an optimum compression ratio based on the speed and load

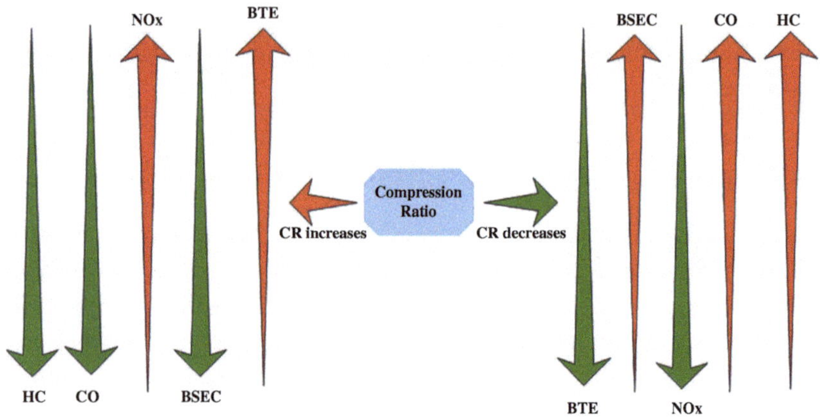

Fig. 4.4 Effects of compression ratios on engine operation.

range to obtain simultaneous particulate matter and NO_x reduction at the exhaust. The significant effect of varying compression ratio is depicted in Fig. 4.4.

In general, renewable fuels pose a considerable issue of augmented viscosity, which makes the injection difficult compared to that of neat diesel fuel. At a higher compression ratio, some biodiesel fuel increases the thermal efficiency of the engine with considerable effect on the NO_x emission characteristics of the engine. The biodiesel-powered engines have a higher fuel spray penetration length, which results in an enhanced probability of the wall impingement effect and NO_x formation. Even though the increase in compression ratio reduces the probability of wall impingement and spray penetration distance owing to the higher charge density, the NO_x emissions are observed to be higher at the exhaust. The magnitude of NO_x at a high compression ratio can be decreased by adopting certain strategies such as retarding injection timing and EGR. Some of the studies providing the optimum compression ratio of biofuel-powered engines in view of NO_x reduction are given in Table 4.2.

4.10 Role of compression ratio in NO_x mitigation for SI engines

The influence of the compression ratio on NO_x formation in SI engines is related to a wide array of engine processes. The increase in instantaneous gas temperature at starting followed by a decreasing trend is observed with respect to compression ratio. In addition, cylinder pressure and surface-

Table 4.2 Optimized compression ratio for some common fuels.

Fuel used	NO_x (ppm)/%/(g/kWh) variation	Optimized compression ratio	Reference
Waste cooking oil biodiesel	Increased by 3.05	Blend B40@CR21	[28]
Jatropha biodiesel	Decreased by 25	B100 @ CR18	[29]
Waste tire pyrolysis oil	Increased by 20	CR18.5	[30]
Calophyllum inophyllum oil	Decreased by 2	CR18	[31]
Tamunu biodiesel	Increased by 201	CR18	[32]
Cotton seed biodiesel	Increased by 32	CR17	[33]

to-volume ratio are increased due to which the turbulence is enhanced during the compression stroke. On the other hand, the increased frictional resistance and pumping work may reduce the mechanical efficiency of the engine. These factors have a clear influence on the burning rate of fuel, flame quenching, heat transfer inside the cylinder, combustion stability, and other engine processes that in turn directly affect the magnitude of NO_x emissions. While using natural gas as fuel in SI engines, it has to be noted that for all spark timings, the NO_x magnitude varies directly concerning the compression ratio. However, at maximum brake torque spark timing, the increase in compression ratio first increases the NO_x, after which the emissions are diminished due to inferior combustion characteristics [34]. The relationship between the compression ratio and NO concerning combustion duration is given in Fig. 4.5.

For typical SI engines, the decrease in the combustion duration augments NO, irrespective of the compression ratio. Considering the least burn duration ($\Theta_b = 40°$), we can observe that NO emissions are augmented rapidly with a compression ratio until 9, whereas the rate of augmentation decreased after 9 until 12. Upon visualizing the highest duration ($\Theta_b = 100°$), a very minute change in NO concentration is observed from the 6 to 9 compression ratio. Meanwhile, a steep decline in the emission trend is observed at a compression ratio from about 9 to 12. The reason for this phenomenon is quite complex, as the change in NO_x concerning the compression ratio is a complex combination of thermodynamic and chemical changes inside the cylinder during combustion. Due to the elevated initial gas temperatures, the primary adiabatic gas temperatures are highest for the peak compression

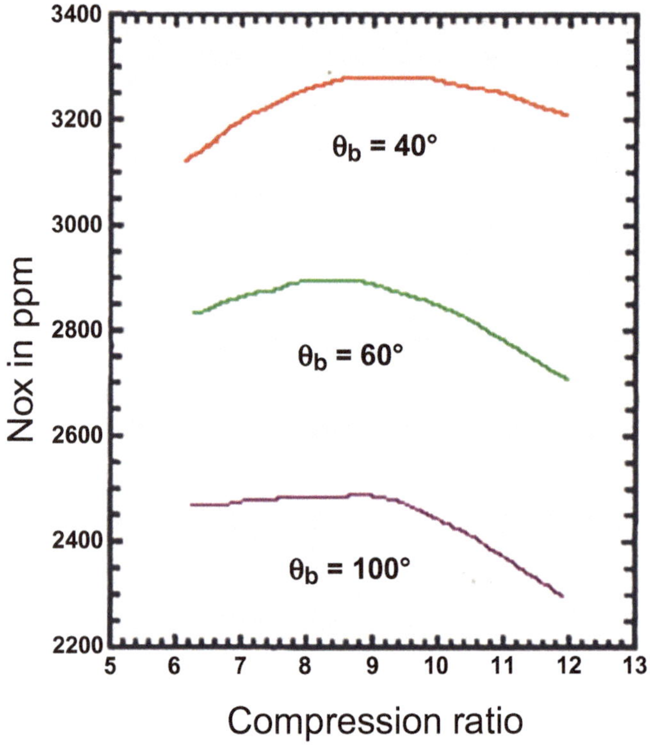

Fig. 4.5 Compression ratio vs. NO (ppm) with different combustion durations.

ratio of the engine. Meanwhile, the gas temperatures of the adiabatic zone are augmented rapidly owing to augmented heat release due to the high mass charge for lower compression ratios. The high heat loss during high compression ratio operation leads to a reduction in peak cylinder temperature. This, in turn, decreases the temperature of the adiabatic zone, which diminishes NO at the exhaust. The oxygen atom concentration during the combustion phase is one of the crucial factors in deciding the magnitude of NO_x. It is clear that the concentration of oxygen atoms rapidly increases with respect to the in-cylinder temperature and reaches a peak around 15° ATDC. From the above discussion, it is clear that the NO_x emission of a typical SI engine predominately depends upon the gas temperature and oxygen concentration. Even though the given discussion is obtained from a standard SI engine fueled with straight isooctane fuel, these results can show variations when compared with engines operated with alternate fuels such as alcohols.

4.11 Effect of valve timing and design on NO_x emissions

In addition to charge compression, the quantity of fresh charge inducted and the extent to which burnt gases are expelled after combustion also affect the emission attributes of IC engines. The quantity of charge intake and burnt gases scavenged is governed by valve timing. The valve timing is specified with correspondence to the position of the pistons. A typical valve timing diagram depicts the breathing process of the engine and is given in Fig. 4.6.

The valve timing is designed based on multiple parameters that influence NO_x magnitude; among these, the pressure created by the piston motion is an important parameter. Next to that, the quantity of intake charge is crucial as it aids the incomplete combustion of injected fuel. Therefore, the intake valve is closed during the compression stroke to obtain the required amount of charge in the cylinder. The exhaust valve is generally engineered in such a way that it opens before the cylinder reaches the BDC on its power stroke. This is done to relieve some of the cylinder pressure so as to minimize piston pumping losses. During the next suction stroke, both inlet and outlet valves are kept open for a short duration to get rid of all the exhaust gases. This is a simultaneous operation called valve overlap, and it is generally helpful at low speeds to obtain increased torque [35]. This scavenging process is the most important as improper scavenging leads to the accumulation of burnt gases in the chamber, which deteriorates combustion in the next cycle. The deteriorated combustion leads to reduced chamber temperature, thereby increasing HC emissions along with lower NO_x [36]. It has to be noted that

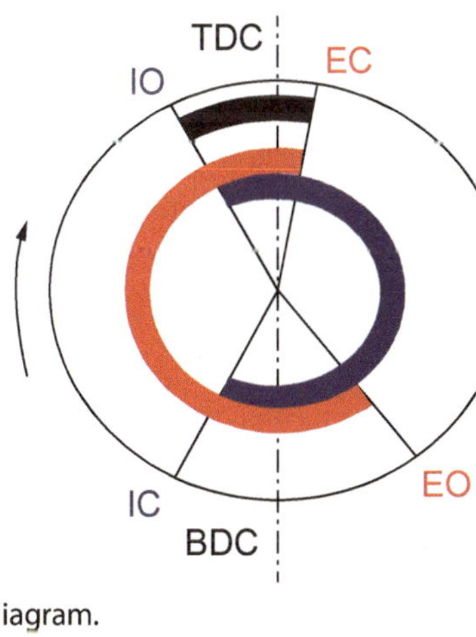

Fig. 4.6 Valve timing diagram.

reducing NO_x in such a manner is not advisable, as a major sacrifice needs to be made on the performance side.

Meanwhile, the multivalve design aims to increase power by enhancing the breathing process of the engine. The addition of more valves increases the valve area, thereby enhancing combustion. This method also assists in low-speed operations. The relationship between NO_x and PM for two valve configurations of a car engine is shown in Fig. 4.7.

An inclined injector at an angle of 20 degrees and 10 degrees to the vertical reduces NO_x emissions due to poor combustion characteristics as compared to a vertical injector. The four-valve configuration showed enhanced NO_x reduction as compared to a two-valve configuration for a fixed load condition. The valves are sometimes released so as to open at different times, which induces turbulence to mix the fuel thoroughly, thereby increasing combustion. Further, the increased number of valves allows positioning spark plugs in SI engines so as to obtain the optimum level of combustion. The manufacturing of such designs is complicated and increases the overall cost. The objective of the multivalve is to provide maximized power output that automatically augments NO_x owing to better fuel combustion. The BSFC of a multivalve engine is lower for the same load compared to a conventionally operated engine. The decrease in BSFC directly shows

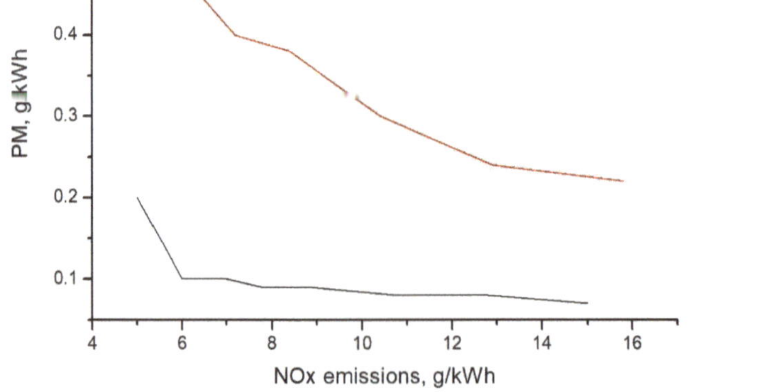

Fig. 4.7 NO_x vs. PM characteristics.

better performance and increased NO_x magnitude. Multivalve operation increases the swirl rate due to the directional nature, which forces the fuel packets to burn completely during the power stroke. Better combustion always assists in decreasing CO emissions and increasing NO_x. Also, the oxidation of hydrocarbons is promoted due to the augmentation of cylinder pressure and temperature. This eliminates the HC and smoke emission at the exhaust, accompanied by increased NO_x as it is completely associated with the high in-cylinder temperature [37]. The higher swirl and turbulence due to multivalve injectors are major reasons for this change in performance and emission aspects. The high heat release rate in the premixed combustion phase leads to increased NO_x emissions in multivalve injection operation. Apart from the negative aspect of NO_x emissions, multivalve operation enhances the performance to a significant extent along with a reduction in smoke. Therefore, the number of valves is crucial for not only the directional attributes but also for the emission and performance characteristics.

4.12 Effect of thermal barrier coating on NO_x emissions

The in-cylinder system accompanied with intake system modifications significantly increases engine power and reduces NO_x emissions. In spite of increasing the combustion temperature, it is also possible to hold the temperature by providing insulation coating over the cylinder wall to obtain better performance. The use of a thermal barrier coating in engines began as a measure to increase the performance of the engine. The engine performance is mainly dependent on the efficiency of combustion, that is, the cylinder temperature and stoichiometry of the combustion. Therefore, different types of thermal coating were employed to check the interaction with combustion and emission. Furthermore, studies showed the effect of thermal coatings in engines to enhance combustion and performance [38]. Materials such as NiCrAl, zirconia, TiO_2, mullite, alumina, spinel $MgO{\cdot}Al_2O_3$, etc., are tested as barrier coatings in engines. Generally, the barrier coats are made up of ceramic materials that act as insulators to heat. This is done to increase the in-cylinder temperature, thereby enhancing combustion efficiency. The coating materials should possess certain material properties that affect the performance of the engine; these are given in Table 4.3. This temperature augmentation inside the cylinder paves the way for certain exhaust emissions such as NO_x. The use of ceramics in this application is justified by their resistance to oxidation, corrosion, and creep resistance.

Table 4.3 Effects and requirement of coating material properties.

Property	Requirement	Effects of property
Melting point	High	Resistance to failure at high temperature application
Oxidation and corrosion resistance	High	Resistance to component damage from working fluids
Thermal expansion	High	It must be close to bond coats
Thermal conductivity	Low	To reduce heat loss
Strain tolerance	High	To withstand strain imposed on components
Phase change	Stable	Resist the phase change as it can deteriorate the coating

The thickness of the coating depends on the application and use of the component. Typical CI engines can have a maximum coating of 1000 μm, causing a high degree of insulation [39]. The increased coating thickness acts as an insulator to cylinder heat, thereby increasing the thermal NO_x. The fuel also plays a role in the selection of barrier material, as higher-octane fuels will produce more heat. The use of ceramics such as partially stabilized zirconia (PSZ) with high thermal resistance tends to increase the thermal efficiency, as the heat stays in the cylinder volume. This high heat inside the cylinder catalyzes the reaction between nitrogen and oxygen molecules inside the chamber to form NO_x emissions. Using ceramics such as TiO_{2-} shows an increase in combustion temperature but not as much as PSZ due to its lower thermal resistance. The use of a thermal barrier coating is beneficial to increase the combustion and performance characteristics of the engine irrespective of fuels and operating conditions.

Generally, complete combustion is desired for better performance, but increased in-cylinder temperature augments certain emissions such as NO_x and CO_2. The higher the thermal resistance of the coating, the lower the carbon-based emissions. The NO_x emissions increase for engines with all types of thermal barrier coatings, irrespective of the coating and stabilizing materials. This is due to the higher in-cylinder temperature due to the retainment of heat, leading to higher temperatures and longer residence times. Ceramics such as PSZ show higher NO_x emissions compared to TiO_{2-} due to the higher thermal resistance [4]. Also, ceramics that have high oxidation properties do not emit NO_x emissions such as a PSZ-coated engine. While using a barrier in a biodiesel-powered engine, the trend of NO_x remains the same. The NO_x emissions are lower for bio-oil compared to neat diesel for coated engines. This may be due to the deprived combustion phenomena observed in bio-oil-based fuels. Therefore, decreasing the NO_x emission is plausible only if the in-cylinder temperatures are lowered. Using a thermal coating sacrifices NO_x characteristics for better performance.

4.13 Low-temperature combustion for NO_x reduction

As the thermal barrier coating is given over the cylinder wall, the increase in core combustion temperature is responsible for improved performance and augmentation of NO_x emissions. All design strategies to reduce NO_x focus on reducing combustion temperature. In this view, an advanced combustion strategy such as low-temperature combustion was developed to reduce combustion temperature. The homogeneous air-fuel lean mixture achieved in low-temperature combustion leads to higher thermal efficiency by attaining uniform temperature distribution, leading to the reduced transfer of heat. This also aids in reducing heat-sensitive emissions such as NO_x. Low-temperature combustion is generally divided into three phases: (i) precombustion, which mainly depends on the characteristics of charge flow, (ii) combustion, which depends on the kinetic behavior of fuel (chemical kinetics), and (iii) postcombustion, which is the final phase that primarily depends on the mixing circumstances. The effect of employing various low combustion strategies on NO_x emissions is illustrated in Fig. 4.8.

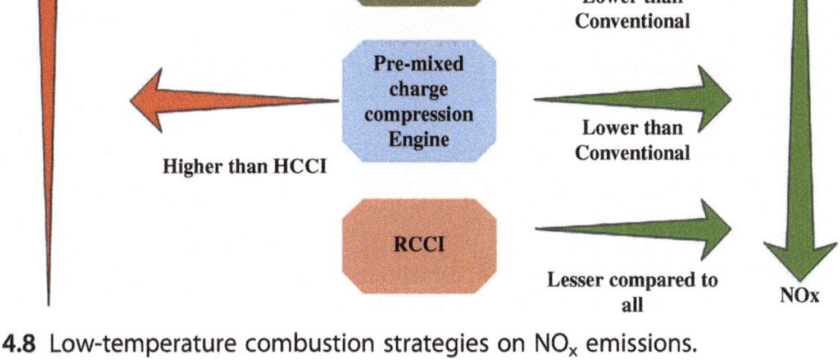

Fig. 4.8 Low-temperature combustion strategies on NO_x emissions.

Low-temperature combustion can be achieved in engines by elevating exhaust gas recirculation by adjusting the air-fuel ratio to a higher level [40]. During combustion, the oxidation of fuel leads to a reduced amount of oxygen at the spray region, which leads to increased soot emissions [41,42]. Nevertheless, attaining high exhaust gas recirculation is difficult in the real-time working of the engine due to the operating range being narrower before the fuel consumption is raised to the required level and an additional charge boost is required to meet the requirements of increased exhaust gas recirculation [43,44]. In recent times to achieve low-temperature combustion, the engines have been equipped with more injection, dual fuel operation, and negative valve overlap. These modern methods require higher cost compared to conventional engines [45]. The pattern commonly observed in low-temperature combustion is the low-temperature heat release, which can be expressed as the difference in time for the initialization of combustion and the lower base of the rate of high-temperature heat release and low-temperature heat release [46]. The lower heat inside the chamber reduces the free radical formation of nitrogen and curbs the oxidation reaction, which in turn decreases NO_x emissions. Various methods to achieve low-temperature combustion are listed below.

Partially premixed low-temperature combustion operates with increased air-fuel mixing with the help of elevated ignition delay, which can be achieved by blending fuel with a lower cetane number with diesel. This mode exhibits reduced NO_x and smoke emissions compared to homogeneous charge compression ignition (HCCI) [47]. The current type can be further divided into two categories based on injection: early injection and late injection. Early injection takes place during halfway compression while late injection takes place in the region of top dead center. The increase in ignition delay causes the engine using this method to create more noise compared to conventional engines. Next to that, HCCI has increased efficiency and reduced NO_x emissions. Low-temperature combustion can be achieved in this approach with the use of lean mixture and a premix of air-fuel ahead of combustion [41]. One of the limitations of this model is the increase in knock due to unforeseen autoignition in the chamber, which leads to increased oscillation and frequency in the cylinder pressure. This mode can be used for all ranges of octane numbers provided a high compression ratio that can lead the fuel to the autoignition temperature. This mode gives the required advantages when the engine is operated at elevated loads, whereas due to the chances of a misfire, this method shows limitations in reduced loads of engines [48]. This mode produces less NO_x as compared to other methods, but the engine operation is difficult relative to other modes.

Premixed charge compression ignition (PCCI) is a mode of achieving low-temperature combustion where the process is done by enriching the air-fuel mixture ahead of combustion initialization by increasing the time consumed between the start of injection and combustion. The air-fuel mixture in this mode can either be lean or leaner provided there is a compression ratio higher than that of a conventional SI engine. This method produces lesser CO and HC emissions but slightly higher NO_x compared to HCCI. Generally, this model is adopted with split injection to avoid wall impingements. In terms of knocking and prediction of the autoignition region, the PCCI method proves to be better than HCCI. Due to the truncated volatility and increased flammability of diesel, the PCCI mode faces certain limitations. Spark-aided combustion can be used for less-volatile fuels to obtain comparatively enhanced engine characteristics. The dual-fuel mode, which is known as PCCI-direct injection, has a higher potential to reduce NO_x and soot emissions with increased thermal efficiency [49]. Meanwhile, reactivity-controlled compression ignition (RCCI) mode utilizes fuels of reactivity at various levels with preplanned injection intervals to achieve the required advantages of combustion. With said different ranges of reactivity, the fuels with lesser reactivity are injected through the port fuel injection method while the fuels with high reactivity are injected with multiinjection strategies ahead of combustion initialization. The advantage of this model is the ability to operate over an extended array of engine loads with reduced NO_x emissions and increased thermal efficiency (relatively less than PCCI). The reactivity is divided into two parts: the first is global reactivity, which is evaluated based on the type of fuel and the injected amount of fuel. The second is known as the reactivity gradient, and it predominantly depends on the fuel injection strategy [50]. The earlier part influences NO_x emissions to a greater extent, as the fuel properties play a vital role in the combustion phase. At higher speed and load, this method exhibits increased pressure at an alarming rate that might damage the engine. Low-temperature combustion along with the above modes demands certain requirements such as fuel properties, operating parameters, charge mixture, and operating modes. The modes listed above such as PCCI, HCCI, and conventional engine have certain limitations that are overcome by RCCI [51]. Table 4.4 shows various modes of low-temperature combustion and their limitations. Thus, low-temperature combustion can be used to reduce NO_x and soot emissions using the above methods; among these, RCCI has a better potential for future work.

Table 4.4 Various modes of low-temperature combustion and their limitations.

S. No.	Mode	Limitations
1	Conventional engine	Increased NO_x and soot emissions
2	PCCI	Reduced power output
3	HCCI	Lesser operating range
4	RCCI	Increased pressure to high speed and load

4.14 Overall engine design requirements and considerations for NO_x mitigation

It can be clearly seen that the NO_x rate is affected by several engine design parameters. To summarize, the in-cylinder temperature variation due to the engine design positively affects the NO_x magnitude inside the cylinder. Being a vital design parameter, an increase in compression ratio enhances the NO_x output, whereas a low compression ratio decreases the performance of the engine. So it is crucial to have an optimum compression ratio for desired power output with reduced emission levels. Next to that, the shape and geometry of the combustion chamber paves the way for better combustion. The thermal barrier coating restricts heat release from the cylinder, which further adds up the heat to form more NO_x. Therefore, the selection of a suitable chamber geometry for the desired application becomes more important to reduce NO_x. Also, the selection of a suitable barrier coating can maintain NO_x with an optimum performance level. Also, designing the proper injection system is crucial for emission reduction in an engine. The geometry of the nozzle hole significantly affects the spray characteristics, which in turn affect the magnitude of NO_x. In addition, employing multi-valves in the engine influences the combustion of fuel, thereby augmenting NO_x. Considering the intake system, the grooves and modifications in the intake manifold increase the rate of swirl, which leads to the complete combustion of fuel. The degree of combustion determines the NO_x level, as it clearly depends on temperature during the combustion phase. The forced induction systems such as turbochargers increase the density of the intake charge and shorten the delay period of the fuel injected. Finally, the advanced low-temperature combustion modes are focused on reducing the in-cylinder temperature, thereby curbing NO_x levels at the exhaust. The overall aspects of engine design parameters on NO_x emissions are illustrated in Fig. 4.9.

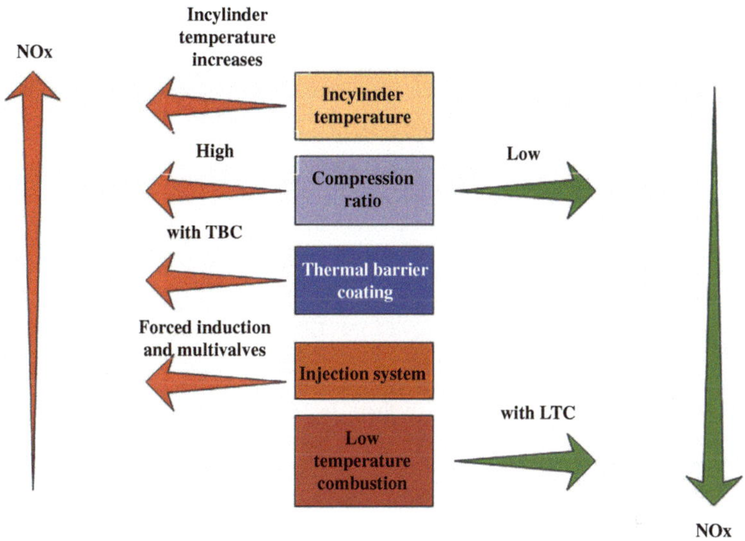

Fig. 4.9 Overall design modifications on NO_x emissions.

4.15 Conclusion

NO_x plays a vital role in environmental degradation and its impact can be observed in all life forms. Being one of the major producers of NO_x, the automotive sector has to be reformed with strict emission norms to curb this dangerous pollutant. The elimination of NO_x at the source level is still possible through varying the salient design parameters discussed in the above topics. Implementation of the design aspects of IC engines described above will have a significant effect on NO_x reduction at the exhaust. On the whole, this chapter provides a detailed view of the influence of engine design changes in NO_x formation and mitigation, whereas the next chapter discusses the effect of the operating parameters on NO_x mitigation.

References

[1] Paramasivam B, Ramesh K, Sakthivel R. Investigation and improvement on storage stability of pyrolysis oil obtained from *Aegle marmelos* de-oiled seed cake. Energy Sources Part A 2021;43(8):953–67.

[2] Devarajan Y, Munuswamy DB, Nagappan B. Emissions analysis on diesel engine fuelled with cashew nut shell biodiesel and pentanol blends. Environ Sci Pollut Res 2017;24(14):13136–41.

[3] Venkateswaran SP, Nagarajan G. Effects of the re-entrant bowl geometry on a DI turbocharged diesel engine performance and emissions—a CFD approach. J Eng Gas Turbines Power 2010;132(12):122803.

[4] Ravikumar V, Senthilkumar D. Reduction of smoke and NOx from the combined effect of thermal barrier coatings and *Raphanus sativus* biodiesel in a diesel engine for transport sectors. Int J Green Energy 2019;16(13):1016–23.
[5] Sakthivel G, Sivaraja CM, Ikua BW. Prediction OF CI engine performance, emission and combustion parameters using fish oil as a biodiesel by fuzzy-GA. Energy 2019;166:287–306.
[6] Wang Y, et al. An analytic study of applying Miller cycle to reduce NOx emission from petrol engine. Appl Therm Eng 2007;27(11):1779–89.
[7] Li C, et al. Application of miller cycle with turbocharger and ethanol to reduce NOx and particulates emissions from diesel engine – a numerical approach with model validations. Appl Therm Eng 2019;150:904–11.
[8] Prabhakara Rao G, Raju VRK, Rao SS. Effect of fuel injection pressure and spray cone angle in DI diesel engine using CONVERGE^{TM} CFD code. Procedia Eng 2015;127:295–300.
[9] Sadeq A, et al. New induction manifold designs for high performance and low emission diesel engine running on alternative fuels. In: 13th international energy conversion engineering conference at Orlando, FL, USA; 2015.
[10] Prasad D, Rangadu V. Reduction of emissions by intensifying air swirl in a single cylinder di diesel engine with modified inlet manifold. Int J Appl Eng Technol 2011;1:18–23.
[11] Harari P, Patil J. A review on effect of nozzle hole geometry on the performance combustion and emission parameters of CI engine Fuelled with diesel and biodiesel; 2016.
[12] Narayanan S, Ramesh K, Sakthivel R. Optimization of nozzle hole number for a diesel engine fueled with kapok methyl ester blend. Energy Sources Part A 2019;1–13.
[13] Pierpont DA, Reitz RD. Effects of injection pressure and nozzle geometry on D.I. diesel emissions and performance. SAE International; 1995.
[14] Som S, et al. Effect of nozzle orifice geometry on spray, combustion, and emission characteristics under diesel engine conditions. Fuel 2011;90(3):1267–76.
[15] Lee B, et al. Effect of the number of fuel injector holes on characteristics of combustion and emissions in a diesel engine. Int J Automot Technol 2010;11:783–91.
[16] Bergstrand P, Denbratt I. Diesel combustion with reduced nozzle orifice diameter; 2001.
[17] Dolan B, et al. Effect of nozzle spacing on nitrogen-oxide emissions and lean operability. AIAA J 2016;54:1–9.
[18] Yin C, et al. Effect of the piston top contour on the tumble flow and combustion features of a GDI engine with a CMCV: a CFD study. Eng Appl Comput Fluid Mech 2016;10(1):311–29.
[19] Dadsetan M, Chitsaz I, Amani E. A study of swirl ratio effects on the NOx formation and mixture stratification in an RCCI engine. Energy 2019;182:1100–14.
[20] Nayak SK, Chandra Mishra P. Combustion characteristics, performances and emissions of a biodiesel-producer gas dual fuel engine with varied combustor geometry. Energy 2019;168:585–600.
[21] Mamilla VR, Mallikarjun MV, Rao GLN. Effect of combustion chamber design on a DI diesel engine fuelled with Jatropha methyl esters blends with diesel. Procedia Eng 2013;64:479–90.
[22] Jaichandar S, Annamalai K. Influences of re-entrant combustion chamber geometry on the performance of Pongamia biodiesel in a DI diesel engine. Energy 2012;44(1):633–40.
[23] Wei S, et al. Numerical analysis on the effect of swirl ratios on swirl chamber combustion system of DI diesel engines. Energ Conver Manage 2013;75:184–90.
[24] Prasad BVVSU, et al. High swirl-inducing piston bowls in small diesel engines for emission reduction. Appl Energy 2011;88(7):2355–67.

[25] Dolak J, Shi Y, Reitz R. A computational investigation of stepped-bowl piston geometry for a light duty engine operating at low load. SAE Technical Papers; 2010.
[26] Lin L, et al. Effects of combustion chamber geometry on in-cylinder air motion and performance in DI diesel engine. SAE International; 2000.
[27] Varun PS, et al. Modification in combustion chamber geometry of CI engines for suitability of biodiesel: a review. Renew Sustain Energy Rev 2017;79:1016–33.
[28] Muralidharan K, Vasudevan D. Performance, emission and combustion characteristics of a variable compression ratio engine using methyl esters of waste cooking oil and diesel blends. Appl Energy 2011;88(11):3959–68.
[29] Jindal S, et al. Experimental investigation of the effect of compression ratio and injection pressure in a direct injection diesel engine running on Jatropha methyl ester. Appl Therm Eng 2010;30(5):442–8.
[30] Sharma A, Murugan S. Potential for using a Tyre pyrolysis oil-biodiesel blend in a diesel engine at different compression ratios. Energ Conver Manage 2015;93:289–97.
[31] Channapattana S, et al. Emissions and performance evaluation of DI CI-VCR engine fuelled with honne oil methyl ester/diesel blends. Energy Proc 2015;74:281–8.
[32] Mohanraj T, Mohan Kumar KM. Operating characteristics of a variable compression ratio engine using esterified tamanu oil. Int J Green Energy 2013;10(3):285–301.
[33] Ramanathan A, et al. The performance and emissions of a variable compression ratio diesel engine fuelled with bio-diesel from cotton seed oil. J Eng Appl Sci 2009;4:72.
[34] Takagaki SS, Raine RR. The effects of compression ratio on nitric oxide and hydrocarbon emissions from a spark-ignition natural gas fuelled engine. SAE International; 1997.
[35] Höckerdal E, Orrling S, Fantenberg E. Cylinder charge estimation in diesel engines with dual independent variable valve timing. This work is funded by the Vinnova FFI project BRAVE. IFAC-PapersOnLine 2019;52(5):636–41.
[36] Kulandaivel D, et al. Effect of retarded injection timing and EGR on performance, combustion and emission characteristics of a CRDi diesel engine fueled with WHDPE oil/diesel blends. Fuel 2020;278:118304.
[37] Rajamohan S, Marshal JJ, Suresh S. Derivation of synthetic fuel from waste plastic: investigation of engine operating characteristics on DI diesel engine. Environ Sci Pollut Res 2021;28:11976–87.
[38] Ravikumar V, Senthilkumar D. Reduction of NOx emission on NiCrAl-titanium oxide coated direct injection diesel engine fuelled with radish (*Raphanus sativus*) biodiesel. J Renew Sustain Energy 2013;5(6):063121.
[39] Shanmugam M, et al. Effect of thermal barrier coating in CI engines fueled with *Citrus Medica* (citron) peel oil biodiesel dosed with cerium oxide nanoparticle. Mater Today Proc 2020. https://doi.org/10.1016/j.matpr.2020.07.485.
[40] Huang H, et al. Combustion performance and emission characteristics of a diesel engine under low-temperature combustion of pine oil–diesel blends. Energ Conver Manage 2016;128:317–26.
[41] Srihari S, Thirumalini S. Investigation on reduction of emission in PCCI-DI engine with biofuel blends. Renew Energy 2017;114:1232–7.
[42] Srihari S, Thirumalini S, Prashanth K. An experimental study on the performance and emission characteristics of PCCI-DI engine fuelled with diethyl ether-biodiesel-diesel blends. Renew Energy 2017;107:440–7.
[43] Imtenan S, et al. Impact of low temperature combustion attaining strategies on diesel engine emissions for diesel and biodiesels: a review. Energ Conver Manage 2014;80:329–56.
[44] Thangaraja J, Kannan C. Effect of exhaust gas recirculation on advanced diesel combustion and alternate fuels – a review. Appl Energy 2016;180:169–84.

[45] Shim E, Park H, Bae C. Comparisons of advanced combustion technologies (HCCI, PCCI, and dual-fuel PCCI) on engine performance and emission characteristics in a heavy-duty diesel engine. Fuel 2020;262:116436.
[46] Drews P, et al. Model-based optimal control for PCCI combustion engines. IFAC Proc 2010;43(7):288–93.
[47] An Y, et al. Homogeneous charge compression ignition (HCCI) and partially premixed combustion (PPC) in compression ignition engine with low octane gasoline. Energy 2018;158:181–91.
[48] Kokjohn S, et al. Experiments and modeling of dual-fuel HCCI and PCCI combustion using in-cylinder fuel blending. SAE Int J Engines 2010;2:24–39.
[49] Wang Y, et al. Study on combustion and emission of a dimethyl ether-diesel dual-fuel premixed charge compression ignition combustion engine with LPG (liquefied petroleum gas) as ignition inhibitor. Energy 2016;96:278–85.
[50] Zhou D, et al. Efficient combustion modelling in RCCI engine with detailed chemistry. Energy Procedia 2017;105:1582–7.
[51] Krishnamoorthi M, et al. A review on low temperature combustion engines: performance, combustion and emission characteristics. Renew Sustain Energy Rev 2019;116:109404.

CHAPTER 5

Effect of engine operating parameters in NO_x reduction

A. Tamilvanan[a], B. Ashok[b], T. Mohanraj[c], P. Jayalakshmi[d], P. Dhamodharan[e], and R. Sakthivel[c]

[a]Department of Mechanical Engineering, Kongu Engineering College, Erode, Tamil Nadu, India
[b]Engine Testing Laboratory, School of Mechanical Engineering, Vellore Institute of Technology, Vellore, Tamil Nadu, India
[c]Department of Mechanical Engineering, Amrita School of Engineering, Amrita Vishwa Vidyapeetham, Coimbatore, Tamil Nadu, India
[d]Hindustan College of Engineering & Technology, Coimbatore, Tamil Nadu, India
[e]SSN College of Engineering, Chennai, Tamil Nadu, India

5.1 Introduction

The reduction of nitrogen oxides (NO_x) from engines has been mandated by government's policies globally. This means that engine manufacturers and researchers must strike the best compromise between engine emissions and performance. There are a number of strategies to diminish NO_x emission levels within boundaries without overly affecting the engine performance, including engine design parameters, operating parameters, and after-treatment systems. The formation of NO_x emissions is eliminated by optimizing the suitable design parameters such as combustion chamber geometry, compression ratio, nozzle position, and manifold design. Optimizing the engine design parameters could lead to lower NO_x emissions with an elevation of other emissions such as carbon monoxide (CO), hydrocarbon (HC), and particulate matter (PM). The influence of engine design parameters on NO_x emissions is feasible under static (specific load and speed) conditions. However, NO_x emissions are uncontrollable under dynamic running (variable load and speed) conditions because the design parameters of the engine are fixed or constant, not varied or optimized, while the engine is in running condition. From that, it is observed that design parameters are not exactly suitable for controlling NO_x emissions under variable loading conditions of the engine. Therefore, varying engine operating parameters is the best way to control NO_x emissions under different running conditions when compared to varying engine design parameters. For the last few decades, the key way to enhance internal combustion (IC) engine combustion and emissions has been through optimizing the different operating

NOx Emission Control Technologies in Stationary and Automotive Internal Combustion Engines
https://doi.org/10.1016/B978-0-12-823955-1.00005-X

strategies. The emissions characteristics of the engine mainly rely on the combustion behavior of the fuel inside the combustion chamber, where combustion can be subject to several factors such as fuel spray pattern, fuel injection pressure, fuel injection timing, fuel quantity injected, engine design such as shape and size of the combustion chamber, compression ratio, location of the fuel injector, the size and number of the injection nozzle hole, air swirl, coolant properties, inlet air properties, fuel properties, etc. [1]. From the NO_x emission point of view, diesel engines produce more NO_x emissions when compared to petrol engines as a result of nonhomogeneous air-fuel mixing, high-temperature combustion, and higher compression ratio. NO_x is produced inside the combustion chamber in the elevated temperature zones in the hot combustible gases, and the formation rate surges exponentially with rising temperature [2]. At elevated temperatures, the enrichment of oxygen and sufficient time for the reaction are the major factors facilitating NO_x emission formation [3].

There are numerous strategies to diminish NO_x emissions from the engine, but they increase the PM emissions and soot deposits on the components of the engine while decreasing engine durability. Recently, a severe drop of NO_x emissions with a minimum level of HC, PM, and CO emissions has been a key subject in a compression ignition (CI) engine to adhere to stringent exhaust emissions. A recent study on vehicle exhaust emissions shows that NO_x is a particular concern in CI engines when compared to other emissions. Several progressive technologies have been established to gradually satisfy stringent emission protocols, including exhaust gas recirculation (EGR), various in-cylinder techniques, and an after-treatment system [4]. EGR is the most promising strategy to diminish NO_x emissions from IC engines, wherever the exhaust gas reduces fresh air from entering the combustion chamber. A lesser quantity of oxygen in the fresh intake mixture decreases the air-fuel ratio which decreases flame temperature during the combustion process. This leads to decrease the NO_x emissions and power loss along with higher PM emissions [5]. Another strategy for NO_x reduction is after-treatment techniques such as a lean NO_x trap (LNT) and a selective catalytic reduction system (SCR), where NO_x emissions are reduced to a greater extent. Compared to in-cylinder techniques and EGR, the trade-off between NO_x and PM is eliminated in after-treatment techniques. However, these after-treatment devices occupy additional space in the vehicles and this technique is highly expensive [6]. Further exhaust emission has been decreased through in-cylinder techniques which resolves the expensive after-treatment arrangement. The in-cylinder combustion techniques are achieved by varying the fuel modifications of the engine. Fuel modification includes blending different forms of fuels such as biodiesel, biogas, and alcohol this could reduce NO_x emissions with a slight penalty in engine performance [7]. Additives such as higher cetane number (CN) and energy density can assist in the elimination of inferior performance. However, this leads to elevated NO_x emissions and deteriorates the life of the engine parts as well as the fuel supply system. On the other hand, variations in engine design parameters and fuel modification may not be adopted for a wide range of loads. Among those, CN variation of operating parameters is the easiest and most economical method to control NO_x emissions without any accessories such as an after-treatment system. The in-cylinder techniques reduce NO_x emissions significantly with a slight increment or decrement of other emissions such as HC, CO, and PM. Also, the same enhancement and curtailment also happen in the performance characteristics of the engine. Also, the operating parameters can be optimized based on the engine operating load range. The impact of the various operating parameters on achieving in-cylinder combustion and the influence of these parameters in the formation of NO_x emissions are discussed in the next section.

5.2 Engine operating factors influencing NO_x emissions in CI and SI engines

The in-cylinder combustion is the foremost reason for the formation of NO_x emissions in the engine. In-cylinder combustion in an IC engine can be obtained simply by varying engine operating parameters such as fuel injection parameters, inlet conditions of air, fuel modifications, coolant flow properties, engine speed, and load. However, the in-cylinder combustion techniques of spark ignition (SI) engines can be obtained by varying the ignition properties (inlet conditions, coolant flow properties, engine speed, and load). The various operating factors that affect NO_x emissions are shown in Fig. 5.1. Here, some of the operating parameters such as fuel injection, fuel ignition, equivalence ratio, angle of overlapping, air inlet pressure, and temperature are directly involved in the combustion process inside the combustion chamber; these are then called in-cylinder operating parameters. However, some other operating parameters such as coolant temperature, engine speed, and load are controlled outside the combustion chamber, and are not directly involved in the in-cylinder combustion processes. These are called non-in-cylinder operating parameters.

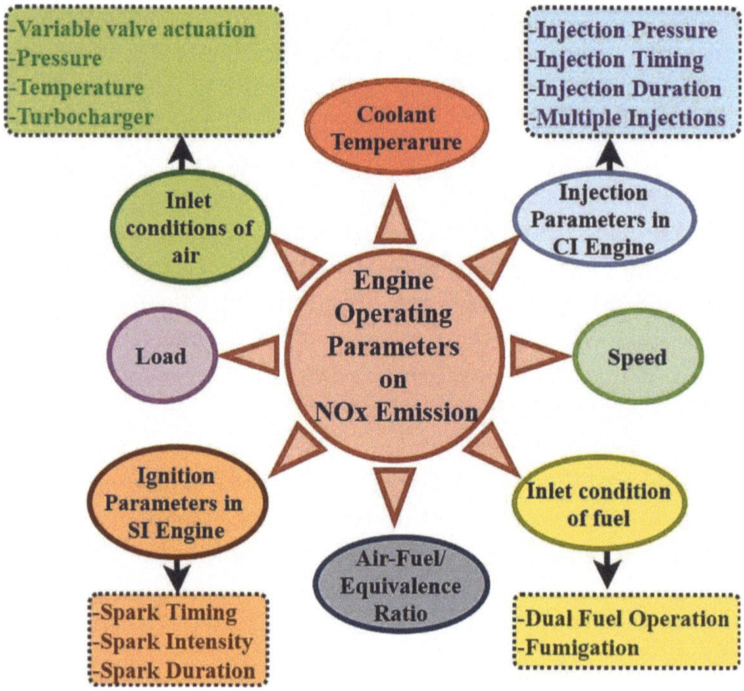

Fig. 5.1 Various engine operating factors that affect NO_x emissions.

Out of several operating parameters influencing NO_x emissions, fuel injection parameters most strongly affect NO_x emissions in CI engines. Fuel injection parameters include a variation of injection timing, injection pressure, injection duration, and multiple injections. An increase in injection pressure, advancement of injection timing, and shorter injection duration resulted in high-temperature combustion, which leads to higher NO_x emissions. A postinjection strategy, which is a shorter injection after the main fuel injection, is another potential method employed in NO_x reduction. For the same amount of fuel injected, NOx emissions decrease significantly with rises of crank angle interval and postponing post-injection timing [3]. In a standard SI engine, combustion originates by a spark generated at the sparkplug by an electrical release. Current developments in the electronic control module in the powertrain make it probable to employ online spark alteration such as spark intensity, spark timing, and spark duration to improve the engine operation in terms of better emissions without compromising fuel economy and power [8]. Meanwhile, feasible ignition parameter limits depend on the working conditions of the engine.

It is useful to extract more power with enhanced combustion and lesser emissions.

Along with other engine operating parameters, the air-fuel/equivalence ratio is a major concern for NO_x formation in IC engines. Oxygen in the air is the foremost factor in the formation of NO_x emissions in IC engines. Generally, most emission control strategies join with the air/fuel ratio control strategy of an engine that regulates and monitors the air/fuel ratio supplied into the engine with the intention of optimizing the emission reduction [9]. Excessive oxygen in the air-fuel mixture makes it tougher for that chemical reaction to happen. Challengingly, not enough oxygen makes it harder to dispose of other pollutants such as carbon monoxide and unburned hydrocarbons in the exhaust [10]. From this view, a gasoline engine typically has a good balance of oxygen when compared to a diesel engine. Apart from regulating the quantity of the air-fuel mixture in controlling NO_x emissions, the quality or condition of the inlet air is a major concern in the development of NO_x emissions. The condition of the inlet is modified by varying the parameters such as pressure, temperature, mass flow rate, and quantity of residual gases inside the cylinder. These parameters are modified by variable valve actuation and operation of the turbocharger. In the use of turbochargers, the quantity of air required for combustion inside the combustion chamber is increased by creating a pressure difference before the manifold and after the manifold. Variable valve actuation (VVA) is accomplished in many ways, but advancing and retarding the opening and closing timing of the inlet and exhaust valve are quite effective and economical. Here, the air required for combustion is adjusted along with residual gases from the previous cycle of combustion. The impacts of various in-cylinder operating parameters on the NO_x emissions of an IC engine are shown in Table 5.1.

Table 5.1 Impacts of various in-cylinder operating parameters on NO_x emissions of an IC engine.

Operating parameters	Variation	Effect on NO_x emissions	Other output parameters
Fuel injection timing	Advance (before TDC)	NO_x ↑ linearly with advance of injection	BTE ↑ up to some extent, HC & CO ↓
	Retarding (after TDC)	NO_x ↓ linearly with retarding of injection	BTE ↓, HC & CO ↑

Continued

Table 5.1 Impacts of various in-cylinder operating parameters on NO_x emissions of an IC engine—cont'd

Operating parameters	Variation	Effect on NO_x emissions	Other output parameters
Fuel injection pressure	↑	NO_x ↑ due to better atomization	BTE ↑, HC & CO ↓
	↓	NO_x ↓ due to long combustion duration	BTE ↓, HC & CO ↑
Fuel injection duration	↑	NO_x ↓ linearly longer combustion duration	BTE ↓, HC & CO ↑
	↓	NO_x ↑ linearly shorter combustion duration	BTE ↑, HC & CO ↓
Start of ignition (SOI)	Advance (before TDC)	NO_x ↑ linearly with advance of SOI	BTE ↑, BSFC ↓, HC & CO ↓
	Retarding (after TDC)	NO_x ↓linearly with retarding of SOI	BTE ↓, BSFC ↑, HC & CO ↑
Spark intensity	↑	NO_x ↑ linearly due to ↑ flame speed	BTE ↑, HC & CO ↓
	↓	NO_x ↓ linearly due to ↓ flame speed	BTE ↓, HC & CO ↑
Equivalence ratio	↓ ($\phi<1$)	NO_x ↑ and then ↓ (peak @ $\phi=0.9$)	Best economy, HC & CO ↓
	↑ ($\phi>1$)	NO_x ↑linearly with $\phi>1$	Best power, HC & CO ↑
Angle of overlap (AOL)	↓	NOx ↑linearly due to ↓ residual gas	BTE ↑, HC & CO ↓
	↑	NO_x ↓ linearly due to ↑ residual gas	BTE ↓, HC & CO ↑
Inlet pressure of the air	↑	NO_x ↑due to ↑ mass of air & (Δp↑)	BTE ↑, HC & CO ↓
	↓	NO_x ↓due to ↓ mass of air & (Δp↓)	BTE ↓, HC & CO ↑
Inlet temperature of the air	↑	NO_x ↑due to ↑ vaporization of fuel	BTE ↑, HC & CO ↓
	↓	NO_x ↓due to ↓ vaporization of fuel	BTE ↓, HC & CO ↑

Apart from controlling various in-cylinder operating parameters, some operating parameters outside the cylinder such as the cooling system, speed, and load are employed to control NO_x emission. One way of reducing NO_x emissions without regulating combustion-related parameters is called an engine cooling system. Here, some percentage of excess heat energy produced by the burning of fuel is taken by the coolant as heat carried away by cooling water because that excess heat energy is responsible for the formation of NO_x emissions. Correspondingly, the heat carried by cooling water mainly relies on the inlet temperature of the coolant, where a lower inlet coolant temperature results in more heat transfer by the coolant. This lowers the amount of heat that takes part in combustion, thereby lowering the combustion temperature, which results in lower NO_x emissions [11]. Engine speed and load are closely connected to fuel consumption and the rate of exhaust emission [12]. The time available for the combustion of the air-fuel mixture inside the cylinder is greatly influenced by speed and load. At high speeds, there is not enough time provided by the engine for the combustion of the air-fuel mixture which affects the engine performance. So, it is necessary to vary the other operating parameters along with a variation of the speed to obtain better performance and emissions. On the other hand, the load has a greater impact on the pressure and temperature of the combustible mixture inside the combustion chamber, which is the major reason for the formation of NO_x emissions. Generally, the load is the greater contributor to NO_x emissions when compared to other operating parameters. It is essential to reduce the NO_x emissions at maximum loads by controlling the other working parameters without compromising performance and fuel economy because most of the engines are operating at maximum loads from an economic perspective The impacts of non-in-cylinder operating parameters on NO_x emissions of an IC engine are shown in Table 5.2.

Table 5.2 Impacts of various non-in-cylinder operating parameters on NO_x emissions of an IC engine.

Operating parameters	Variation	Effect on NO_x emission	Other output parameters
Coolant temperature	↑	NO_x ↑ linearly due to ↓ heat rejection, (ΔT↓)	BTE ↑ up to some extent, HC & CO ↓
	↓	NOx ↓ linearly due to ↑ heat rejection, (ΔT↑)	BTE ↓, HC & CO ↑
Engine speed	↑	NO_x ↓ due to ↓ time needed for combustion (Δt↓)	BTE ↓, HC & CO ↑
	↓	NO_x ↑ due to long combustion duration (Δt↑)	BTE ↑, HC & CO ↓
Engine load	↑	NO_x ↑ linearly due to ↑ in pressure & temperature	BTE ↑, HC & CO ↓
	↓	NO_x ↓ linearly due to ↓ in pressure & temperature	BTE ↓, HC & CO ↑

5.3 Effect of fuel injection parameters on NO_x emissions in CI engines

The various injection parameters that affect NO_x emissions in the CI engine are injection timing, injection pressure, and injection duration, which are shown in Fig. 5.2. Fuel injection parameters such as injection pressure, injection timing, and number of injections can control the engine's overall efficiency among several other factors such as compression ratio, air swirl level, etc. [13]. The detailed effect of these above parameters is discussed below.

5.3.1 Injection pressure

NO_x emissions are significantly influenced by the atomization of fuel droplets in the combustion chamber, which is mainly dependent on the injection pressure of the fuel in the nozzle. Generally, NO_x emissions increase with an increase in injection pressure. The greater injection pressure of fuel generates a reduced size of fuel droplets that results in improved atomization, enhanced fuel surface area, better air-fuel mixing, improved evaporation rate of the fuel, and improved combustion [1]. Compared to that of lesser fuel injection pressures, the larger injection pressure of fuel could result in an elongated spray tip penetration and higher spray area at the equivalent

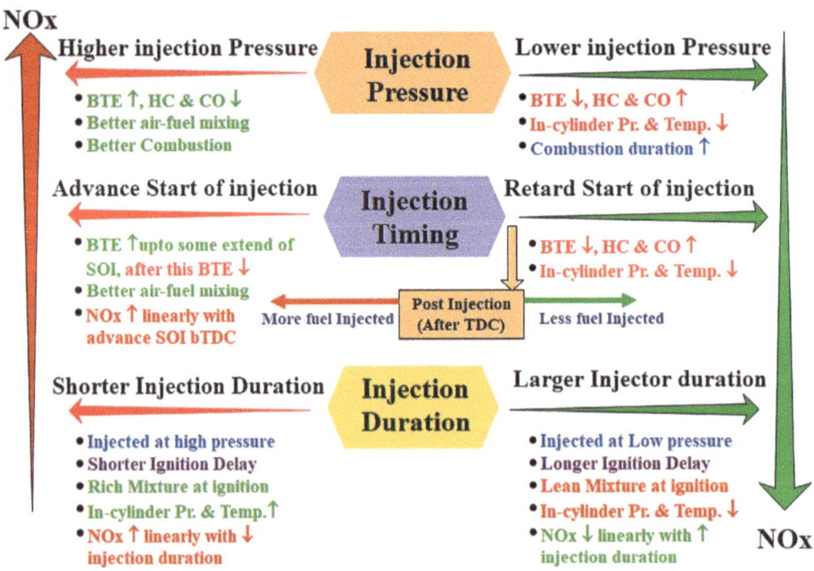

Fig. 5.2 Effect of injection parameters on NO_x emissions.

elapsed time after the start of injection (SOI). This would generate a better air-fuel mixture inside the combustion chamber, resulting in better combustion as well as elevated temperature [14]. The above factors increase the mean gas temperature inside the cylinder, which is the foremost cause for the formation of NO_x emissions. On the other hand, lower injection pressure could produce lower NO_x emissions, but there is a slight penalty in brake thermal efficiency (BTE) and HC and CO emissions. The reduction of injection pressure is not suitable for the engine running at a higher speed. Here, the lower injection pressure leads to lengthier combustion where the time required for the fuel and air mixture preparation is larger. This results in most of the mixture burning on the late combustion stroke, which causes inferior combustion and lower efficiency.

5.3.2 Injection timing

Fuel injection pressure along with injection timing have a significant role in ignition delay (ID) which affects engine combustion characteristics. As a result of injection timing, the pressure and temperature changes significantly nearer to top dead center (TDC) will be the reasons for the formation of NO_x emissions. Advances in the SOI with high injection pressure resulted in higher NO_x emissions. For the same fuel under similar load and speed, the NO_x emissions increase linearly with the advance of injection before TDC [13]. NO_x emissions are lower for retarded SOI with low fuel injection pressure; however, it was higher for the same retarded SOI with high injection pressure. Finally, retard SOI significantly decreases the NO_x emissions. Generally, biodiesel has higher NO_x emissions compared to diesel, but retarding the SOI of biodiesel fuel blends could result in lower NO_x emissions compared to diesel under similar injection pressure and load [15]. NO_x emissions are minimum when SOI timings nearer to TDC. The maximum pressure and occurrence of maximum pressure (in crank angle) inside the combustion chamber are decreased when SOI is retarded to before TDC. The lower maximum pressure could result in the creation of lower NO_x emissions [16].

Further, NO_x emissions can be decreased by post fuel injection (PFI) technique with different intervals of CA (crank angle). For similar PFI quantity, NO_x emissions have a declining tendency as CA interval rises. Contrastingly, NO_x emissions are increased for the same crank interval postinjection timings with more fuel injected. This may raise in-cylinder temperature, enabling the NO_x development to some amount. In the meantime, the formation of NO_x emissions is predisposed by the time separation of post-injection from the main injection. A short time separation between the main injection resulted in an inadequate chemical reaction, which limits NO_x formation [3]. A substantial drop of peak temperature is required for lowering the NO_x emissions, which can be achieved by late injection timing. This is mainly for transferring the main combustion phase into expansion stroke, i.e., after TDC [17].

5.3.3 Injection duration

The injection duration rate is gradually reduced with rising fuel injection pressure. The difference in injection duration is larger for comparatively lower pressures of 350 bar and 500 bar, but the duration difference is very small for the two injection pressures of 850 bar and 1000 bar [16]. This is mainly because a pressure difference exists between the injector and ambient cylinder conditions. A larger pressure difference results in a shorter injection duration, where the fuel particles travel quickly in the air inside the cylinder. In shorter injection duration, more fuel is accrued in the ID period, which leads to a slightly rich fuel-air mixture. This leads to shortening ID and higher cylinder pressure, which results in higher NO_x emissions. Generally, biodiesel blends have higher NO_x emissions when compared to diesel because of the shortening injection duration of biodiesel fuel blends. This is attributed to a higher density of biodiesel when compared to diesel fuel. Because of the higher density difference that exists between injected biodiesel and air inside the cylinder, more fuel is injected with lesser time. This leads to the shorter injection duration of biodiesel blends. A maximum reduction of NO_x is possible for late fuel injection with the intention of shifting the main combustion phase into the expansion stroke, which also results in higher PM exhaust.

5.4 Effect of fuel ignition parameters on NO_x emissions in SI engines

In this section, fuel ignition parameters such as spark timing, spark intensity, and spark duration are explored to assess the NO_x exhaust levels in the SI engine. The variation of these operating parameters is a feasible way to reduce NO_x emissions and enhance the engine process in terms of fuel economy, efficiency, and emissions of the SI engines.

5.4.1 Spark timing

Adjusting spark timing also has a crucial role in increasing or decreasing NO_x emissions. Spark timing is an alternative significant parameter for regulating combustion and dropping exhaust emissions. The disparity of NO_x emissions with respect to spark timing is shown in Fig. 5.3. In regular engine operation, the advanced start of ignition is expected for increased output power. At all loads, advances of spark timing before TDC resulted in a linear increase of NO_x emissions, which was mainly due to the increase of the average of in-cylinder temperature and elevated burned gas temperatures. However, advances in spark timing resulted in linear decreases of BSFC; therefore, spark timing is typically a trade-off between NO_x emissions and BSFC [18]. Through advancing the timing of spark ignition, the flame development period is enhanced, and the fast combustion period as well as total combustion duration are reduced. Advances of ignition timing also resulted in knocking due to the uneven burning that takes place inside the cylinder. This then resulted in elevated NO_x emissions. On other hand, NO_x emissions can be reduced to a greater extent by retarding the ignition timing; this is also accomplished through a reduction in BTE [19]. Moreover, the retardation of ignition timing often results in combustion instability such as a misfire and partial burn, which lead to incomplete combustion. The maximum in-cylinder pressure declines linearly with the retardation of spark timing from the maximum brake torque position. Also, postponing spark timing diminishes the quantity of mass burned at the end of

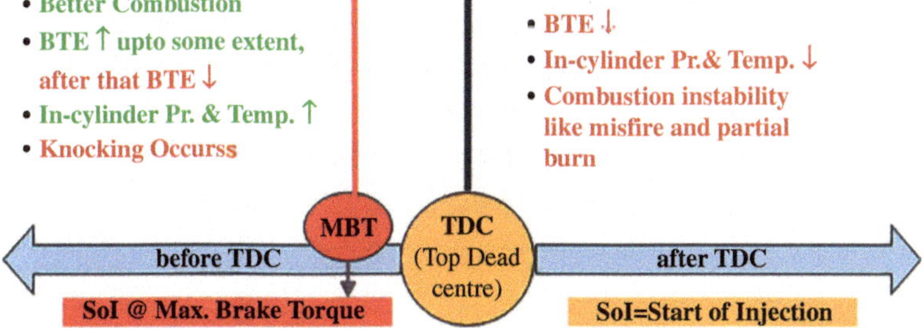

Fig. 5.3 The variations of NO_x emissions with respect to various spark timings.

the combustion stroke where larger masses are burned during the late expansion stroke, which could lead to lower NO_x emissions along with lower BTE. The expansion stroke gets decreased during retarded spark timing; part of the combustion occurs nearer the exhaust valve opening and produces minimum expansion work [20]. Anyway, the adjustable limit of ignition timing is often restricted in practical cases, because advanced and retard ignition path leads to knocking and instability in combustion. However, the NO_x emissions accompanied by advanced injection timing can be eliminated with the help of EGR, but it has a slight penalty in BTE. Finally, the NO_x emissions can be diminished in the SI engine by altering the best possible spark timing along with an appropriate percentage of EGR while confirming the economy of the engine.

5.4.2 Spark intensity

Spark intensity is another significant parameter in SI engine combustion, and it determines the speed of the combustion that takes place inside the cylinder [21]. The disparity of NO_x emissions with respect to variable spark intensity is shown in Fig. 5.4. Generally, high spark intensity is needed for the higher

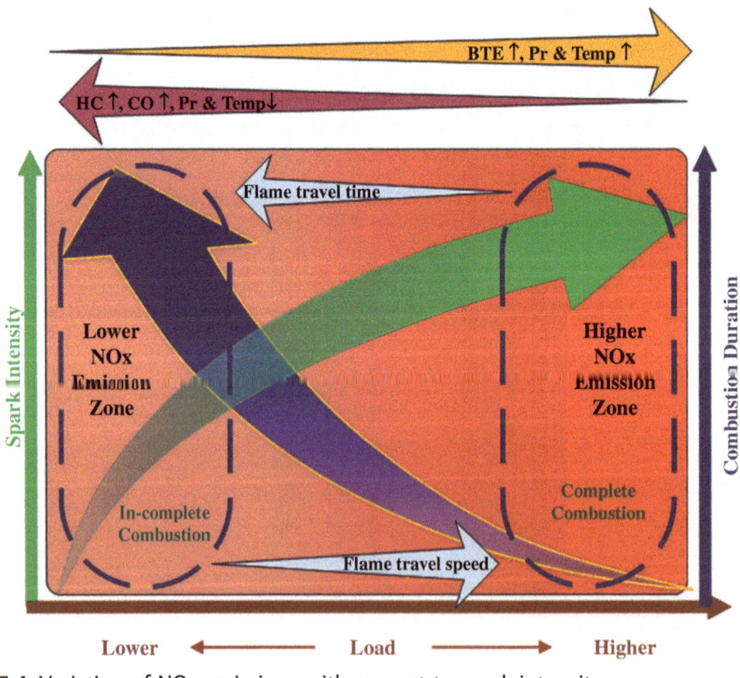

Fig. 5.4 Variation of NO_x emissions with respect to spark intensity.

load and speed of any SI engine to attain higher efficiency because the combustion duration is drastically reduced when it moves to a higher speed. Increasing the spark intensity may increase the area of the flame front with time, which could result in shorter combustion duration that leads to higher temperature and pressure inside the combustion chamber. This would probably increase the flame travel speed and rate of combustion where the time required to ignite the adjacent layer of the mixture is minimized, which leads to higher BTE and higher NO_x emissions. Here, the increase in spark intensity also led to complete combustion because the air-fuel mixture nearer the cylinder walls also ignited as a result of strong flame stretch. The spark intensity can be increased by providing the coil structure, type of sparkplugs, and number of anodes. On other hand, reducing the spark intensity resulted in lower NO_x emissions with a slight penalty in BTE with slightly higher HC and CO emissions. Lower spark intensity can cause reduced flame speed and diminished area of the flame front with respect to time, where the time required to burn the air-fuel mixture of the adjacent layer and nearer the cylinder walls is increased. This would result in lower in-cylinder pressure and temperature as well as most of the mixture burning during the late combustion stroke. This may be a reason for lower NO_x emissions along with lower BTE. Producing slightly lower NO_x emissions without compromising the efficiency by providing lower spark intensity can be achieved by providing more sparkplugs and anodes. Finally, the higher spark intensity with a suitable spark position that is nearer to before the TDC is optimum to produce slightly lower NO_x emissions without compromising the efficiency of the engine.

5.4.3 Flame travel distance

For engines with larger combustion space, the flame travel distance is the major factor for the combustion of the mixture. Mostly, all SI engines are operated in lean mixture conditions for better fuel economy. At lean operating conditions, BTE and acceleration are lowered because the spark travel distance is minimum, which results in incomplete combustion. Normally, this happens with the engine running at high speeds because the reduction of spark travel distance is needed linearly with an increase in speed where combustion is faster at this condition. In order to achieve complete burning of the mixture, increasing spark intensity and the advanced start of ignition are needed, but this results in higher NO_x emissions accompanied by power loss. In that condition, the pressure acting on the piston is uneven and the temperature variation is inconsistent, resulting in the elevation of local pressure and temperature. This problem in lean mixture conditions can be eliminated by providing an increasing number of sparkplugs as well as anodes without altering spark intensity and spark timing. By providing more than one sparkplug or more than one anode in a single sparkplug in a combustion chamber helps to ignite the lean mixture uniformly which results in stable pressure and temperature rise. Here, the average local temperature decreases result in lower NO_x emissions with better performance. Moreover, this is the most suitable technique for over square engine to achieving toward complete combustion with enhanced speed and fuel economy.

5.5 Effect of air-fuel/equivalence ratio on NO_x emissions

Generally, NO_x emissions greatly depend on how much air and fuel take part during combustion inside the cylinder. The relation between the air and fuel proportion is termed the equivalence ratio (ϕ), which is the ratio of the actual fuel-air ratio to the stoichiometric fuel-air ratio. Another term is the air-fuel ratio (λ), which is the ratio of the mass of air to the mass of fuel. During combustion with $\phi < 1$ or $\lambda > 14$ is termed a lean mixture; otherwise, with $\phi > 1$ or $\lambda < 14$, the mixture is called a rich mixture when more amount of fuel is taking part in combustion when compared to stoichiometric combustion ($\phi = 1$ or $\lambda = 14.7$). The variation of NO_x emissions with respect to different air-fuel/equivalence ratios is shown in Fig. 5.5. Most of the IC engines work under leaner mixture or stochiometric conditions because of high fuel economy as well as lower HC and CO emissions. However, it resulted in higher NO_x emissions, which is mainly due to the presence of relatively larger oxygen content when compared to fuel content at lean mixture conditions. With the engine operating at stoichiometric conditions, the flame propagation time is very low. Therefore, the heat generated inside the combustion chamber will not get time to disperse as an adequate quantity of oxygen exists for complete combustion. This results in higher combustion temperature inside the cylinder, which is the key reason for NO_x formation. Hence, it remains desirable for the SI engine to run in a little rich zone ($\phi > 1$) with EGR. Here, EGR reduces the fraction of oxygen to take part in the combustion that overwhelms the combustion temperature, which brings about lesser NO_x emissions. On other hand, engines with a rich air-fuel mixture or $\phi > 1$ resulted in lower NO_x emissions along with greater emissions of HC and CO. At an equivalence ratio greater than one, more fuel is injected relative to the quantity of oxygen for

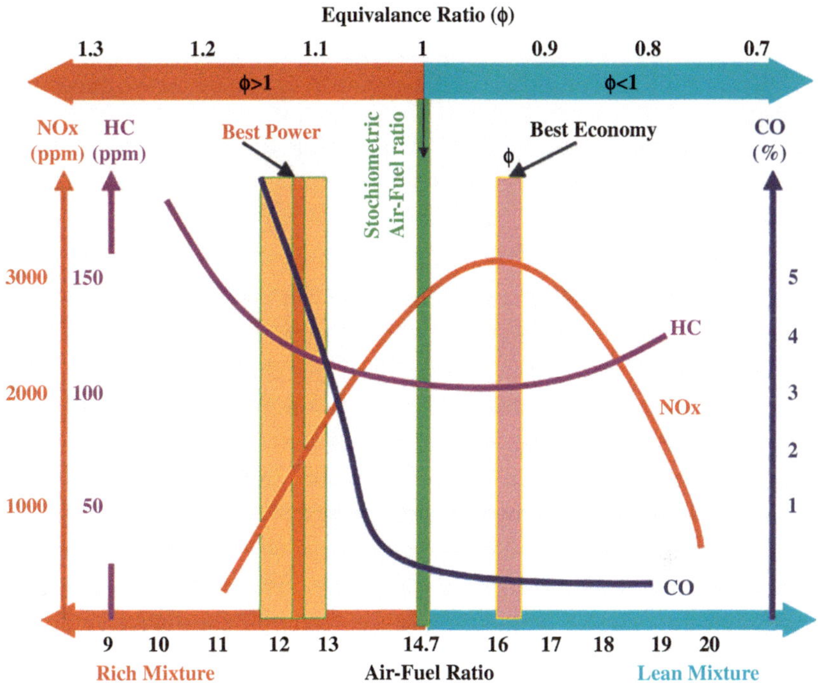

Fig. 5.5 Variation of NO_x emissions with respect to different air-fuel/equivalence ratios.

stoichiometric combustion. At that condition, the fuel should cool the combustion chamber environment owing to the latent heat of fuel, which reduces the flame speed as well as combustion temperature that caused lesser NO_x emission. At high speed and high loads, the temperature of the engine is so high, which is the foremost reason for the formation of NO_x emissions because more fuel takes part in the combustion. In this situation, the formation of NO_x is reduced by operating the engine at a slightly rich fuel mixture. Here, the fuel automatically cools the combustion environment produced by the previous cycle, which results in a reduction of maximum combustion temperature that results in lower NO_x emissions. The counterpart effect resulted in greater emissions of HC and CO due to incomplete combustion as a result of a relatively higher equivalence ratio.

5.6 Effect of inlet conditions on NO_x emissions

The condition of air at the inlet is also one of the major concerns in forming NO_x emissions apart from controlling the air-fuel/equivalence ratio of the air-fuel mixture needed for combustion. Different inlet conditions of air are achieved by varying some of the important parameters such as pressure temperature, mass, and volume flow rate of air. These key factors are optimized for different running conditions of the engine to obtain better working characteristics. These parameters can be varied with the help of a turbocharger and valve actuation techniques; this is discussed in the below subsection.

5.6.1 Variable valve actuation

Variable valve timing has a noteworthy effect on a drop in NO_x emissions compared to other emission controls. It has a promising effect of reducing fuel consumption by a gain in thermal efficiency throughout all speeds and loads. It is economical and the valve timing is optimized based on the running condition of the engine. Here, not only does the exhaust valve timing change, but the exhaust valve's lift profile is also variable. In this technique, NO_x formation is regulated by a fraction of mass of residual gas (RG) as well as dilution of fresh charge with residual gases, which is often called internal exhaust gas recirculation [22]. The quantity of residual gas is regulated by fine-tuning the opening and closing of the inlet and exhaust valve timing. However, controlling the intake valve opening is often quite effective when compared to controlling the exhaust valve. Early opening of the intake valve has been more effective in reducing both HC and NO_x emissions. Advanced opening of the inlet valve resulted in more residual gases entering the intake manifold because of high pressure in the combustion chamber. After the opening of the exhaust valve, a sudden reduction of pressure takes place in the combustion chamber. At this momentum, the fresh air/fuel charge along with the residual gas present in the intake manifold enter the combustion chamber, which reduces the amount of oxygen content in the fresh charge. This results in lowering the combustion temperature and pressure inside the cylinder that leads to lower NO_x emissions. On the other hand, advancements in the closing of the intake valve will result in an increased compression ratio, which leads to slightly higher NO_x emissions and lower HC emissions. The counter valve called the exhaust valve also plays a major role in controlling NO_x emissions by varying residual gases inside the cylinder. Advancing the closing of the exhaust valve has also been effective in reducing NO_x as well as HC emissions. This is because of the trapping of quench deposits of gases retained in the exhaust stroke, which has higher concentrations of hydrocarbons. Therefore, the left-out gases from the exhaust stroke have lower HC emissions. Moreover, the quench gas has produced an internal EGR effect in the combustion process, which reduces in-cylinder pressure and temperatures that leads to lower NO_x emissions. Delaying the exhaust valve closing after the TDC also reduces NO_x emissions by retaining residual gases during the suction stroke from the exhaust manifold [23]. In addition, to control inlet and exhaust valves individually, both valves are controlled simultaneously to regulate NO_x emissions. Here, the piston moves nearer to TDC, and both valves are open during a certain period called the valve overlapping period, which has a significant influence on NO_x emissions. Increasing valve overlapping results in a linear decrease in NO_x emissions because of higher quantity of residual gases takes part in combustion, as shown in Fig. 5.6. As overlapping period increases, opening of intake valve and closing of exhaust valve are advanced and retarded respectively. Therefore a large quantity of residual gases accumulated in the inlet and exhaust manifold attributed to the previous exhaust stroke entered into the cylinder during suction stroke. Hence, the flow of residual gas reduces the quantity of fresh oxygen contributed during combustion, which results in lower pressure and temperature inside the cylinder. Furthermore, overlapping has a significant influence on CO emissions; increasing the overlapping period results in a linear decrease of CO and HC emissions [24].

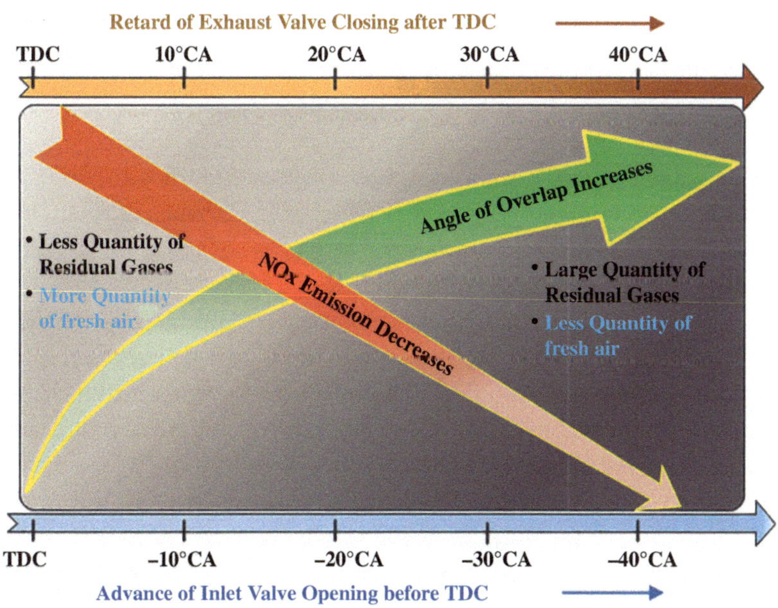

Fig. 5.6 Impact of valve overlap on the variation of NO_x emissions.

5.6.2 Turbocharger

Nowadays, a turbocharger is preferable in the IC engine due to its consistency and outstanding fuel efficiency. Here, the energy transferred along with exhaust gas is utilized by a turbocharger and transformed into valuable work. Actually, the turbocharger is typically present in existing diesel engines. In a turbocharger, the turbine is driven by the energy in the exhaust gas, where the turbine is connected to compressor blades with a shaft that upsurges the inlet pressure of the air provided to the engine. This results in a larger quantity of incoming air entering the engine cylinder; similarly, it upsurges the temperature of the air [25]. Meanwhile, it is also used for generating turbulences such as swirl and tumble motion inside the combustion chamber for better mixing of fuel-air. The velocity of the flame propagation is primarily affected by the turbulence motion of the mixture, which is the major reason for the rate of pressure as well as the formation of abnormal combustion inside the cylinder [26]. Generally, the rate of the reaction is slower for the nonturbocharged engines because of less turbulence inside the combustion chamber, where advanced spark and injection are needed for SI and CI engines, respectively, for better fuel economy and performance. This advancement may result in a longer delay period, which may affect the emission characteristics of the engine such as higher NO_x emissions. But the turbocharger induces larger turbulence, which accelerates the rate of a chemical reaction by the close mixing of fuel and air. Therefore, spark and injection advance may be limited in SI and CI engines, respectively. Here, the flame speed is increased due to turbulence, which shortens the combustion duration and thus a tendency for the occurrence of abnormal combustion is reduced at larger loads.

Conversely, the air-fuel ratio is rich at higher loads, where it results in the deficiency of oxygen that promotes the formation of black smoke. But in turbocharged engines, it is eliminated due to the availability of oxygen even at the full load. At a leaner air-fuel ratio at lower and medium loads, the burning may be affected by the improper mixing of fuel and air. This is mainly due to the lack of turbulence created inside the combustion chamber in nonturbocharged engines, which results in more unburned hydrocarbon emissions at engine idling and light load operation than at full load. But this can be eliminated by creating turbulence with the assist of a turbocharger, which also supports the burning of the lean air-fuel mixture. But extreme turbulence may quench the flame, which results in noisy and rough engine operation. Generally, turbocharger operation may reduce the combustion duration, which results in a higher premixed combustion phase and elevated rate of pressure rise. Therefore, it resulted in improved BTE accompanied by higher NO_x emissions. So, it is essential to diminish the NO_x emissions of turbocharged engines without compromising for the performance and fuel economy by varying other operating parameters such as the spark, injection, dilution of residual gas, and fuel-air ratio [27].

5.6.3 Inlet air temperature

Intake air temperature and pressure play significant roles in altering the combustion efficiency, stability, and exhaust emissions. When the air inlet temperature (AIT) increases, fuel droplets break up into finer droplets, which accelerate the molecular collision and reaction rate and promote preflame reactions. The above factors contributed to the lessened delay period, slightly higher premixed combustion, increased combustion efficiency, and enhanced BTE. Because combustion efficiency increases, a higher rate of oxidation has occurred that causes elevated temperature inside the combustion chamber, which leads to the formation of greater NO_x emissions. Beyond a certain limit of the inlet air temperature, BTE decreases because of the dilution of fuel due to the higher air-fuel ratio. This may be due to the reduced volumetric efficiency and mass flow rate of air sucked by the engine and it reflects on the engine performance as well as emissions. The very high inlet temperature of air makes the combustion more rapid, which causes abnormal oxidation processes as well as uncontrolled combustion processes. At higher AIT, the fuel is vaporized before the air-fuel charge arrives at the engine cylinder and initiates the combustion earlier, before the piston reaches the TDC position. This advanced start of combustion (SOC) increases the knocking strength and diminishes the engine power output along with increased NO_x emissions [28]. Due to the shorter ID and better evaporation rate, carbon monoxide and hydrocarbon emissions decrease for elevated air intake pressure and temperature. At the same time, at higher AIT, the flame front burns the soot particles faster and decreases the smoke emissions. Here, a rise in the air inlet temperature may help in forming a better homogenous air-fuel mixture, which increases flame speed.

When the AIT decreases, the ID period is lengthened due to the poor mixing of charge that results in the probability of accumulation of fuel particles and finally results in knocking. On the other side, due to the lower combustion temperature, the NO_x emissions decrease and the poor oxidation of carbon particles results in increased HC, CO, and smoke emissions. When the AIT is increased to an optimum level, the effect of fuel viscosity and surface tension will be overcome by the improved fuel characteristics. As a result, the fuel undergoes faster evaporation and better atomization, and the cylinder wall impingement of the fuel particles is also reduced eventually. The performance of the lubricating oil can be enhanced by reduced blow-by problems. If the AIT is too low, engine starting creates problems, and the exhaust emissions during warm-up can become excessive. Maintaining the air inlet pressure is also to be monitored carefully for optimum power.

5.7 Effect of inlet condition of fuel on engine NO_x emissions

Apart from controlling the inlet condition of air, the inlet condition of fuel is another strategy to diminish the formation of NO_x emissions inside the cylinder. The inlet condition of fuel can be varied by two methods: dual-fuel operation and fumigation. The effect of the these two parameters on NO_x emissions is discussed in the following section.

5.7.1 Dual fuel operation

Dual fuel operation can be obtained in two ways: adding or blending a suitable additive liquid fuel along with the main fuel and adding gaseous fuel along with the inlet air. In the first case, the additive fuel has a substantial part in the reduction of NO_x emissions. The additive fuel has properties such as ahigh boiling point, higher latent heat of vaporization, and lower CN. The additive fuel should absorb the heat of the cylinder, thereby reducing the temperature inside the combustion chamber that in turn reduces the formation of NO_x emissions. Moreover, the addition of additive fuel would lead to an unfavorable condition for the reaction of oxygen and nitrogen inside the combustion chamber because of insufficient oxygen, where the hydrocarbons present in the additive fuel should react with the excess oxygen in the air [29]. Also, the quantity of additive fuel may be varied with different operating conditions such as the speed and load of the engine. The additive fuel should result in enhancing the BTE and lowering the specific fuel consumption of the engine. This way, dual fuel operation is efficient to diminish NO_x emissions of the engine without compromising performance. A second case is gaseous dual fuel operation, which decreases NO_x emissions in contrast to pure diesel mode operation, where the maximum quantity of the fuel is burned under lean premixed conditions which consequence in lesser local temperature [30]. Here, oxygen in the incoming air is utilized by HC present in the gaseous fuel, thereby less oxygen takes part in the reaction with oxygen. The emissions of unburned HC under dual-fuel mode operation are noticeably larger than regular diesel mode operation, specifically at low and medium loads [31]. However, the other emissions can be controlled by varying the quantity of pilot fuel.

5.7.2 Fumigation

In fumigation, a secondary fuel is added by injecting liquid fuel at an inlet manifold along with inlet air. The fumigation of alcohols has a greater impact on engine performance and emissions. Generally, fumigation of alcohols in a diesel engine could cause a lessening of both NO_x and particulate emissions [32]. A dramatic reduction of NO_x emissions is possible in fumigation, which is the most effective technique to control NO_x emissions. The particulate and NO_x emission drop is greater with an increasing quantity of fumigation. But overall, fumigation results in larger CO and HC emissions. During fumigation, the heat inside the cylinder due to the previous cycle is absorbed by the fumigated fuel during the suction and compression stroke because of the high latent heat of vaporization. This results in reduced in-cylinder combustion pressure and temperature, which may lead to a longer ignition delay period attributed to lower NO_x formation accompanied by larger HC and smoke emissions. An increasing quantity of fumigation may lead to a linear decrease in NO_x emissions where the maximum heat release rate and cylinder peak pressure occur after some degrees of TDC during the expansion stroke. This may also result in lower BTE where a major part of combustion occurs during the expansion stroke, which is possibly high at low loads. This fumigation method is more suitable for medium and high loads where the combustion temperature is slightly higher, which is required for the atomization of pilot injected fuel [33].

5.8 Effect of coolant temperature on NO_x emissions in CI and SI engines

The inlet coolant temperature has a substantial consequence on the performance, volumetric efficiency, and emission characteristics of the engine. The volumetric efficiency of the engine decreases with an increase in the coolant temperature of the engine. The impact of the coolant inlet temperature on performance and emissions characteristics is represented in Fig. 5.7. It also shows that an increase in the temperature of the coolant could result in elevated cylinder wall temperature, which leads to an upsurge in the heat transferred from the wall to the cold fresh charge during the suction stroke.

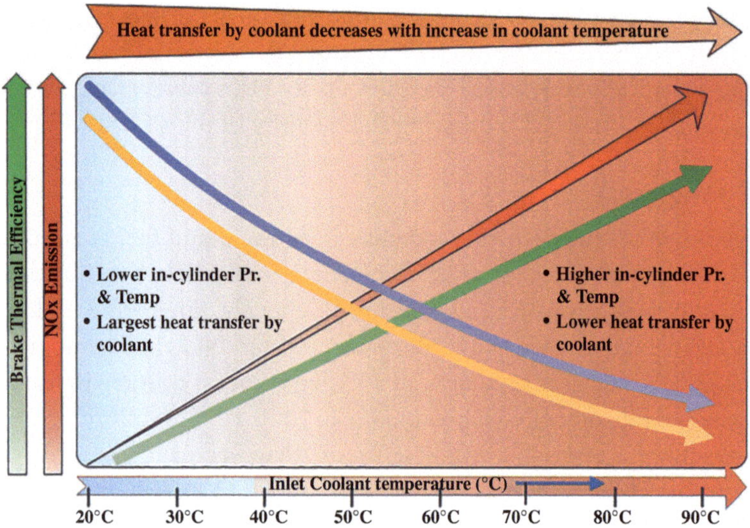

Fig. 5.7 Impact of inlet coolant temperature on the variation of NO_x emissions.

This would probably increase the fresh charge temperature, which diminishes the mass flow rate and density of air; this can subsequently decrease the volumetric efficiency. Also, the equivalence ratio declines with upsurging the engine coolant temperature [34]. In that condition, the mass of the fuel flow rate is reduced by a greater rate than the equivalent diminished mass of airflow rate. Due to this effect, the equivalence ratio of the air-fuel mixture decreases by means of increasing the engine coolant temperature. Also, increasing the coolant temperature decreases the coolant loss, which in turn elevates the conversion of heat into work. As an outcome, an increase in BTE and a reduction of the brake-specific fuel consumption (BSFC) are obtained. From the emission point of view, the coolant temperature has a considerable consequence on NO_x emissions and slight significance on carbon monoxide, carbon dioxide, and oxygen volumetric percentages. Increases in coolant temperature reduce heat loss from the cylinder wall to the coolant, which in turn increases the heat transfer to the air-fuel mixture. This would result in elevated combustion of reaction, which upsurges the in-cylinder gas temperature that leads to the formation of greater NO_x emissions. Also, it seems that NO_x emissions increase linearly with increasing coolant temperature because loss of heat through the cooling system gets reduced while moving to a higher inlet coolant temperature [35].

Generally, raise in coolant inlet temperature can decreases equivalence ratio, fuel mass flow rate, cooling water loss, and volumetric efficiency.

These circumstances can develop a high temperature oxidation zone inside the combustion chamber that results in the formation of higher NO_x emissions accompanied by higher BTE and CO_2 emissions. However, the emissions of CO and HC are reduced to a greater extent as a result of superior oxidation. The decreasing trend of the above parameters gets reversed by reducing the inlet coolant temperature. Here, NO_x emissions decrease linearly with reducing inlet coolant temperature along with a slight reduction in BTE and CO_2 emissions. But HC and CO emissions are slightly increased due to a little inferior combustion. As the coolant temperature decreases, the heat taken by the cooling water increases, and heat contributed to the combustion decreases. As a result of this heat transfer, the peak combustion temperature becomes reduced, which brings about lower NO_x emissions [36]. In modern systems, engines are equipped with engine thermal management (ETM) and active coolant control (ACC) to regulate heat rejection as well as exhaust formations without compromising power loss. This can be achieved by an electrical pump with variable speed and a smart valve with variable position, which regulates the distribution of coolant flow rate between the radiator and its bypass line to control heat rejection [37]. Along with coolant inlet temperature, the mass flow rate of coolant has a great influence on NO_x emissions. For the same inlet temperature and different flow rate conditions, NO_x emissions decrease linearly with increasing mass flow rate of the coolant accompanied by slightly higher CO and HC emissions. This is because the higher mass of the coolant makes the engine cylinder temperature low, resulting in poorer oxidation [38].

5.9 Effect of engine speed on NO_x emissions

Generally, the engine should be operated at high engine torque speed to reduce BSFC. However, at lower and higher engine speeds, the stability of in-cylinder combustion degrades, which results in higher CO, HC, and NO_x emissions. Engine speed is one of the major factors for estimating the duration of combustion, which has a major impact on the formation of NO_x emissions [39]. Commonly, the duration needed for combustion decreases linearly with an increase in engine speed where the duration of cycles has been increased while running with high speeds. During high speeds, the time required for the combustion of the air-fuel mixture is not sufficient when compared to low speeds at the same fuel-air ratio and compression ratio. Otherwise, the time required for the completion of a working cycle is reduced,; similarly, the time of the combustion process is also reduced. This means that the working fluid residence time in the cylinder at high temperature is reduced. The above has a substantial influence on the development of NO_x emission because the critical time period for the formation of NO_x is the time between the SOC and the time at which the peak cylinder pressure occurs. Thus, NO_x emissions decrease with the increase in speed of diesel engines. At maximum speed conditions, the fuel given to the engine is greater, and more fuel is burned through the diffusive combustion phase. This is because insufficient combustion time at premixed combustion also resulted in a lower premixed combustion phase. This resulted in lower NO_x emissions accompanied by lower BTE [12]. To overcome the reduced BTE, the fuel injection is advanced to some extent when compared to low-speed conditions. Thus, it resulted in higher efficiency along with higher NO_x emissions and lower HC emissions. Contrarily, NO_x emissions are higher at higher speed along with higher loads because at higher loads, CI engines are operated at a lean mixture to attain better fuel economy. This upsurge in NO_x emissions is mainly due to the higher pressure and temperature of the air, which make the fuel vaporize easily at high loads and greater turbulence is obtained as a result of the sudden pressure difference due to high speeds. These two factors created a better air-fuel mixture inside the cylinder, thus resulting in improved combustion efficiency. However, the escalation of these NO_x emissions is reduced by operating engines with a slightly higher equivalence ratio.

In SI engines, the NO_x emissions are lower for engines running at higher speeds at constant spark timing and mixture strength, which is accompanied by lower BTE. This is because of the lesser time available for combustion, which results in reduced peak pressure and lower combustion efficiency. In order to avoid performance losses, advancing spark ignition timing is needed. For instance, the speed is increased to 2N, then the spark is advanced to 2d to obtain the same peak pressure, where "N" and "d" are speed and degree of advance before TDC, respectively. This is because the flame propagation speed is halved when the engine speed is doubled. However, at high loading conditions, the high pressure and temperature inside the combustion chamber make the delay period shorter. This resulted in slightly higher premixed conditions leading to higher NO_x emissions when compared to lower load and higher speeds. Subsequently, the duration of the combustion is reduced by increasing the flame speed at higher speeds. This occurred by creating turbulence inside combustion inside the combustion chamber with the assist of a turbocharger operation and by varying the design of the combustion chamber and inlet manifold. Here, the flame speed upsurges linearly with rising engine speed because it elevates the rate of turbulence inside the combustion chamber.

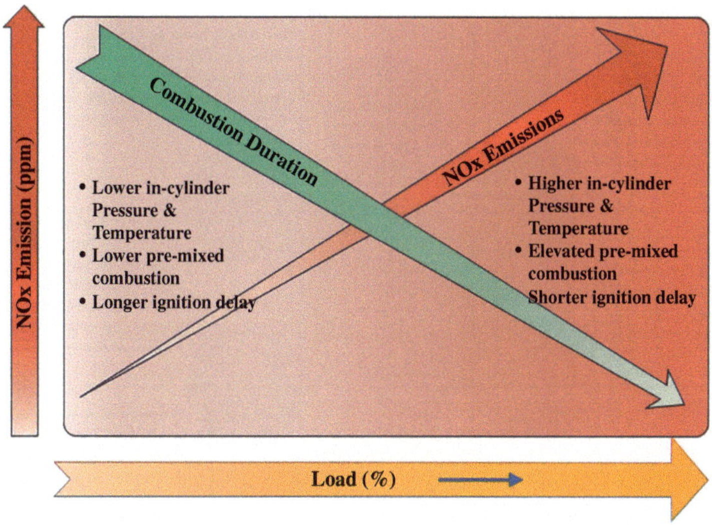

Fig. 5.8 Impact of engine load on the variation of NO_x emissions.

5.10 Effect of engine load on NO_x emissions

The formation of NO_x emissions during combustion fluctuates with peak combustion temperature where it mainly relies on engine load. The impact of engine load on the variation of NO_x emissions is shown in Fig. 5.8. The combustion temperature tends to upsurge as the engine load increases, which results in an increased rate of NO_x formation [40]. NO_x emissions of both CI and SI engines increase linearly with increases in engine load. By increasing the engine load, the flame speed may increase due to a diminished ID period, which results in elevated premixed combustion. This is attributed to increased cycle operating pressure and temperature, and the combustion chamber is occupied with a slightly rich air-fuel mixture when compared to lower loads [39].

5.11 Comparison of different operating parameters

NO_x formation does not rely on a single operating parameter but depends on different operating parameters. The way to reduce NO_x formation may affect the engine performance and other emissions such as CO and HC; it also affects fuel economy. This can be overcome by operating the engine with optimum operating parameters with a slight effect on performance as well as fuel economy, which is shown in Fig. 5.9. The operating parameter can be optimized over the entire load range to obtain lower

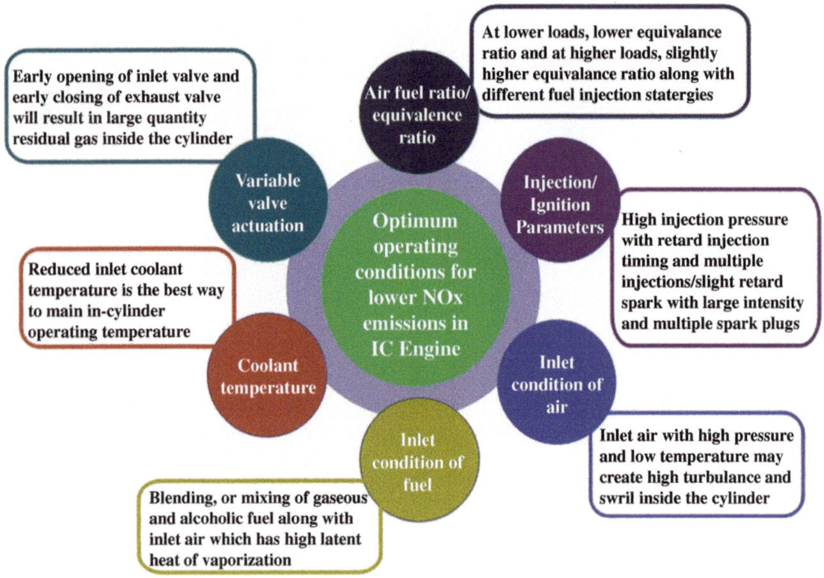

Fig. 5.9 Optimum operating parameters for lower NO_x emissions in IC engines.

NO_x emissions without compromising engine performance and fuel economy. The coolant inlet parameter is best to control NO_x formation because it doesn't directly contribute to the in-cylinder combustion process. Also, it doesn't affect the engine performance. Following that, the inlet conditions of the air and fuel are significant aspects in NO_x formation. Also, the amount of residual gas during combustion is a significant factor in NO_x reduction, but it results in lower performance.

5.12 Conclusion

The reduction of NO_x emissions from IC engines has attracted toward the researches who are working in the development of advanced combustion strategies as well as after treatment techniques. Commonly, in-cylinder combustion strategies are more feasible because the formation of NO_x is controlled inside the combustion chamber when compared to after-treatment techniques, where the generated NO_x is reduced. The key benefit of varying different operating parameters is controlling NO_x formation at different speeds and load conditions with a slight or negligible effect on engine performance as well as other exhaust emissions. Generally, every engine should operate to produce better performance with the best fuel economy, which would lead to higher NO_x emissions as a result of enhanced combustion and oxidation inside the cylinder. Moreover, a large amount of NO_x is produced during high engine operating temperature because NO_x emissions primarily rely on operating temperature. However, this higher NO_x could be reduced by running the engine with different operating parameters to reduce engine operating temperature, resulting in a negligible or slight loss in engine performance and fuel economy. Varying different engine parameters is the most economical method to reduce the formation of NO_x emissions when compared to other NO_x control techniques.

References

[1] Jaichandar S, Annamalai K. Combined impact of injection pressure and combustion chamber geometry on the performance of a biodiesel fueled diesel engine. Energy 2013;55:330–9.

[2] Mohan B, Yang W, Kiang Chou S. Fuel injection strategies for performance improvement and emissions reduction in compression ignition engines—a review. Renew Sust Energ Rev 2013;28:664–76.

[3] Liu W, Song C. Effect of post injection strategy on regulated exhaust emissions and particulate matter in a HSDI diesel engine. Fuel 2016;185:1–9.

[4] Ashok B, Nanthagopal K, Chyuan OH, Le PTK, Raje N, Raj A, et al. Multi-functional fuel additive as a combustion catalyst for diesel and biodiesel in CI engine characteristics. Fuel 2020;278:118250.

[5] Agarwal D, Singh SK, Agarwal AK. Effect of exhaust gas recirculation (EGR) on performance, emissions, deposits and durability of a constant speed compression ignition engine. Appl Energy 2011;88(8):2900–7.

[6] Pandiaraj S, Subbaiyan D, Ayyasamy T, Nagarajan S. Design, development and analysis of mullite catalytic converter for CI engines. SAE Technical Paper; 2019. https://doi.org/10.4271/2019-28-0017. Report No.: 0148-7191.

[7] Tamilvanan A, Balamurugan K, Ashok B, Selvakumar P, Dhamotharan S, Bharathiraja M, et al. Effect of diethyl ether and ethanol as an oxygenated additive on Calophyllum inophyllum biodiesel in CI engine. Environ Sci Pollut Res 2020;1–19. https://doi.org/10.1007/s11356-020-10624-3.

[8] Haskara I, Zhu GG, Winkelman J. IC engine retard ignition timing limit detection and control using in-cylinder lonization signal. SAE Trans 2004;1537–45.

[9] Lewis DJ. System and method for controlling the air/fuel ratio in an internal combustion engine. Google Patents; 2003.

[10] Sakthivel R, Mohanraj T, Baranitharan P, Tamilvanan A, Gomathi K. Emission aspects of biomass-based advanced second generation bio-fuels in IC engines. Recent technologies for enhancing performance and reducing emissions in diesel engines. IGI Global; 2020. p. 44–64. https://doi.org/10.4018/978-1-7998-2539-5.ch003.

[11] Abdelghaffar WA, Osman MM, Saeed MN, Abdelfatteh AI, editors. Effects of coolant temperature on the performance and emissions of a diesel engine. Internal combustion engine division spring technical conference; 2002.

[12] Kean AJ, Harley RA, Kendall GR. Effects of vehicle speed and engine load on motor vehicle emissions. Environ Sci Technol 2003;37(17):3739–46.

[13] Kannan G, Anand R. Effect of injection pressure and injection timing on DI diesel engine fuelled with biodiesel from waste cooking oil. Biomass Bioenergy 2012;46:343–52.

[14] Agarwal AK, Dhar A, Gupta JG, Kim WI, Lee CS, Park S. Effect of fuel injection pressure and injection timing on spray characteristics and particulate size–number distribution in a biodiesel fuelled common rail direct injection diesel engine. Appl Energy 2014;130:212–21.

[15] Ye P, Boehman AL. An investigation of the impact of injection strategy and biodiesel on engine NOx and particulate matter emissions with a common-rail turbocharged DI diesel engine. Fuel 2012;97:476–88.

[16] Agarwal AK, Dhar A, Gupta JG, Kim WI, Choi K, Lee CS, et al. Effect of fuel injection pressure and injection timing of Karanja biodiesel blends on fuel spray, engine performance, emissions and combustion characteristics. Energy Convers Manag 2015;91:302–14.

[17] Suh HK. Investigations of multiple injection strategies for the improvement of combustion and exhaust emissions characteristics in a low compression ratio (CR) engine. Appl Energy 2011;88(12):5013–9.

[18] Li Y, Wang P, Wang S, Liu J, Xie Y, Li W. Quantitative investigation of the effects of CR, EGR and spark timing strategies on performance, combustion and NOx emissions characteristics of a heavy-duty natural gas engine fueled with 99% methane content. Fuel 2019;255:115803.

[19] Van Blarigan A, Kozarac D, Seiser R, Chen J, Cattolica R, Dibble R. Spark-ignited engine NOx emissions in a low-nitrogen oxycombustion environment. Appl Energy 2014;118:22–31.

[20] Russ S, Lavoie G, Dai W. SI engine operation with retarded ignition: part 1-cyclic variations. SAE Trans 1999;1522–31.

[21] Brequigny P, Halter F, Mounaïm-Rousselle C, Dubois T. Fuel performances in spark-ignition (SI) engines: impact of flame stretch. Combust Flame 2016;166:98–112.

[22] Bechtold R, Marshall W. Valve timing: its effect on emissions and fuel economy. Bartlesville, OK: Energy Research and Development Administration; 1977.

[23] Pourkhesalian AM, Shamekhi AH, Salimi F. NOx control using variable exhaust valve timing and duration. SAE Technical Paper; 2010. Report No.: 0148-7191.

[24] Tomoda T, Ogawa T, Ohki H, Kogo T, Nakatani K, Hashimoto E. Improvement of diesel engine performance by variable valve train system. Int J Engine Res 2010;11 (5):331–44.

[25] Karabektas M. The effects of turbocharger on the performance and exhaust emissions of a diesel engine fuelled with biodiesel. Renew Energy 2009;34(4):989–93.

[26] Jiaqiang E, Zhao X, Qiu L, Wei K, Zhang Z, Deng Y, et al. Experimental investigation on performance and economy characteristics of a diesel engine with variable nozzle turbocharger and its application in urban bus. Energy Convers Manag 2019;193:149–61.

[27] Jung C, Park J, Song S. Performance and NOx emissions of a biogas-fueled turbocharged internal combustion engine. Energy 2015;86:186–95.

[28] Torregrosa A, Olmeda P, Martin J, Degraeuwe B. Experiments on the influence of inlet charge and coolant temperature on performance and emissions of a DI diesel engine. Exp Therm Fluid Sci 2006;30(7):633–41.

[29] Tamilvanan A, Balamurugan K, Mohanraj T, Selvakumar P, Ashok B, Sakthivel R. Influence of nano-particle additives on bio-diesel-fuelled CI engines: a review. Recent technologies for enhancing performance reducing emissions in diesel engines. IGI Global; 2020. p. 85–104. https://doi.org/10.4018/978-1-7998-2539-5.ch005.

[30] Liu J, Yang F, Wang H, Ouyang M, Hao S. Effects of pilot fuel quantity on the emissions characteristics of a CNG/diesel dual fuel engine with optimized pilot injection timing. Appl Energy 2013;110:201–6.

[31] Ayyasamy T, Balamurugan K, Duraisamy S. Production, performance and emission analysis of Tamanu oil-diesel blends along with biogas in a diesel engine in dual cycle mode. Int J Energy Technol Pol 2018;14(1):4–19.
[32] Zhang Z, Tsang K, Cheung CS, Chan TL, Yao CD. Effect of fumigation methanol and ethanol on the gaseous and particulate emissions of a direct-injection diesel engine. Atmos Environ 2011;45(11):2001–8.
[33] Jiang Q, Ottikkutti P, VanGerpen J, VanMeter D. The effect of alcohol fumigation on diesel flame temperature and emissions. SAE Trans 1990;892–915.
[34] Ganesan V. Internal combustion engines. McGraw Hill Education (India) Pvt Ltd; 2012.
[35] Thomas S, Saroop A, Rajak R, Muthiah S. Investigation on the effect of coolant temperature on the performance and emissions of naturally aspirated gasoline engine. SAE Technical Paper; 2011. Report No.: 0148-7191.
[36] Heywood JB. Internal combustion engine fundamentals. McGraw-Hill Education; 2018.
[37] Mohamed ES, Allam EM. Technology. Effect of active cooling control on internal combustion engine exhaust emissions and instantaneous performance enhancement. Brit J Appl Sci 2016;12(2):1–16.
[38] Pang HH, Brace C, Akehurst S. Potential of a controllable engine cooling system to reduce NOx emissions in diesel engines. SAE Technical Paper; 2004. Report No.: 0148-7191.
[39] Sun Z, Wang X, Wang X, Zhou J, editors. Combustion and emission analysis of heavy-duty vehicle diesel engine. AIP Conference Proceedings. AIP Publishing LLC; 2017.
[40] Tamilvanan A, Balamurugan K, Vijayakumar M. Effects of nano-copper additive on performance, combustion and emission characteristics of Calophyllum inophyllum biodiesel in CI engine. J Therm Anal Calorim 2019;136(1):317–30.

CHAPTER 6

Application of exhaust gas recirculation of NOx reduction in SI engines

Dhinesh Balasubramanian[a,b,c], Inbanaathan Papla Venugopal[a], Rajarajan Amudhan[a,d], Tanakorn Wongwuttanasatian[b,c], and Kasianantham Nanthagopal[e]

[a]Department of Mechanical Engineering, Mepco Schlenk Engineering College, Sivakasi, India
[b]Mechanical Engineering, Faculty of Engineering, Khon Kaen University, Khon Kaen, Thailand
[c]Center for Alternative Energy Research and Development, Khon Kaen University, Khon Kaen, Thailand
[d]Department of Mechanical Engineering, CK College of Engineering and Technology, Cuddalore, India
[e]Engine Testing Laboratory, School of Mechanical Engineering, Vellore Institute of Technology, Vellore, Tamil Nadu, India

6.1 Introduction

In this chapter, we will discuss the control of NOx using exhaust gas recirculation (EGR, as presented in Box 6.1) in petrol engines. The effect of the NOx trade-off with hydrocarbons and other emissions in the SI engine while varying the EGR percentage has been presented. The various opportunities of stratified exhaust gas such as lateral stratification, radial stratification, and axial stratification have been discussed while the characteristic comparison of hot and cooled EGR has also been studied. The rate of enhancement of cooled EGR systems with a turbocharged SI engine for NOx reduction and knocking as well as dissociation with possible technologies has also been reviewed. The implementation of EGR in advanced SI engine techniques such as gasoline direct injection, multi-point fuel injection, and lean-burn combustion has been discussed. This chapter will also discuss the implementation of dedicated EGR powered by natural gas and EGR implementation powered with hydrogen-assisted jet ignition in SI engines.

The exhaust gas of an engine is recirculated back into the combustion chamber to mix with the fresh inlet charge, and also, it does not take part in a combustion reaction. The application of exhaust gas still manages to reduce the emissions of NO_X that too significantly. EGR means exhaust gas recirculation, and as the name suggests, the exhaust gas of an engine is recirculated back into its combustion chamber in some amount with fresh

NOx Emission Control Technologies in Stationary and Automotive Internal Combustion Engines
https://doi.org/10.1016/B978-0-12-823955-1.00006-1

BOX 6.1 Abbreviation.

BDC: bottom dead center
LP: low pressure
CI: compression ignition
MBT: maximum brake torque
HC: hydrocarbon
N_2: nitrogen
H_2O: water
RPM: revolutions per minute
EGR: exhaust gas recirculation
O_2: oxygen
CO: carbon monoxide
NOx: nitrogen oxides
CO_2: carbon dioxide
SI: spark ignition
HP: high pressure

air or charge. It is used in both CI and SI engines to decrease the formation of NO_X emissions. The mass of the exhaust introduced in the engine depends upon the conditions of load, engine design, RPM, type of combustion that takes place, and cylinder temperature. The maximum percentage of recirculation depends upon the type of ignition. EGR is effected in the CI engine at the intake stroke when the entire cylinder is filled with air, and then it is compressed. As compression increases, the temperature and the pressure in the cylinder rise, after which the fuel is injected; it autoignites and starts burning. Now, the oxygen in the air completes the combustion in the cylinder. The burning of the fuel increases the temperature, and the pressure and the piston get pushed down to the BDC of the cylinder; hence, the power is generated. At higher load conditions, the entire oxygen available inside the combustion chamber would be used for burning the fuel. But when the engine is at part load condition, the power requirement is less. Hence, it requires less fuel, which is evident as there is no need to waste more fuel. Inside the cylinder, there is enough oxygen available, sufficient to burn a large amount of fuel. But at partial load conditions, the amount of fuel quantity is less and so the entire oxygen available inside the cylinder will not be utilized. So, the combustion temperature is high, and more vacant oxygen atoms are available [1]. These atoms react with nitrogen and form nitrogen oxides, NO_X, a major hybrid pollutant produced as per the pollution norms. But it is impossible to stop the entry of nitrogen inside the combustion chamber, as 71% of the air consists of it. We need to allow less air with a limited amount of oxygen to enter, which would be sufficient to burn the fuel at 50% load condition. If there is no excess air supply, then there are no extra vacant oxygen molecules inside the cylinder. But at the compression stroke, due to a lack of mass, the rise in the pressure and the temperature will be less. So when the fuel is injected, it cannot get autoignited, as the pressure and the temperature are not sufficient for combustion to take place. Here, the application of EGR exists, as some of the exhaust gas from the earlier cycle is diverted, and then it is cooled to some extent. The temperature of the gas is relatively lower than the exhaust gas, but it is more than the atmospheric temperature. Now, the exhaust gases are recirculated again to the inlet through the EGR valve. According to the load condition, the opening of the EGR valve varies. Lowering the load will make more exhaust enter the inlet manifold while as the load rises, more fresh air is allowed, and exhaust gas recirculation is decreased.

The combustion chamber is filled with a mixture of fresh air and recirculated exhaust gas. So when it gets compressed, the temperature and the pressure rise would be high. The exhaust gas recirculation also adds up to the temperature level. When the fuel is injected now, it autoignites quickly, decreases the ignition lag, and gives controlled combustion. After combustion completion, there is no oxygen left inside the cylinder, and only a certain mass of fresh air is allowed to enter, which is sufficient enough to burn the exact quantity of injected fuel. All the gas used to fill the combustion chamber is recirculated exhaust as it is already burned up gas; it does not have any vacant oxygen molecules, and now it can burn again. This recirculation of inert exhaust gas serves the purpose of increased mass required for compression and controlling the NO_X formation by limiting the supply of oxygen atoms. The decrease in temperature decreases the chances of knocking and NOx formation if some vacant oxygen atoms are still left. The recirculated exhaust is inert, and hence it does not take part in the combustion reaction. After combustion, the combustible gases rise in temperature but not in recirculated exhaust temperature. The recirculated exhaust will absorb heat from the combustible gas, and this heat absorption will result in a decrease in the combustion chamber's internal temperature, and it cools. For SI engines, the use of EGR is to decrease the internal temperature and NO_X formation. Due to EGR application in SI engine, the inlet mixture with more mass introduction may lead to autoignition, which is not desired as it may lead to knocking in SI engines. Apart from cooling, there is another advantage of a reduction in pumping losses.

In the SI engine, the flow of mass of air is restricted to restrict the oxygen supply and hence limit the NO_X production directly. This reduction in air-flow also helps in decreasing the knocking tendency. But this restriction in air supply increases the pumping losses. At suction, the piston has to suck air from a smaller opening of the throttle, which is reduced to restrict the mass of air with EGR. EGR can allow more air to enter, decreasing the pumping losses while controlling the oxygen supply at the same time. For SI engines at partial load conditions, the opening of EGR is inversely proportional to the engine's load. Due to the rise in load, more fuel is needed to burn. Hence the amount of oxygen requirement increases by the reduction in input quantity EGR. More amount of fresh air enters into the inlet manifold at heavy load conditions. At the same time, EGR valve is closed to ensure a high abundance of oxygen to burn more fuel thus producing more torque and power. Even at an idle condition, the EGR valve is closed. The simple reason behind this activity is the engine idling when it stops at signals, and in case the EGR valve is open, the RPM of the engine is low to save the fuel, but it also means each cycle takes more time to complete. If accelerated, the EGR valve will close, but the already sucking air is just sufficient to burn less fuel. So, the whole cycle ends with less power, and in the next cycle, it consumes the exhaust mixed in the air, which is present in the inlet manifold. As a result, the whole cycle is wasted again, and after that, the fresh air will come as the engine is at the lower speed at idle, and each cycle takes more time to complete. Hence this will give a lag in acceleration, which is not desired at all, and therefore the EGR valve is closed at idle also. This is the way EGR works in SI engines and helps to decrease NO_X and engine temperature. It also has the advantage of reducing pumping losses in SI engines, and this is shown in Table 6.1.

6.2 Different types of EGR set-up

The supply of exhaust gas back to the combustion chamber can be done using two different methods. One is supplying the exhaust gas externally

Table 6.1 Advantages of EGR use in CI and SI engines.

In CI engine	In SI engine
Decrease in NO_X	Decrease in NO_X
Decrease in engine temperature	Decrease in engine temperature
Decrease in ignition lag	Decrease in pumping loss

to the inlet manifold while the other is to decrease the timing of the opening of the exhaust valve, making the exhaust gas remain inside the cylinder. Significant modifications are needed in cam profiles to achieve variable valve timing to control the exhaust gas flow effectively. Due to such complications, this method has a substantial disadvantage in its usage in the SI engines. The external supply of exhaust gas is a widely used method among many researchers because of its low-cost implementation. In this methodology, the exhaust gas flow is controlled effectively by a separate EGR valve [2–4]. A dedicated stainless steel pipe system along with the EGR control valve and EGR intercooler are used. The exhaust gas passes through this system and enters into the inlet manifold. The EGR control valve varies the amount of exhaust gas entering into the inlet side. The EGR intercooler cools the exhaust gas, which would be at a high temperature when moving out from the combustion chamber [2].

In the case of multicylinder engines, the EGR design is based on the acceptable dynamic response and homogeneous distribution of exhaust gas to all the cylinders. In general, there are two ways of supplying the exhaust gas to the cylinder, centralized and decentralized EGR. Fig. 6.1 shows the centralized EGR system in which the EGR control valve would be at the starting of the inlet manifold. This allows the proper mixing of the exhaust gases over the inlet air, thus entering the cylinders. Fig. 6.2 represents the decentralized EGR, which supplies the exhaust gases to the respective cylinders through the inlet valves. This can be divided into two different methods: one method is fitting an EGR valve for each cylinder, and another way is the use of single multiple valves that supply exhaust gas equally to all

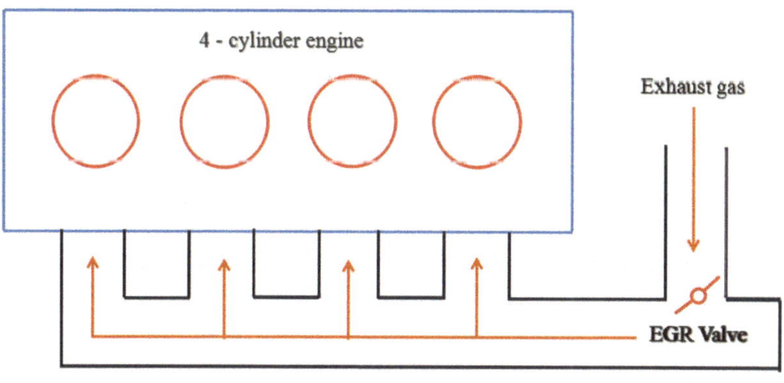

Fig. 6.1 Centralized EGR with a single EGR valve at the end of the inlet manifold.

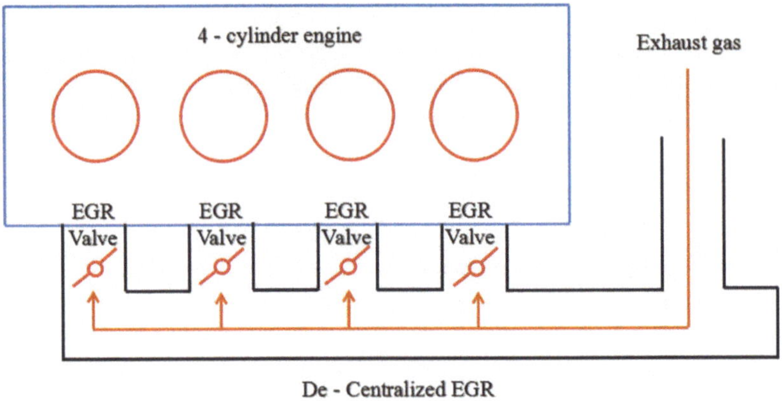

Fig. 6.2 Decentralized EGR with an EGR valve for each cylinder in the inlet manifold.

Table 6.2 Comparison between decentralized and centralized EGR systems.

Decentralized EGR system	Centralized EGR system
Lesser thermal load on the components	Smaller in size
Components of the throttle valve and intake manifold are less contaminated	Lower costs
Distribution of EGR is controlled	Less in weight

the cylinders. The advantage of decentralized EGR over centralized EGR is an excellent dynamic response. The extra flow of exhaust gases produces the stratification of fuel inside the combustion chamber. This flow of exhaust gas is done by proper EGR distribution inside the cylinder because of the appropriate arrangement of EGR control valves. But the still type of method is not used widely due to some constraints in the EGR valve assembly [5]. Table 6.2 shows the comparison between the centralized and decentralized EGR systems for a four-cylinder diesel engine.

6.3 Stratified form of EGR

The homogeneous exhaust gas recirculation system reduces laminar flame speed, burning speed, cycle-to-cycle variations, and rise in hydrocarbon emission. It also makes it difficult to achieve steady combustion [6, 7]. During the suction and compression strokes, highly stratified EGR will diversify the air-fuel mixture, and it will quickly overwhelm the above complications. In a stratified EGR system, it separates fresh air and EGR inside the cylinder.

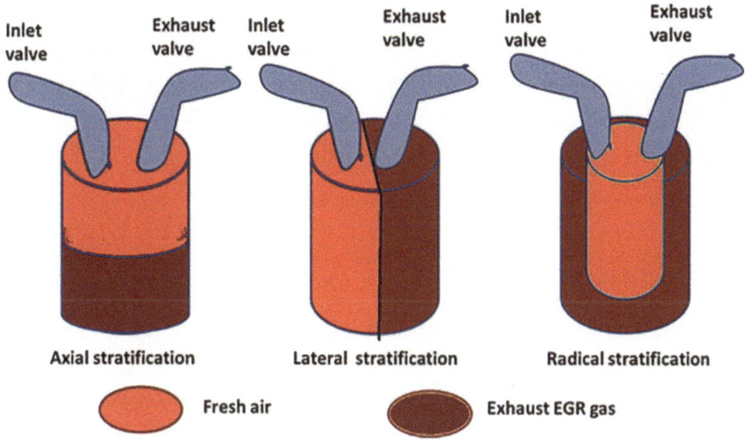

Fig. 6.3 Stratified EGR techniques.

Flame propagation is improved when a lower concentration of exhaust gas exists near the spark plug, and the compatibility of EGR is also increased. Due to the complex flow inside the cylinder, it is difficult to separate the exhaust gas and air. The second challenging task is to achieve EGR stratification at the suction stroke and maintain it during the compression stroke.

The various opportunities of stratified exhaust gas, such as lateral stratification, radial stratification, and axial stratification, are shown in Fig. 6.3. In lateral stratification, the separation happens as a mixture of air and fuel exists at the combustion chamber's intake valve and EGR exists at the exhaust valve. This system requires a large intensity of solid tumble to keep the momentum in the vertical flow direction [8]. In radial stratification, air and EGR would swirl in the same direction to reach the required force balance due to the conservation of angular momentum at the location of the interface. In axial stratification, the combustion cylinder is split into two different regions–one at the top with air and the other at the bottom with EGR. The injected air first pierces into the pure air zone, then fuses with EGR, and then again would be ignited and burned. A flow in the vertical direction, such as squeeze, mixes the gases thoroughly.

The tumble flow is simply collapsed by turbulent flow when the piston transports to the top dead center. But in radial stratification, the arrangement seems to easily maintain the tumble flow. In this type, the mixture of air and fuel exists at the center of the cylinder, and EGR exists in the cylinder's outer area. In the compression stroke, the piston moves upward, and the cylinders are compressed in the axial direction. To reach a new equilibrium stage, the two regions would rearrange their location boundary by maintaining the angular momentum when both EGR and air are swirling in the same direction. Radial stratification will withstand considerably extended time toward the end of the compression stroke [9]. Due to this, radial stratification is the most suitable technique for intake and exhaust stratification.

6.4 Hot and cooled EGR

The main difference between cooled and hot EGR is that if the exhaust gas is supplied directly to the inlet manifold, it would be hot EGR. If the exhaust gas is supplied after cooling using an intercooler, it is known as cooled EGR. Hot EGR increases the thermal efficiency while cooled EGR increases the volumetric efficiency of the gasoline engine. The use of cooled EGR reduces NOx emissions while it slightly increases the HC emissions due to the dilution effect, which is created inside the combustion chamber. A characteristic comparison of hot and cooled EGR is shown in Table 6.3.

Both types of EGR supply have significant effects on HC and NOx emissions. When the cooled EGR supply rate is increased at the inlet manifold, NOx emission decreases while an increase in the hot EGR supply rate raises NOx emissions. The reason behind this is the reduction in the stability of combustion inside the cylinder along with the increased cooled EGR rate. Simultaneously, the increase in hot EGR raises the combustion temperature and facilitates the better mixing of fuel and air, leading to an increase in NOx emissions. The amount of EGR addition into the fresh inlet charge has a specific limit since the excess supply of EGR might quench the flame, which would ultimately increase the HC emission. In HC emissions, hot EGR would produce less HC while cooled EGR dilutes the fresh charge, increasing the HC emission values. Table 6.4 shows the various strategies used in cooled EGR and their merits.

Table 6.3 Hot and cooled EGR characteristic comparison.

Characteristics	Hot EGR	Cooled EGR
NOx	High	Low
HC	Low	High
Knock	High	Less
Structure	Simple	Complex
Cost	Low	High

Table 6.4 Various strategies used in cooled EGR.

Investigators	EGR strategy	Merits	Demerits
Potteau et al. [10]	Cooled EGR in turbocharged SI engine	Reduction in BSFC, NO_X, and CO	Soot formation
Kaiser et al. [11]	Cooled EGR indirect injection SI boosted engine	Improvement in thermal efficiency and reduction in NO_X	–
Liu et al. [12]	Cooled EGR in downsized direct-injection SI Engine	Improvement in fuel economy and reduction in NO_X	–

6.5 Correlation between knock and NOx emissions

In diesel engines, EGR has been used widely to decrease the cylinder's temperature, thereby reducing NOx formation inside the combustion chamber [13]. The mixing of exhaust gas with air intake minimizes the appearance of very high temperatures inside the cylinder. EGR also affects the heat release rate and rate of pressure rise inside the combustion chamber, which would ultimately reduce the combustion noise produced from the engine [14, 15]. Due to higher NOx emissions and fuel consumption, nowadays various countries around the world have introduced several norms on NOx emission values. While moving toward this objective, gasoline engines are developed to drive the future world without the disadvantages of diesel engines. To obtain high power density, SI engines are attached with superchargers and turbochargers. The former receives power from the crankshaft while the latter takes power from the exhaust gas. Turbocharged SI engines are more popular in NOx reduction nowadays despite their problems at higher loads due to knocking and high exhaust gas temperatures [16].

Knocking is produced due to the compressed charge's autoignition effect before the flame formed by the sparkplug reaches it. The preignition of the compressed charge has to be controlled inside the combustion chamber. Many researchers have conducted various studies on reducing the formation of knocking inside the cylinder [17–20]. Knock formation is primarily due to the high-end gas pressure and temperature. The reaction rate of the end gas reaches the spontaneous ignition level due to the uncontrolled expansion of the combusted gases. Many researchers [10, 21] have found that knock occurs due to the acceleration of flame inside the homogeneous compressed charge, which is on par with the flame propagation theory. Knock occurs due to the increased speed of flame rather than the combustion noise [21]. The increased rate of pressure rise and temperature increase the knocking tendency sharply, which would cause more engine damage. The knock tendency would be measured by using a band-pass filter. The filter senses the signals from the pressure of the knock produced. The amount of knocking cycles would also be determined by using a band-pass filter. A reference voltage would be set in the filter to compare with the signals generated in the filter concerning the cylinder pressure. For every cycle, the voltage generated from the pressure variation would be compared with the reference. When the induced voltage is higher than the reference, knocking occurs inside the combustion chamber [10, 21].

In this chapter, more concentration is given on the discussion of knock suppression methods in comparison with the variation in NOx emissions. The above-mentioned methods can be divided into three different types, which are a decrease in compression ratio, an air-fuel ratio reduction, and supply of cooled EGR. The decline in compression ratio will result in a decrease in in-cylinder pressure. Air-fuel ratio reduction results in an excess amount of fuel injection, which decreases the temperature combustion inside the cylinder. In this method, HC and CO emissions would increase because of the reduction in in-cylinder temperature.

Also, the working of a three-way converter at the exhaust would be affected because it operates well only in stoichiometric conditions. This operation also increases the fuel consumption inside the cylinder. While comparing the above two methods, the supply of cooled EGR in the fresh intake charge would be the most efficient alternative in reducing knock formation. Also, the EGR would allow the prevailing of overall stoichiometric conditions inside the combustion chamber. In SI engines, cooled EGR use would be far better than the air-fuel ratio reduction to suppress knock effectively [22, 23]. Table 6.5 shows the fuel economy improvements in cooled EGR

Table 6.5 Fuel economy improvements in cooled EGR.

EGR type	Benefit	Reason
Cooled EGR	Knock suppression	Advanced ignition timing Increased compression ratio
	Increase in relative specific heat	Improved thermal efficiency
	Lowering exhaust gas temperature	Reduced fuel enrichment

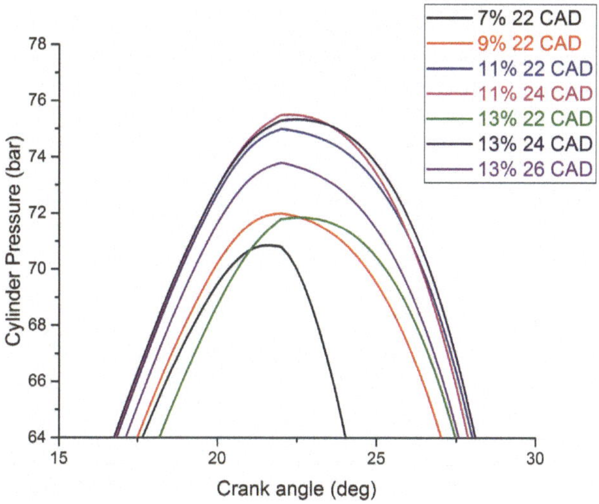

Fig. 6.4 Crank angle vs. cylinder pressure.

Fig. 6.4 depicts the cylinder pressure variation and the various crank angle degrees at different EGR rates and spark timings. In Fig. 6.4, the legends with 7% 22 CAD represent the EGR rate and respective spark timing. The spark timing in the legends would define the crank angle before TDC while the x-axis in Fig. 6.4 shows the crank angle degrees after TDC. We can see that the cylinder peak pressure rises as the EGR rate increases while the advance in spark timing along with the EGR rate rise restrict the rise in peak cylinder pressure. An optimum spark timing with an increased EGR rate would be preferred for the maximum cylinder pressure inside the combustion chamber. The reason behind the rise in in-cylinder pressure along with the increase in EGR rate would be the dilution effect caused inside the cylinder, thus reducing the peak combustion temperature, which induces the formation of knock. This knock formation increases the peak cylinder pressure.

Fig. 6.5 represents the variation of average generated amplitude concerning variations in EGR temperature. In this study, the EGR level was set at a 10% rate, and the angle of ignition was set at 20 crank angle degrees before top dead center. The BMEP was maintained at 1.64 bar ±0.1 bar. In the test, the EGR temperature varied from 80 to 200°C with a 20°C rise at each step. The intake air temperature was maintained at 40°C before EGR mixing, and so the inlet manifold temperature dropped down to 37°C. The results from Fig. 6.5 indicate that the engine at low EGR temperatures produced more knocking combustion.

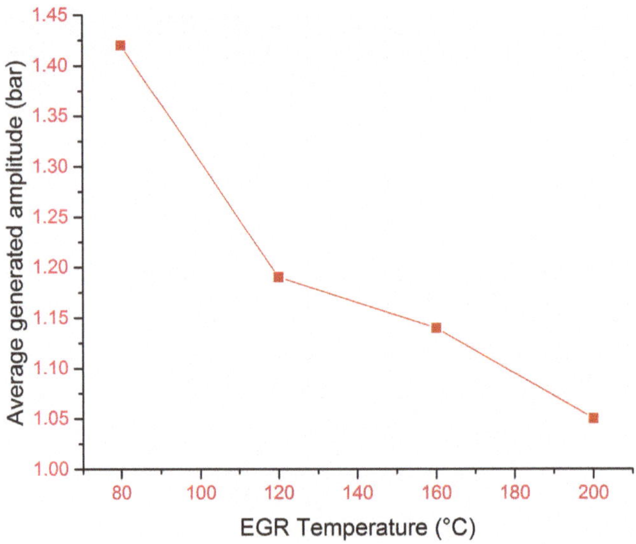

Fig. 6.5 EGR Temperature vs. average generated amplitude.

6.6 EGR vs. NOx and soot emissions

The formation of NOx emissions is because of the increased temperature inside the combustion chamber and the high oxygen content. NO is the main constituent in the emission of NOx in spark ignition and diesel engines because oxidation occurs while reacting with air to form NO_2. The increase in temperature will mainly cause NOx to increase. The NOx formation rate is meager at 1800 K. It will increase sharply after reaching 2500 K. After reaching that level, the rate of formation of NOx would double. NOx formation varies from CO and HC. NOx would be formed due to the complete combustion of the mixture. The NOx formation reaction is extremely complex. Mainly, the nitrogen content is rich in fossil fuels. At high-temperature combustion, N_2 and O_2 in the air thermally react to form NOx. In the postflame combustion process, NOx is formed. The creation of nitrogen oxides during combustion could be derived by the Zeldovich mechanism [9, 24, 25]. The major reactions that take place for the conversion of molecular nitrogen to NO at a stoichiometric air-fuel mixture are:

$$N + O2 \Leftrightarrow NO + O \tag{6.1}$$

$$N + OH \Leftrightarrow NO + H \tag{6.2}$$

$$N2 + O \Leftrightarrow NO + N \tag{6.3}$$

Eqs. (6.1), (6.2), and (6.3) denote the initial formation of NO at a controlled rate. In this derivation, NO depicts molar concentration while $[N_2]e$ and $[O_2]e$ denote concentrations at equilibrium [16]. The influence of oxygen and temperature concentrations in NO formation is shown in the above equations. The reduction of NO in the combustion chamber is the only means of reducing temperature and oxygen.

$$\frac{d[\mathrm{NO}]}{dt} = \left[\frac{6\times 10^{16}}{T^{0.5}}\right] \exp\left[\frac{-0.69.096}{T}\right] [\mathrm{O_2}]_e^{0.5} [\mathrm{N_2}]_e \ \mathrm{mol\,s/cm^3} \tag{6.4}$$

In the spark-ignition engine, EGR is the highly used method for reducing NO_X emissions. H_2O, N_2, CO_2, and O_2 are the major constituents of EGR. The CO_2 exerts three effects in the combustion chamber:

1. Thermal effect: The various emissions released from the exhaust contain two and three atom molecules. The capacity of heat of the three atom molecules in the combustion reaction rises rapidly. The flame temperature will reduce by the effect of the improved heat capacity of the oxidizer.
2. Chemical effect: In the combustion process, CO2 becomes an active molecule, and chemically involves in the reaction.
3. Dilution effect: This effect is created due to the oxidizer content reduction and reduced reactive species in the combustion reaction. Then, it decreases the collision frequency.

These three effects have been referred to from the articles [26–28]. From various tests carried out in the engines, many have determined that NOx is highly reduced when the concentration of oxygen is concentrated in the combustion chamber [1]. The use of EGR in petrol engines replaces O_2 molecules by the exhaust gas, due to which the O_2 content inside the combustion chamber reduces. When the oxygen content in the cylinder is decreased, a specified volume of fuel will diffuse before ultimately attaining sufficient oxygen to form a stoichiometric mixture. But for a prescribed volume of fuel, the combustion cylinder not only contains a stoichiometric mixture but also includes an additional amount of N_2, CO_2, and H_2O. Other gases absorb the surplus energy released during combustion. This leads to NOx formation, and the flame temperature will be lower [29, 30]. EGR effects in the SI engine and its pollutant formation are shown in Fig. 6.6.

Fig. 6.7 represents the variation of CO, HC, and NOx emissions concerning the change in the EGR rates of 0–10%. For the fuel enrichment case, the emissions obtained are 200 ppm, 1300 ppm, and 40,000 ppm of NOx, HC, and CO emissions, respectively. The maximum emission values in

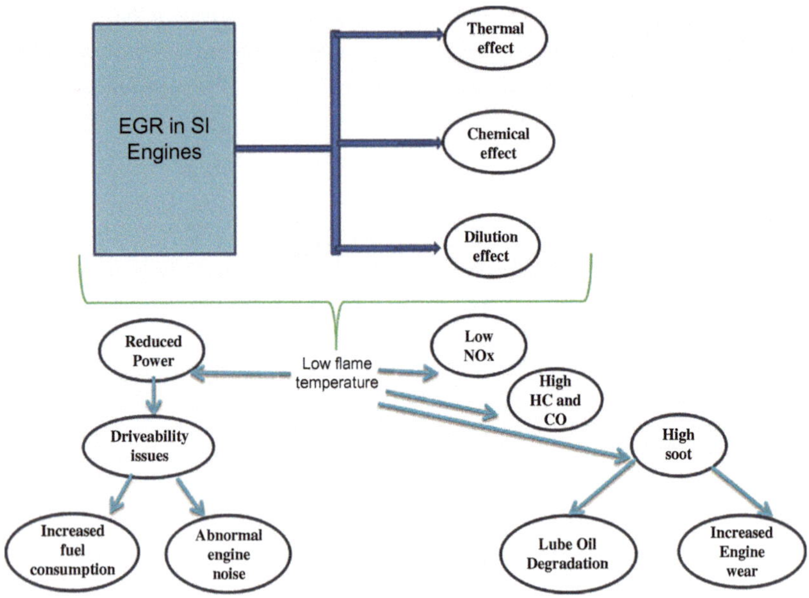

Fig. 6.6 EGR effects in the SI engine and its pollutant formation.

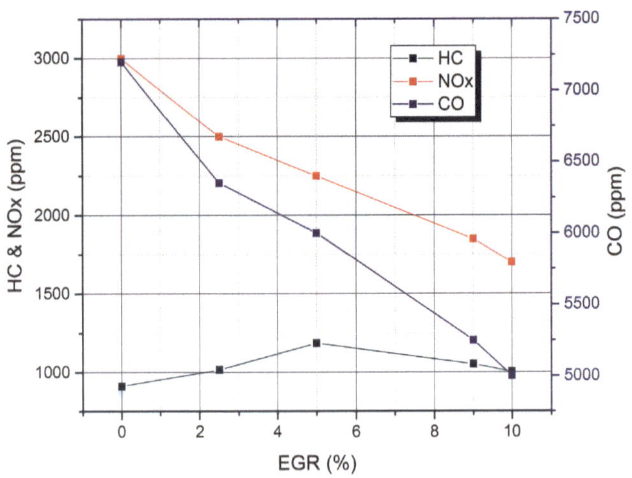

Fig. 6.7 Variation of HC, CO, and NOx emissions with various EGR rates.

EGR addition are 3000 ppm, 1200 ppm, and 7250 ppm of NOx, HC, and CO, respectively, as shown in Fig. 6.7. Using EGR over fuel enrichment is due to the decreased CO emission values of about 4%. If the mixture is stoichiometric, then the decrease in CO is about 0.8%. HC emissions are also reduced by about 0.08% while NOx emissions increase for the stoichiometric mixture.

The significant benefit of SI engines over diesel engines is that operation in stoichiometric combustion promises the three-way catalyst's optimal functionality. All the significant pollutants (CO, HC, and NO_X) have a high conversion rate achieved using a catalyst in SI engines. SI engines with unleaded gasoline do not have major difficulties concerning particulate matter emissions or soot [31]. The soot reduction in diesel engines has been extensively done by many researchers, which made SI engines also to concentrate more on reduction of soot particles [32, 33]. For soot formation, the blend preparation approach in engines plays a significant part [34]. Through a three-way catalyst, the ultrafine particle emissions pass through it. They need the installation of filters such as gasoline particulate filters, which would increase the cost and make it difficult for the after-treatment arrangements.

Despite the widespread investigation of soot formation, explicit knowledge of the essential phenomenon and a complete chemical reaction for the increase of soot nuclei have not been defined [31, 35]. Modern researchers developed a soot formation model for gas turbine simulation to characterize soot emissions [36]. It contains all the chemical and related physical methods of forming soot, and it is confirmed for partially premixed and diffusion flames with dissimilar fuels.

Soot emissions depend upon the stability between the nonequilibrium process of oxidation and formation. This development depends on pressure, temperature, fuel type, and oxygenated additives [31, 35]. Much research for nonaromatic and aromatic fuels has reported that particulate matter emissions show a bell-shaped trend as a function of temperature. The soot volume increases with temperature at lower temperatures. At higher temperatures, the relation is inverted [37–40]. The extreme soot yield temperature is a function of fuel and differs over diverse experimental formations. The studies on simulation and the practical part denote steadied flames in a burner. The extreme temperature is simply a function of the heat losses by radiation and conduction as well as heat release from combustion. The same studies concerning temperature have been investigated for diesel through the T-φ maps [41, 42].

The equivalence ratio is the crucial parameter affecting soot formation in a spark-ignition engine. During a stoichiometric or lean mixture, most fuels have minimum soot formation [43]. The influence of EGR on soot formation has been investigated on direct injection engines using gasoline fuel. Due to cooled EGR, the reduced combustion temperature minimizes the particulate matter and soot emissions [44–46]. At lower loads, the effect of combustion temperature influences the rate of particulate formation.

At greater loads, the eradication of fuel enrichment is the main reason for the reduction of soot [44, 45]. The lesser exhaust temperature with cooled EGR is attained by working closer to the maximum brake torque owing to knock mitigation with the capability of absorbing extra heat. Therefore the conversion from enriched fuel to the stoichiometric mixture considerably decreases soot emissions. But a further rise of EGR results in improved soot [46]. For those cases, the formation of soot emissions is less due to concentrated soot oxidation owing to lower oxygen concentration and lower temperatures [47]. As a result, cooled EGR has the advantage over soot for the SI engine's functioning regime.

6.6.1 Fuel/air ratio on NOx emissions

The fuel-air ratio is an essential parameter in combustion because it significantly impacts the variation in NOx emissions. A rich fuel-air mixture will reduce NOx emissions while a lean fuel-air mixture would increase NOx emissions. However, too lean a mixture would deteriorate the quality of combustion and would also lead to a decrease in NOx emission values. In this study, the advantage of EGR over fuel enrichment has been shown clearly. An experiment was conducted with 0.9 bar EGR boost pressure and a rich fuel-air ratio of about 0.9. The ignition timing was set close to TDC at 6.6 crank angle degrees before TDC to avoid knocking. Fig. 6.8 represents the variation of cylinder pressure along with the change in crank angle at the prescribed air-fuel ratio of 0.9 and the EGR rate of 10%. The delayed

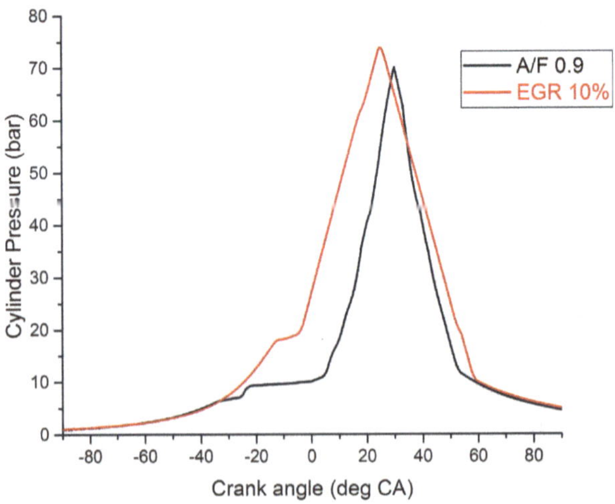

Fig. 6.8 Crank angle vs. cylinder pressure (A/F and EGR comparison).

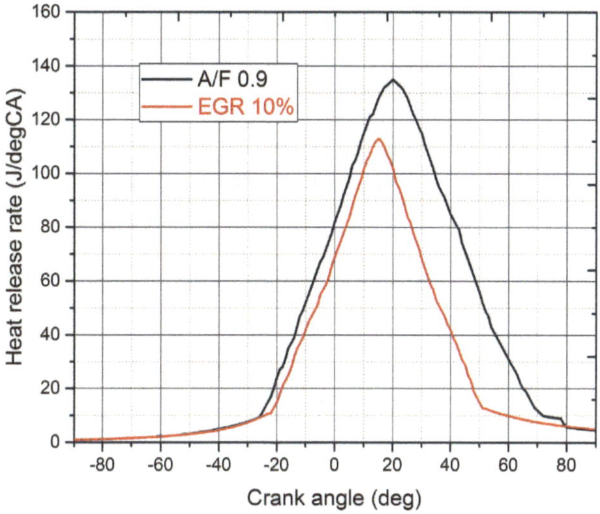

Fig. 6.9 Crank angle vs. heat release rate (A/F and EGR comparison).

ignition angle resulted in lower peak cylinder pressure. Fig. 6.9 shows the heat release rate variation along with the change in the crank angle. The addition of EGR has been achieved by using a longer duration of burning. Here, the combustion is divided into two different parts. In the first part, the maximum amount of burn duration would take place. In the air-fuel ratio of 0.9, 10% burn duration took place at 16.6 crank angle degrees while in the case of the EGR 10% rate, 10% burn duration took place at 24.5 crank angle degrees. Thus, the development of flame in the case of EGR is affected more than the propagation of turbulent flame. Fig. 6.10 shows that the increase in EGR rate would reduce NOx emissions along with a drastic reduction in peak heat release rate and peak cylinder pressure. Because the exhaust gas led inside the inlet charge dilutes it and reduces the temperature and pressure inside the cylinder, it reduces NOx emissions.

6.6.2 Effect of ignition timing on NOx emission

Concerning the initial ignition timing, its effect on NOx emissions would also differ in the case of GDI engines. When the engine runs at the maximum brake torque condition, an early ignition will increase the NOx emissions because more time would be available for combustion. Retarded ignition timing from maximum brake torque would increase the exhaust temperature and also the better postflame characteristics. This would reduce the formation of NOx with the reduction in peak temperature inside the cylinder.

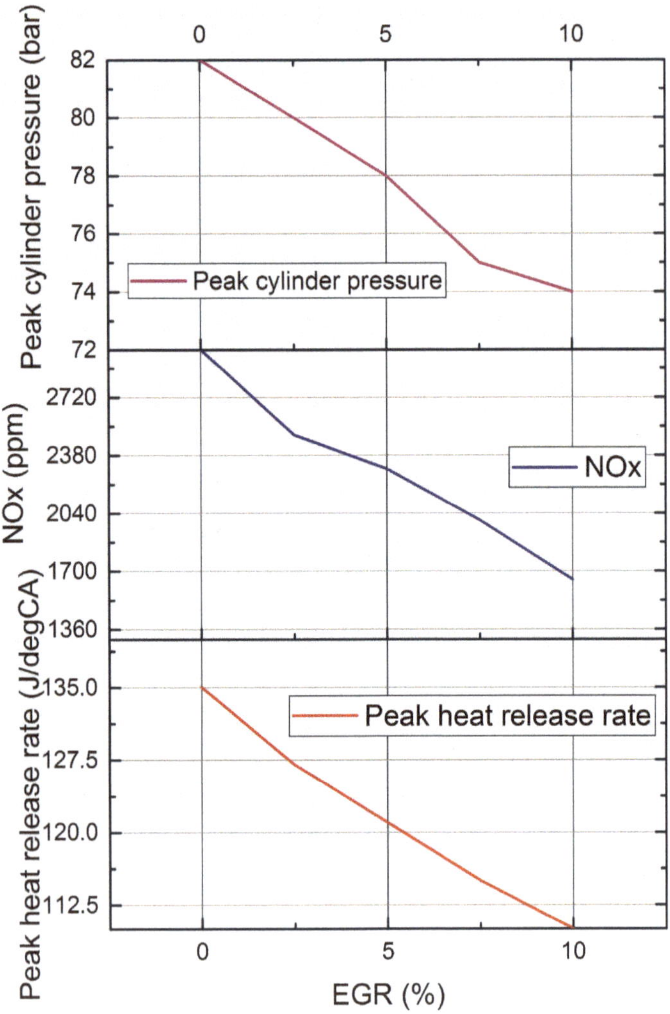

Fig. 6.10 EGR vs. NOx, peak cylinder pressure and peak heat release rate.

Fig. 6.11 depicts the variation in the consumption of fuel concerning the change in ignition timing. Here, the comparison between fuel enrichment and EGR rate has been shown in terms of fuel consumption. At higher loads, a rich mixture is needed to withstand the load condition and higher fuel consumption. With EGR at a 13% rate, fuel consumption decreases about 10% with a proper knocking combustion inside the cylinder.

Fig. 6.12 represents the variation of brake mean adequate pressure concerning the various spark timings at different EGR rates supplied to the inlet

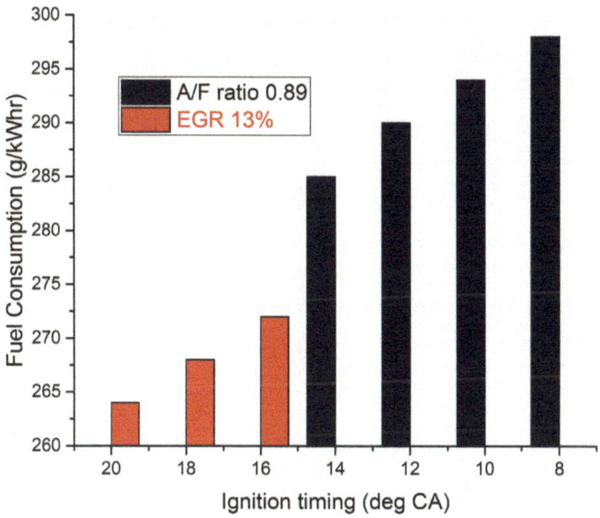

Fig. 6.11 Fuel consumption vs. ignition timing.

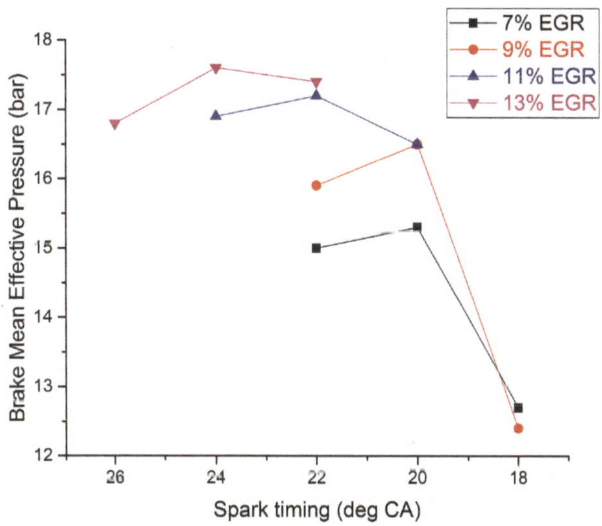

Fig. 6.12 Brake mean effective pressure vs. spark timing.

charge. In this study, maximum brake mean sufficient pressure is restricted by the angle of ignition and the EGR rate. The detection of knock by the band-pass filter has been set above 2.5 bar cylinder pressure and 960°C of exhaust gas temperature [17]. From Fig. 6.12, concerning the increase in EGR rate (7% 13%), maximum BMEP would increase along with the increase in the angle of ignition. The increase would reduce the restriction of the rise in BMEP in the EGR rate. The mass of the compressed charge inside the combustion chamber would be increased, thereby lowering the combusted gas temperature by diluting it before it reaches the high temperature. Also, more fuel can be combusted within this limit. With the same EGR rate of 13% when the ignition angle is advanced, the BMEP would be steady with no further rise because it is restricted by the formation of knock inside the cylinder. From Fig. 6.12, concerning the increase in EGR rate, NOx emissions decrease as the exhaust gases dilute the fresh charge, thus affecting the combustion taking place inside the cylinder.

6.7 EGR in advanced SI engines

Various advancements have been made in the case of spark-ignition engines. Some of the latest developments are the implementation of gasoline direct injection (GDI engines), multipoint fuel injection (MPFI engines), and lean-burn combustion technique (lean burn engines).

6.7.1 EGR in MPFI engines

EGR is primarily involved in MPFI engines, which often work under stoichiometric conditions because of reduced NOx emissions, fuel consumption, and throttling loss inside the engine. A wide-open throttle condition is needed in this type of engine to compensate for the pumping loss. Fuel economy could be improved due to the reduction in pump work and reduced heat transfer to the cylinder walls. Due to the application of EGR, the inlet charge density reduces. Thus, the use of a supercharger to increase the inlet pressure becomes inevitable to compensate for the power and torque output of the engine. The period of combustion and ignition delay would be raised in MPFI engines with EGR because of the reduction in laminar flame speed. The EGR rate would be restricted to 25% because higher EGR rates in MPFI engines would produce a drastic decrease in the power output of the machine.

6.7.2 EGR in GDI engines

The implementation of EGR in gasoline direct injection (GDI) engines [48] is a crucial enabler for reducing NOx emissions. Usually, GDI technology reduces CO_2 emissions from spark-ignition engines. Unlike MPFI engines, GDI delivers lower specific fuel consumption, high compression ratio, and a lower temperature of the charge inside the engine. The application of cooled EGR provides a better reduction in NOx emissions than the application of hot EGR. Cooled EGR reduces postflame oxidation as well as the exhaust temperature.

6.7.3 EGR in lean-burn engines

A lean burn is a kind of technique followed in spark-ignition engines that would operate in a more skeletal mixture mode, thus producing lower NOx emissions. This technique would work under the limit of misfire. So, HC and CO emissions might increase with a relative decrease in engine efficiency due to the deterioration in the engine's stability. From the above condition, one could understand that the reduction in NOx emissions would always increase HC and CO emissions invariably, and it would also reduce the engine's thermal efficiency. To meet the latest emission norms, the use of after-treatment techniques in lean-burn mode becomes inevitable. Selective catalytic reduction (SCR) devices are used at the exhaust of the SI engine running in lean-burn mode [49]. They consist of a catalyst, ammonia storage, an injection system, and feed. Here, the stored ammonia is fed at the starting of the catalyst in the exhaust gas, which is maintained at a specific temperature. In addition, a two-way catalytic converter is also used to reduce the HC and CO emissions from the exhaust gases. A three-way catalytic converter (TWC) could be implemented to minimize all three emissions simultaneously. TWC is less expensive than SCR in lean-burn engines. The excess supply of air has to be reduced to maintain good efficiency of TWC. The engine must Always operate under the stoichiometric air-fuel ratio condition. Under this condition, the tendency of knocking would increase along with the rise in in-cylinder temperature. This would restrict the use of turbochargers, thus resulting in less thermal efficiency. The use of EGR and the TWC technique would result in lower emissions than a lean-burn approach. Some engine parameters such as fuel consumption would not be specific. But still, the engine efficiency and emissions are relatively better with the use of EGR and TWC devices in the lean-burn engine technique.

6.8 EGR implementation in advanced SI engines

The effects of EGR in advanced engine technologies such as turbochargers as well as the utilization of natural gas and hydrogen in engine operation are discussed in the current section.

6.8.1 Turbocharged SI engine with EGR

Two different configurations enable exhaust gas recirculation into the engine inlet. In the low-pressure (LP) configuration, exhaust gases are removed from the turbocharger downstream and injected into the compressor upstream. In diesel engines, the high-pressure (HP) configuration is used widely. In the present configuration, exhaust gases are removed from the turbine upstream and introduced in the inlet port. The schematic diagrams of the HP and LP configurations are shown in Figs. 6.13 and 6.14, respectively.

The HP layout does not have long air paths and transport delays. So, it is suitable for transient operation. However, the LP EGR configuration is ideal for spark-ignition engines. Cooled(LP EGR is highly appropriate for eliminating the high load enrichment of fuel and mitigating knock. Cooled LP EGR is attained by working closer to maximum brake torque (MBT) due to lower knock propensity and improved heat capacity. At lower load conditions, uncooled HP EGR is an effective method for thermally dethrottling the engine. To drive HP EGR in a downsized turbocharged engine, there would be no positive gradient from the inlet manifold to the exhaust

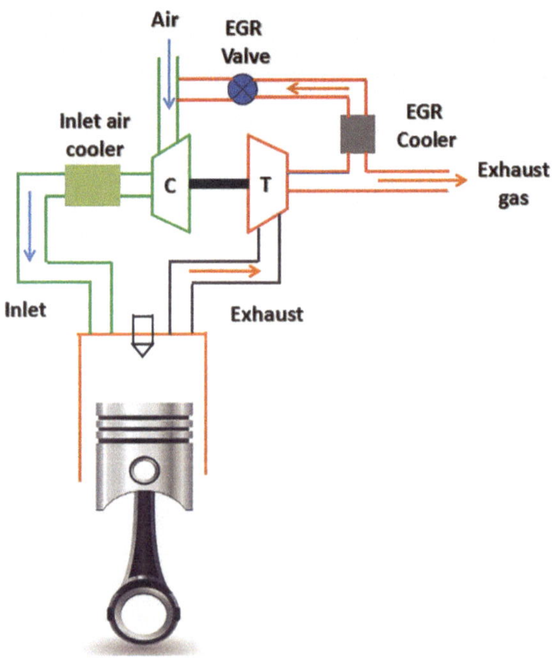

Fig. 6.13 High-pressure EGR loop.

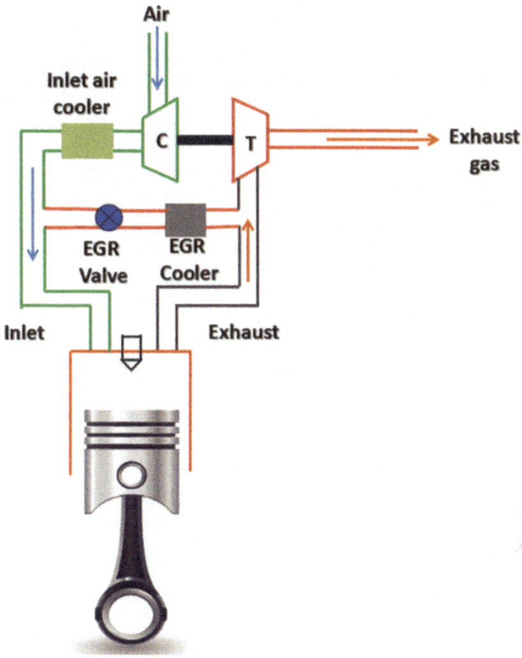

Fig. 6.14 Low-pressure EGR loop.

manifold at higher loads and lower speeds [10, 21]. The inadequate mixing between EGR and air causing cylinder-to-cylinder variations is another limitation for HP EGR [50, 51]. The new inlet charge would be mixed well before moving into the cylinder because of the long path of the LP layout.

In downsized engines, the critical part is the extraction of the EGR post-turbine LP configuration, which would cause less intervention with the turbocharger [50, 52]. After expansion through the turbine in the LP EGR layout, the requirements needed for the cooling of exhaust gases are minimal. Also, the intercooler gives further cooling of exhaust gases, and so EGR is supplied to the inlet at a lower temperature than the HP configuration [52]. This method of temperature reduction is suitable for both the elimination of fuel enrichment and the mitigation of knock intensity. Using LP EGR, the researchers evaluated the difference in pressure to run the flow using both catalyzed and noncatalyzed EGR and compared it [53, 54]. The LP EGR delivery range would be expanded by noncatalyzed EGR, which gives a higher pressure difference. Because of the influence of H_2 and CO on noncatalyzed EGR combustion, the laminar flame speed is greater than that of the catalyzed EGR. However, the main benefit of catalyzed EGR is the

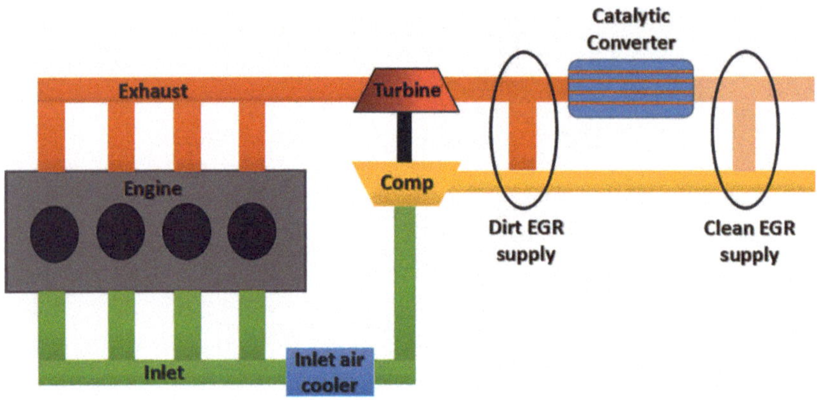

Fig. 6.15 Clean vs. dirty EGR circuits.

reduction in NOx concentration. The elimination of NOx-concentrated EGR significantly enhances knock suppression [55–57]. However, fouling the compressor and EGR circuit could be prevented by clean EGR because HC emissions are reoxidized inside the catalyst. The differences between the supply of clean EGR and dirty EGR are shown in Fig. 6.15. Another advantage of LP EGR over its alternatives is the extraction of EGR from the three-way catalyst downstream.

Various researchers have investigated two different fuels, 9E30 and E850, for their advantage on fuel efficiency and the use of cooled EGR. For both fuels, the fuel economy is between 3% and 5% [11]. Also, a researcher has evaluated the volumetric efficiency of the use of cooled EGR on a turbocharged SI engine with different compression ratios. The effects of cooled EGR resulted in increase in volumetric efficiency due to the mitigation of knock intensity [12]. So, LP cooled EGR is suitable for use in downsized turbocharged spark-ignition engines.

Fig. 6.16 represents the variation of inlet and exhaust manifold pressure concerning a change in the engine speed. At higher load and lower speed conditions, the backpressure created at the exhaust is lower than that of the inlet manifold pressure. Thus, the exhaust gas would not move to the intake because of the difference in pressure created between the inlet and exhaust. This could be prevented by raising the turbine operating pressure and also by lowering the boost pressure. The conditions mentioned above can be achieved by installing a throttle valve near the compressor or the turbine. This can also be achieved by installing a variable geometry turbocharger or waste-gated turbocharger to give the required EGR pressure without affecting the engine's performance.

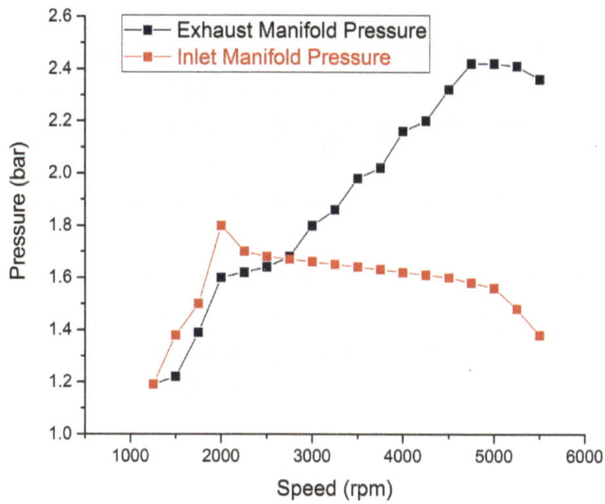

Fig. 6.16 Inlet and exhaust manifold pressure vs. engine speed.

6.8.2 Natural gas-powered SI engine with dedicated EGR

In the SI engine, the thermal efficiency is increased by the lean combustion method. However, the rise in thermal energy by the lean combustion method and the great quantity of nitrogen oxide emissions are created simultaneously. Therefore, it is necessary to improve the rise in thermal efficiency and the reduction of NO_X discharge. Both will be achieved by the use of a dedicated EGR. In this set-up, the first cylinder is used as a dedicated EGR cylinder. The exhaust gas from the first cylinder is premixed with fresh natural gas with various equivalence ratios recirculated into other cylinders. The flame and the laminar burning velocity are taken into account to evaluate the NO_X and thermal efficiency. The dedicated EGR set-up powered by natural gas in the SI engine is shown in Fig. 6.17. Table 6.6 shows the properties of natural gas used in the SI engines.

To distinguish the characteristics of the nitrogen emissions and thermal efficiency, the flame temperature and the laminar burning velocity are measured. The degree of constant volume and heat loss are estimated. The dedicated EGR cylinder functioning at $\phi_{\#1} = 2.1$ and delivering EGR to the remaining cylinders contains a significant quantity of CO and H_2. The chemical species perform progressive parts for more incredible velocity in laminar burning and nearly a reduced impact on the flame temperature.

Due to increased combustion in the dedicated EGR cylinder in the spark-ignition engine, thermal efficiency is comparable with the conventional spark-ignition engine. Even with a higher dedicated EGR ratio,

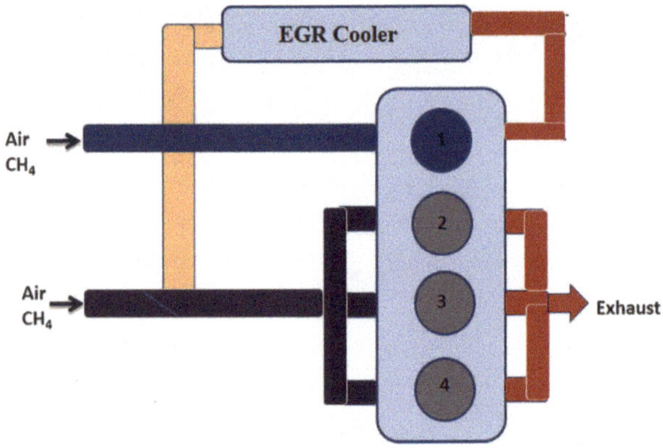

Fig. 6.17 Dedicated EGR set-up in an SI engine.

Table 6.6 Properties of natural gas.

Properties	Units	Natural gas
Calorific value	kJ/kg	38.1
Limit of flammability	Volume %	5–15
Autoignition temperature	°C	450
Fuel density	kg/m^3	187.2
Energy density	MJ/m^3	7132
Octane rating	RON	107
Latent heat of vaporization	kJ/kg	104.8

the degree of volume control can be improved by the combustion effects of carbon monoxide and hydrogen. Due to the enormous quantity of H_2O and CO_2, the flame temperature drops. Thus, the amount of NO_X discharge is lowered intensely. In SI engines, the dedicated EGR ratio varies as 0.5, 0.33, 0.25, and 0.2 with a change in the number of cylinders from three to six. If the number of cylinders is higher than four, the thermal efficiency will be higher than 40%, and nitrogen oxide emissions in three cylinders will be the lowest. So, the engine will operate stably due to the lower flame temperature.

In the spark-ignition engine with a dedicated EGR arrangement, the degree of constant volume of the remaining cylinders is reduced because it works on a lean combustion technique. Hence, once a dedicated EGR arrangement is arranged, the remaining cylinders work in stoichiometric combustion. For increased thermal efficiency in the reliable EGR arrangement, the dedicated EGR ratio will be less than 0.33 with more than four cylinders.

6.8.3 Hydrogen powered SI engine with dedicated EGR

Hydrogen-assisted jet ignition is a progressive ignition method that uses chemically energetic hydrogen fuel, which provides turbulent to start the combustion of ultralean mixtures in a regular gasoline-powered engine. The hydrogen-assisted jet ignition component is installed into the hydrogen (H_2) injector and spark plug, as shown in Fig. 6.18. Tables 6.7 and 6.8 indicate the thermodynamic properties and combustion properties of hydrogen in SI engines.

In this method, a smaller quantity of H_2 is injected near the spark plug electrodes inside the chamber so that a creamy mixture is formed. When hydrogen in the chamber gets ignited, the combusting gas moves through

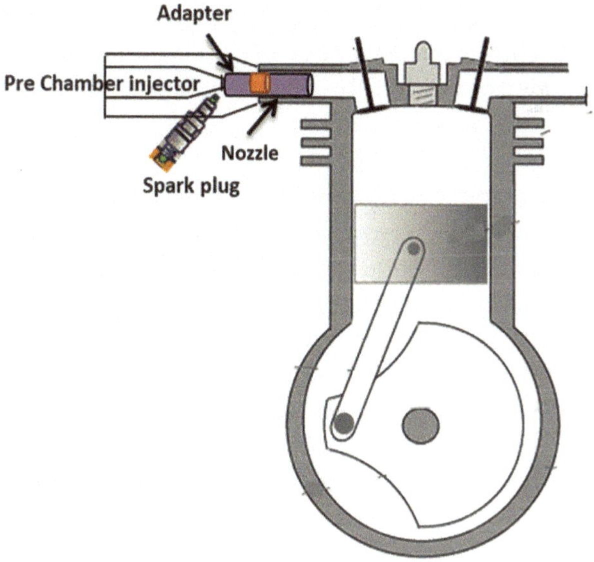

Fig. 6.18 Sectional view of hydrogen=assisted jet injection in an SI engine.

Table 6.7 Thermodynamic properties of hydrogen.

Thermodynamic property	Hydrogen
Molecular weight	2.016
Viscosity at NTP ($g\,cm^{-1}\,s^{-1}$)	8.75×10^{-5}
Specific heat at NTP ($J\,g^{-1}\,K^{-1}$)	14.89
Specific heat ratio at NTP	1.383
Diffusion coefficient at NTP ($cm^2\,s^{-1}$)	0.61
Gas constant, R (cm^2 atm $g^{-1}\,K^{-1}$)	40.7030

Table 6.8 Combustion properties of hydrogen.

Combustion property	Hydrogen
Flame temperature (K)	2318
Autoignition temperature (K)	858
Burning velocity in NTP air	265–325
Limits of flammability in air (vol%)	4–75
Diffusivity in air ($cm^2\ s^{-1}$)	0.63

several orifices into the combustion chamber at more incredible speed and burns the lean chamber charge. The presence of chemically sensitive intermediate reactants such as hydrogen and hydrogen oxides provides a greater extent of turbulence inside the chamber. Thus the energy levels are higher in magnitude, which is established using a spark plug. In hydrogen-aided jet ignition, the primary chamber combustion is dependable over a significantly wider choice of air-fuel ratios. When gasoline acts as the primary fuel, the lean boundary can be prolonged to $\lambda = 5$ at extensive open throttle situations due to which the strong jets of chemically reactive combustion products ignites the main charge quickly with lesser combustion variability.

Fewer nitrogen oxides were produced due to the thinner operation, and proper temperature combustion in the hydrogen-assisted jet ignition system combustion occurs at more moderate temperatures, which result in lower NO_X formation. Lower temperatures also develop efficiency by reducing dissociation, such as carbon dioxide to carbon monoxide, which usually happens at higher temperatures. The process obtains better engine efficiency with additional air, which reduces fuel consumption and permits the engine to be under less throttled conditions while significantly decreasing pumping losses. This has the potential to run the engine in an utterly unthrottled mode in the limits of $\lambda = 2.5$. In these circumstances, hydrocarbon emissions become more critical due to the burning mixtures at low temperatures and growing wall quenched hydrocarbon with decreased crevice sourced hydrocarbons.

6.9 Conclusion

Implementation of exhaust gas recirculation (EGR) in gasoline engines would result in a decrease in NOx and soot emissions with reduced fuel consumption. The use of hot EGR promotes combustion in the cylinder, ultimately improving the engine's thermal efficiency. Cooled EGR use in gasoline engines raises the density of the charge at the inlet by the exhaust gas mixing with the inlet air. It also improves the volumetric efficiency of the engine by reducing the fuel consumption. A decrease in temperature inside the cylinder with the application of cooled EGR decreases NOx emission while it raises the HC emission and knock intensity in comparison with hot EGR. The brake means adequate pressure of the gasoline engine could be improved by limiting the formation of knock inside the combustion chamber. The method adopted to determine the intensity of knocking is fuel enrichment. Fuel enrichment in the engine results in increased fuel consumption and emissions. Cooled EGR on a par with fuel enrichment can also control the temperature of exhaust gases in gasoline engines. In contrast, the rate of cooled EGR supplied to the fresh inlet charge has to be limited because an increased rate of EGR addition to the new inlet charge would lead to the following effects: The speed of the turbulent flame inside the combustion chamber reduces, thereby decreasing the rate of combustion, cylinder peak pressure, and heat release rate. Thus the intensity of knocking inside the combustion chamber is reduced. But the combusted end gas at the end of the power stroke would be at a high temperature, which could initiate knocking inside the cylinder.

- The increased mass of the exhaust gas entering the cylinder would decrease the combustion temperature inside the cylinder, thus reducing the intensity of knocking.
- The specific heat value of the compressed charge would reduce with a rise in the EGR rate, due to which the temperature of the compressed charge increases, ultimately increasing the formation of knock.
- The use of cooled EGR rather than fuel enrichment at higher loads leads to the use of stoichiometric charge, thus helping in reducing the CO and HC emissions.
- The use of cooled EGR would effectively reduce the knock intensity without compromising the fuel consumption and emissions. EGR implementation in turbocharged gasoline engines has become more difficult in the world market because turbo matching has been faced by many researchers, which has to be sorted out shortly.

The various techniques in the EGR method to control NOx in CI engines were reviewed. The possible advanced EGR techniques to control NO_X in CI engines will be discussed in the next chapter. The design and optimization of a long route, short route, and hybrid EGR system will be addressed in steady-state and transient conditions. The significance of EGR implementation on advanced combustion techniques in CI engines such as low-temperature combustion, homogeneous charge compression ignition, partially premixed charge compression ignition, and reactivity controlled compression ignition will also be reviewed. The in situ strategy of EGR to the engine inlet for NOx control and the EGR system requirement for the alternative liquid and gaseous fueled engines will be presented. The next chapter will also discuss the EGR system effect with preheating of blended fuels on the performance, combustion, emission, engine wear, and engine oil contamination.

Acknowledgment

The authors would like to acknowledge the management of Mepco Schlenk Engineering College for encouraging the research work.

References

[1] Inbanaathan PV, Dhinesh B, Tamilarasan U. Experimental investigation of performance and emission characteristics of diesel blended with palm methyl ester along with alumina nano-additive using D.I. diesel engine. Springer Science and Business Media LLC; 2020. p. 151–66. https://doi.org/10.1007/978-981-15-1607-8_15.
[2] Abd-Alla GH. Using exhaust gas recirculation in internal combustion engines: a review. Energy Convers Manag 2002;43(8):1027–42. https://doi.org/10.1016/S0196-8904(01)00091-7.
[3] Fatih S, Maurice K, Amin V, Ulrich S. Potential of reducing the NOX emission in a spray guided DI gasoline engine by stratified exhaust gas recirculation (EGR). SAE paper 2006-01-1261; 2006.
[4] Fontana G, Galloni E. Variable valve timing for fuel economy improvement in a small spark-ignition engine. Appl Energy 2009;86(1):96–105. https://doi.org/10.1016/j.apenergy.2008.04.009.
[5] Yun-long B, Zhi W. Wang Jian-xin Part-load characteristics of direct injection spark ignition engine using exhaust gas trap. Appl Energy 2010;87:2640.
[6] Huang B, Hu E, Huang Z, Zheng J, Liu B, Jiang D. Cycle-by-cycle variations in a spark ignition engine fueled with natural gas-hydrogen blends combined with EGR. Int J Hydrog Energy 2009;34(19):8405–14. https://doi.org/10.1016/j.ijhydene.2009.08.002.
[7] Ladommatos N, Abdelhalim S, Zhao H, Hu Z. The dilution, chemical and thermal effects of exhaust gas recirculation on diesel engine emission—part 4: effect of carbon dioxide and water vapour. SAE technical paper 971660; 1997.
[8] Xu M, Chen G, Daniels C, Dong M. Numerical study on swirl-type high-dilution stratified EGR combustion system. SAE transactions 1689-99; 2000. Jan 1.
[9] Agarwal D, Singh SK, Agarwal AK. Effect of exhaust gas recirculation (EGR) on performance, emissions, deposits and durability of a constant speed compression ignition engine. Appl Energy 2011;88(8):2900–7. https://doi.org/10.1016/j.apenergy.2011.01.066.
[10] Potteau S, Lutz P, Leroux S, Moroz S, Tomas E. Cooled EGR for a turbo SI engine to reduce knocking and fuel consumption. SAE paper 2007-01-3978; 2007.

[11] Kaiser M, Krueger U, Harris R, Cruff L. "Doing more with less"—the fuel economy benefits of cooled EGR on a direct injected spark ignited boosted engine. SAE technical paper; 2010.
[12] Liu Z, Cleary D. Fuel consumption evaluation of cooled external egr for a downsized boosted SIDI DICP engine. In: SAE technical papers, vol. 1. SAE International; 2014. https://doi.org/10.4271/2014-01-1235.
[13] Dhinesh B, Tuan HA, Inbanaathan PV, Arunprasad S, Jianbing G, Tanakorn W. Numerical and experimental evaluation on the pooled effect of waste cooking oil biodiesel/diesel blends and exhaust gas recirculation in a twin-cylinder diesel engine. Fuel 2021;287:119815. https://doi.org/10.1016/j.fuel.2020.119815.
[14] Li ZW, Shu GQ, Wei HQ, Liang XY. Experimental investigation on effect of EGR on combustion noise in transient conditions of DI-diesel engine. Chin Intern Combust Engine Eng 2010;31(3):17–22.
[15] Wei H, Li Z, Liang X, Shu G. Influence of the accelerating operation mechanism on the combustion noise in DI-diesel engines. Int J Automot Technol 2012;13 (3):373–88. https://doi.org/10.1007/s12239-012-0035-2.
[16] Levendis YA, Pavlatos I, Abrams RF. Control of diesel soot, hydrocarbon and NOx emissions with a particulate trap and EGR. In: SAE technical papers. SAE International; 1994. https://doi.org/10.4271/940460.
[17] Livengood JC, Wu PC. Correlation of autoignition phenomena in internal combustion engines and rapid compression machines. Symp Combust 1955;347–56. https://doi.org/10.1016/s0082-0784(55)80047-1.
[18] Leppard WR. Individual-cylinder knock occurrence and intensity in multicylinder engines. In: SAE Preprints. SAE; 1982.
[19] Merola SS, Vaglieco BM. Knock investigation by flame and radical species detection in spark ignition engine for different fuels. Energy Convers Manag 2007;48 (11):2897–910. https://doi.org/10.1016/j.enconman.2007.07.011.
[20] Brunt Michael FJ. Gasoline engine knock analysis using cylinder pressure data. SAE technical paper 980896980896; 1998.
[21] Curry S. A three dimensional study of flame propagation in a spark ignition engine. SAE technical paper; 1963.
[22] Börje G, Hans-Erik Å. Replacing fuel enrichment in a turbo charged SI engine: lean burn or cooled EGR; 1999. SAE paper 1999-01-3505.
[23] Börje G, Hans-Erik Å, Per S, Eric O. Knock suppression in a turbocharged SI engine by using cooled EGR; 1998. SAE paper 982476.
[24] Benson RS, Whitehouse ND. A detailed introduction to the thermodynamics of spark and compression ignition engines, their design and development. Internal combustion engines. Pergamon; 1979. p. 1–216.
[25] Lavole GA, Heywood JB, Keck JC. Experimental and theoretical study of nitric oxide formation in internal combustion engines. Combust Sci Technol 1970;1 (4):313–26. https://doi.org/10.1080/00102206908952211.
[26] Al-Qurashi K, Lueking AD, Boehman AL. The deconvolution of the thermal, dilution, and chemical effects of exhaust gas recirculation (EGR) on the reactivity of engine and flame soot. Combust Flame 2011;158(9):1696–704. https://doi.org/10.1016/j.combustflame.2011.02.006.
[27] Ghazikhani M, Feyz ME, Joharchi A. Experimental investigation of the exhaust gas recirculation effects on irreversibility and brake specific fuel consumption of indirect injection diesel engines. Appl Therm Eng 2010;30(13):1711–8. https://doi.org/10.1016/j.applthermaleng.2010.03.030.
[28] Ladommatos N, Abdelhalim SM, Zhao H, Hu Z. Effects of EGR on heat release in diesel combustion. SAE technical paper; 1998.

[29] Balasubramanian D, Papla Venugopal I, Viswanathan K. Characteristics investigation on di diesel engine with nano-particles as an additive in lemon grass oil. In: SAE technical papers (Issue October). SAE International; 2019. https://doi.org/10.4271/2019-28-0081.
[30] Balasubramanian D, Kamaraj S, Krishnamoorthy R. Synthesis of biofuel from waste cooking oil by alkali doped calcinated waste egg shell powder catalyst and optimization of process parameters to improve biofuel conversion. SAE technical paper 2020-01-0341; 2020.
[31] Heywood JB. Internal combustion engine fundamentals. New York: McGraw-Hill; 1988.
[32] Johnson T. Vehicular emissions in review. SAE Int J Engines 2016;9(2). https://doi.org/10.4271/2016-01-0919.
[33] Kittleson D, Kraft M. Particle formation and models. Encyclopedia of automotive engineering; 2014. https://doi.org/10.1002/9781118354179.auto161.
[34] Price P, Stone R, Collier T, Davies M. Particulate matter and hydrocarbon emissions measurements: comparing first and second generation DISI with PFI in single cylinder optical engines. SAE technical paper; 2006.
[35] Flagan RC, Richard C, Seinfeld JH, John H. Fundamentals of air pollution engineering. Englewood Cliffs, New Jersey: Prentice-Hall, Inc.; 2012.
[36] Di Domenico M, Gerlinger P, Aigner M. Development and validation of a new soot formation model for gas turbine combustor simulations. Combust Flame 2010;157(2):246–58. https://doi.org/10.1016/j.combustflame.2009.10.015.
[37] Blanquart G, Pitsch H. Analyzing the effects of temperature on soot formation with a joint volume-surface-hydrogen model. Combust Flame 2009;156(8):1614–26. https://doi.org/10.1016/j.combustflame.2009.04.010.
[38] Böhm H, Hesse D, Jander H, Lüers B, Pietscher J, Wagner HGG, Weiss M. The influence of pressure and temperature on soot formation in premixed flames. Symp Combust 1989;22(1):403–11. https://doi.org/10.1016/S0082-0784(89)80047-5.
[39] Frenklach M, Taki S, Durgaprasad MB, Matula RA. Soot formation in shock-tube pyrolysis of acetylene, allene, and 1,3-butadiene. Combust Flame 1983;54(1–3):81–101. https://doi.org/10.1016/0010-2180(83)90024-X.
[40] Graham S, Homer J, Rosenfeld JL. The formation and coagulation of soot aerosols generated by the pyrolysis of aromatic hydrocarbons. Proc R Soc Lond A Math Phys Sci 1637;344:259–85.
[41] Kamimoto T, Bae M. High combustion temperature for the reduction of particulate in diesel engines. SAE technical paper 880423; 1988. p. 692–701.
[42] Akihama K, Takatori Y, Inagaki K, Sasaki S, et al. Mechanism of the smokeless rich diesel combustion by reducing temperature. SAE technical paper 2001-01-0655; 2001. p. 648–62.
[43] Kayes D, Hochgreb S. Mechanisms of particulate matter formation in spark-ignition engines. 1. Effect of engine operating conditions. Environ Sci Technol 1999;33(22):3957–67. https://doi.org/10.1021/es9810991.
[44] Alger T, Gingrich J, Khalek IA, Mangold B. The role of EGR in PM emissions from gasoline engines. SAE Int J Fuels Lubr 2010;3(1):85–98. https://doi.org/10.4271/2010-01-0353.
[45] Hedge M, Weber P, Gingrich J, Alger T, Khalek I. Effect of EGR on particle emissions from a GDI engine. SAE Int J Engines 2011;4(1):650–66. https://doi.org/10.4271/2011-01-0636.
[46] Luján JM, Climent H, Novella R, Rivas-Perea ME. Influence of a low pressure EGR loop on a gasoline turbocharged direct injection engine. Appl Therm Eng 2015;89:432–43. https://doi.org/10.1016/j.applthermaleng.2015.06.039.

[47] Thanh BT, Dhinesh B, Tuan HA, Ozcan K, Chan ND, Nam TV. Characteristics of PM and soot emissions of internal combustion engines running on biomass-derived DMF biofuel: a review. In: Energy sources, part a: recovery, utilization, and environmental effects; 2020. p. 1–22. https://doi.org/10.1080/15567036.2020.1869868.
[48] Zhao F, Harrington DL, Ming-Chia L. Automotive gasoline direct injection engines. SAE book; 2002.
[49] Parthasarathy M, Ramkumar S, Lalvani JIJ, Elumalai PV, Dhinesh B, Krishnamoorthy R, Thiyagarajan S. Performance analysis of HCCI engine powered by tamanu methyl ester with various inlet air temperature and exhaust gas recirculation ratios. Fuel 2020;118833. https://doi.org/10.1016/j.fuel.2020.118833.
[50] Lakhlani H, Barman J, Rajput K, Goswami A. Experimental study of EGR mixture design and its influence on EGR distribution across the cylinder for NOx—PM trade-off. In: SAE technical papers, vol. 12. SAE International; 2013. https://doi.org/10.4271/2013-01-2743.
[51] Sjöberg M, Dec JE. Influence of EGR quality and unmixedness on the high-load limits of HCCI engines. SAE Int J Engines 2009;2(1):492–510. https://doi.org/10.4271/2009-01-0666.
[52] Nam K, Yu J, Cho S. Improvement of fuel economy and transient control in a passenger diesel engine using LP(low pressure)-EGR. In: SAE 2011 world congress and exhibition; 2011. https://doi.org/10.4271/2011-01-0400.
[53] Lewis A, Akehurst S, Turner J, Patel R, Popplewell A. Observations on the measurement and performance impact of catalyzed vs. non catalyzed EGR on a heavily downsized DISI engine. SAE Int J Engines 2014;7(1):458–67. https://doi.org/10.4271/2014-01-1196.
[54] Roth DB, Keller P, Becker M. Requirements of external EGR systems for dual cam phaser turbo GDI engines. In: SAE technical papers. SAE International; 2010. https://doi.org/10.4271/2010-01-0588.
[55] Hoffmeyer H, Montefrancesco E, Beck L, Willand J, Ziebart F, Mauss F. CARE—catalytic reformated exhaust gases in turbocharged DISI-engines. SAE Int J Fuels Lubr 2009;2(1):139–48. https://doi.org/10.4271/2009-01-0503.
[56] Kawabata Y, Sakonji T, Amano T. The effect of NOx on knock in spark-ignition engines. SAE technical paper 1999-01-0572; 1999.
[57] Takaki D, Tsuchida H, Kobara T, Akagi M, Tsuyuki T, Nagamine M. Study of an egr system for downsizing turbocharged gasoline engine to improve fuel economy. In: SAE technical papers, vol. 1. SAE International; 2014. https://doi.org/10.4271/2014-01-1199.

CHAPTER 7

Application of exhaust gas recirculation for NOx reduction in CI engines

C. Kannan and T. Vijayakumar
Department of Automotive Engineering, School of Mechanical Engineering, Vellore Institute of Technology, Vellore, Tamil Nadu, India

7.1 Introduction

The impact of implementing conventional/stratified EGR in SI engines in terms of NO_x emission reduction, fuel economy enhancement, and knock suppression was elaborately discussed in the previous chapter. Chapter 6 also sheds light on the challenges and opportunities during EGR implementation in advanced SI engine concepts such as MPFI, GDI, and lean-burn engines. The negative impact of EGR, especially on the after-treatment device, is highlighted in a section. The prospects of EGR in SI engines are also provided at the end of the chapter to accelerate the research in this domain. In this chapter, the implications of EGR implementation in CI engines are elaborated in detail. EGR is an established and effective technique to inhibit NO_x emissions in diesel engines and as much as 50% of the exhaust gas can be recycled. In recent times, the revival of EGR has received sudden interest due to the development of a new generation of electronic controlled valves, which ensure better accuracy and shorter response in transient conditions. The urban traffic density, which makes the passenger car diesel engine operate at partial loads for most of the time, also suggests EGR implementation to get over the ill effects of higher oxygen content and high temperature.

Stringent emission norms could be satisfied through numerous ways such as engine modifications, implementation of after-treatment devices, and adoption of high-quality alternative fuels. The engine modifications needed to meet the emission norms, on the other hand, impose an undesirable impact on the overall cost of a vehicle. Although alternative fuel adoption seems to be relatively simple, determining the suitable proportion of substitute fuels and their blends based on engine geometry is a cumbersome task that requires a lot of experiments and optimization. Three-way

NOx Emission Control Technologies in Stationary and Automotive Internal Combustion Engines
https://doi.org/10.1016/B978-0-12-823955-1.00007-3

catalyst-based after-treatment devices have their maximum conversion efficiency ($\geq$ 95%) when the engine is operated with a stoichiometric air-fuel ratio. Under this mixture condition, they could effectively convert HC, CO, and NO_x emissions into unharmful gases. As diesel engines are operated with excess air in the majority of their operating region, the effectiveness of conventional after-treatment devices on NO_x conversion is partial. This, in turn, demands either selective catalytic reduction and/or NO_x storage reduction to be used along with other techniques. In this context, EGR has become an essential control strategy for both advanced combustion and alternative-fueled engines. This chapter discusses the different EGR designs, configurations, and operating window significance in addition to the implication of EGR under steady and transient state operation of conventional diesel engines, which are elaborately detailed. Different EGR control strategies with a significant emphasis on electronic and hybrid control are presented. The performance, combustion, and emission characteristics of advanced combustion and alternative liquid/gaseous fueled engines with EGR implementation are discussed. The negative aspects of EGR on engine wear, oil contamination, and soot formation are analyzed and presented. This chapter ends with the prospects of EGR. Even though the EGR technique dates back to 1940, its potential to reduce NO_x emissions was first investigated using an SI engine in late 1950. While EGR was considered a vital NOx control technique for light-duty diesel engines during 1970, its implementation in heavy-duty diesel engines happened only during 2000 [1]. To meet the ever-demanding emission norms, it is essential to use EGR in combination with NO_x reduction catalysts. At present, the heavy-duty engines fulfill the Euro VI NO_x limits through EGR adoption along with urea-based selective catalytic reduction (SCR) while nonroad engines alternatively use EGR and ammonia-based SCR [2].

7.2 Exhaust gas recirculation

EGR is a method in which part of the exhaust gas is redirected back to the intake system through a control valve. The working principle of EGR in a diesel engine is schematically presented in Fig. 7.1. The introduction of EGR replaces some of the fresh air in the case of diesel engines. The injected fuel quantity remains constant for a given torque and power output, which eventually results in a lower overall air-fuel ratio. This, in turn, ensures a substantial reduction of exhaust emissions. A relatively high temperature of the recirculated exhaust gases reduces the mass flow rate of inlet charge

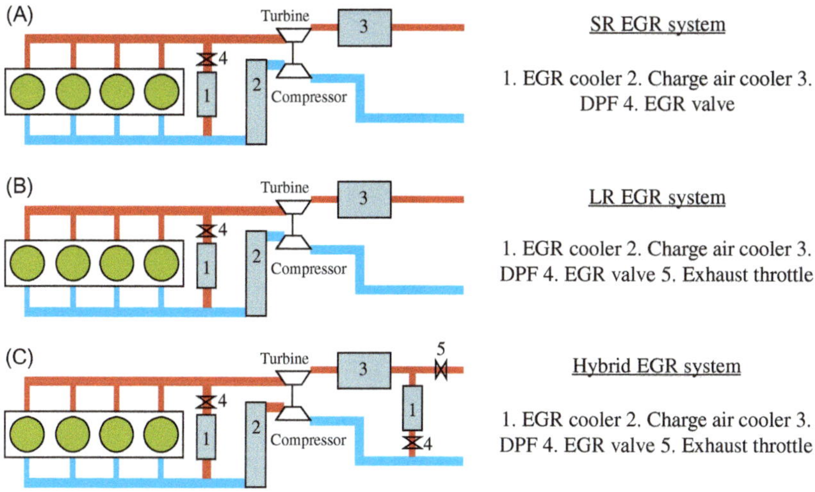

Fig. 7.3 Different EGR design configurations.

drawbacks such as matching the boost pressure and turbocharger, enhancing the kinetic energy by placing venturi, etc. Another system referred to as hybrid EGR also exists. It has a switching capability between SR and LR or can operate both systems at the same time, depending upon the driving situation. This leads to the establishment of an engine delivering its most beneficial efficiency [8]. The schematic of different EGR design configurations is presented in Fig. 7.3.

There are controversial statements over the supremacy of one EGR architecture over the other. However, the effectiveness of LR/LPL EGR can be clearly understood by its superior NO_x reduction (more than 15%) compared to SR/HPL EGR for the same EGR rate. This is correlated to reduced inlet charge temperature. Under the same brake-specific NO_x limit of 5 g/kWh, a passenger car engine could reduce the soot by about 16% and the inlet charge temperature by about 12% as well as nearly a 7.5% drop in the pumping loop losses and about 2% savings in brake-specific fuel consumption (BSFC) by adopting the LR EGR design [9]. This can be inferred from Fig. 7.4. The poor transient performance of LR EGR limits its applications. At present, manufacturers adopt the hybrid EGR strategy that follows a short route during the transient condition and a long route during steady-state operating condition.

Another debate regards the benefits attained by hot and cold EGR. The exhaust gases are cooled and recirculated back to the inlet manifold in the case of cold EGR. This is found to be more effective than hot EGR in

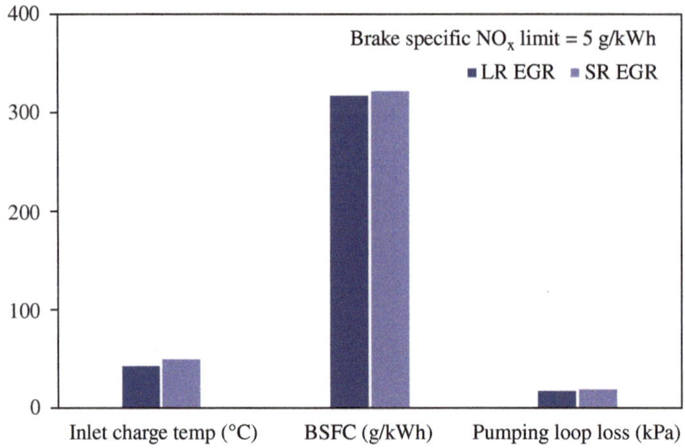

Fig. 7.4 Effectiveness comparison of LR and SR EGR designs under fixed brake -NOx limit in a passenger car diesel engine.

reducing NO_x emissions, as it results in a lower intake air temperature. This, in turn, reduces the in-cylinder peak temperature. Low in-cylinder temperature and oxygen availability eventually reduce NO_x emissions. Generally, a high rate of pressure rise was observed under all loads with different EGR ratios in hot EGR over cold EGR.

7.4 EGR operating window and significance

The percentage of EGR adopted in an engine can be calculated on a mass or volume basis using the Eq. (7.1):

$$\mathrm{EGR}(\%) = \frac{\text{Mass or volume of air without EGR} - \text{Mass or volume of air with EGR}}{\text{Mass or volumen of air without EGR}} \tag{7.1}$$

Another method based on carbon dioxide concentration in the inlet and exhaust manifold can also be used to determine the percentage of EGR using Eq. (7.2):

$$\mathrm{EGR}\,(\%) = \frac{[CO_2]_{\text{inlet}}}{[CO_2]_{\text{exhaust}}} \tag{7.2}$$

There are other parameters such as oxygen mass fraction, dilution ratio, and combustible oxygen mass fraction that have been embraced by researchers [10–12] to replace the EGR ratio to increase the reliability of EGR transient measurement for better control. Generally, under the low load condition (<25%) of diesel engines, the NO_x concentration is lower, which can be satisfied with the lower EGR rate. On the other hand, the high load condition (>75%) of diesel engines necessitates the adoption of a high EGR rate due to the higher NO_x concentration. The tendency of NO_x formation increases gradually during the engine's medium load condition (>25% and<75%), so it is necessary to improve the EGR rate. The EGR rates of 2.4%–4.5%, 7.5%–7.9%, and 9.1%–9.7% are found to be optimal for the low, medium, and high load conditions of turbocharged diesel engines, respectively [13]. The EGR operating window is different for different engine operating conditions. The EGR operating windows for different engine categories and their significance are presented in Table 7.1.

7.5 EGR control strategies

7.5.1 Mechanical control

The most common mechanical method of controlling the EGR valve is utilizing the port vacuum switch. It connects the engine vacuum to the EGR valve when the engine reaches the warm-up temperature. Until then, the EGR valve is closed and thus shuts off the recirculation of exhaust gases [22]. The working of the ported vacuum switch can be interpreted from Fig. 7.5. The recycled exhaust gases are introduced into the downstream of a throttle valve of the inlet manifold.

To improve EGR control during the speed/load transients of an engine, a new system has been devised, which uses the throttle plate as a secondary controller [23]. In this layout, the EGR port is positioned on the upstream of the throttle plate located in the inlet manifold. The schematic of the throttle body ported EGR is shown in Fig. 7.6. The new position of the EGR port helps in achieving: (i) thorough mixing of exhaust gas and fresh air, (ii) instantaneous response, (iii) better performance under steady-state conditions, and (iv) prevention of contamination.

The mechanical EGR control is also comprised of residual gas retention in the cylinder altering the timing of the inlet and exhaust valves. The limited actuation force available for opening and the possible nonclosing of the EGR valve either by weaker springs or soot deposition are the other drawbacks. Ford harmonic drive, Elrod and Nelson cam lobe phasing, Fiat 3 dimensional cam, Renold camshaft phase change were gaining popularity due to their cost-effectiveness [24]. The mechanical controlled EGR systems were in production until the EURO III emission norms and were later migrated to electrical and electronic control.

Table 7.1 EGR operating windows and their significance.

Application	Engine type	EGR type	Operating window	Significance
Passenger cars [14]	2.2 L, four-cylinder direct injection diesel engine	SR and LR EGR	0%–29% by mass	Better NO_x control was achieved with LR EGR. Improved NOx–BSFC was observed with LR EGR under steady-state operating conditions.
Light duty trucks [15]	2.5 L high speed, direct injection, turbocharged diesel engine	SR and cooled EGR	0%–49% by volume	By adopting a high EGR %, the NO_x level in the combustion chamber was reduced to less than half of that without EGR.
Heavy duty trucks [16, 17]	12 L, six cylinder, four valves/cylinder, turbocharged and aftercooled diesel engine	SR and cooled EGR	0%–20% by mass	The implementation of a variable nozzle turbine with a high EGR rate resulted in maximum NO_x reduction with minimum penalties in BSFC and PM.
		LR, SR, and uncooled EGR	0%–40% by mass	40% EGR was limited to 20% of maximum load. LR EGR was effective for NO_x reduction in high load region. To avoid compressor and intercooler problems, SR EGR was recommended.
Marine [18]	MAN B&W 4T50ME-X, four-cylinder, uniflow scavenged, two-stroke, low speed diesel engine	External EGR	0%–40% by mass	By adopting 40% EGR, the brake-specific NO_x was reduced by about 88% and BSFC increased by about 4%.
Locomotive [19]	127 L, 12 cylinder, two-stroke, low speed diesel engine	LR, cooled EGR	0%–10% by volume.	About 42% reduction in NO_x emissions was achieved with EGR.
Power generator [20]	3.9 L, 44 kW, four-cylinder, four-stroke DI diesel engine	LR, cooled EGR	0%–10% by volume.	30% reduced NO_x was reported for 10% EGR. 7.5% EGR was recommended for low and medium loads.
Stationary engine [21]	Single cylinder, 4.4 kW, air cooled DI diesel engine	LR, cooled EGR	0%–15% by volume.	The combustion duration was increased by about 5%. Significant NO_x reductions were reported.

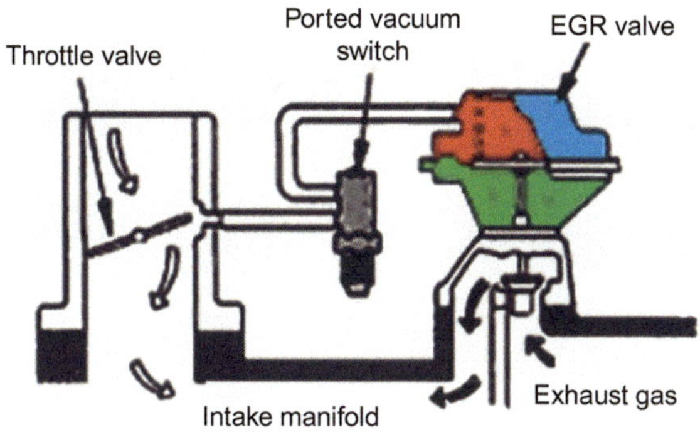

Fig. 7.5 EGR control with a ported vacuum switch.

7.5.2 Electrical control

The electric EGR valves are replacing the pneumatic (vacuum-operated) valves due to their quick response and precision. A linear solenoid, stepper motor, torque motor, and direct current (DC) motor may be used as an actuator to open and close the EGR valve. The actuation force of a solenoid is proportional to the square of its magnetic induction. A DC motor can also be used to actuate the EGR valve, in which the torque generated by the motor is converted into linear actuation force through a system of mechanical gears and levers. The actuation by the DC motor is achieved by utilizing a position sensor. These two types of EGR valve electrical actuation are shown in Fig. 7.7.

To overcome the problems associated with the DC motor brushes, a torque motor that employs contactless rotary actuators is being used. It is economical due to a reduced number of components and is also compact as it avoids mechanical gear transmission. The EGR valves are also controlled by stepper motors. Reversing a stepper motor without gear transmission is an

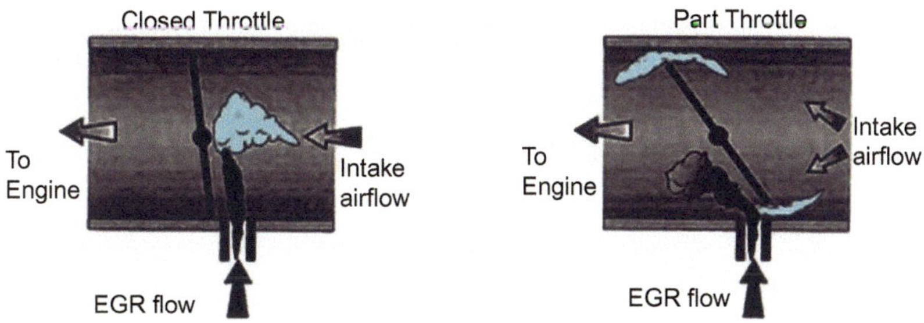

Fig. 7.6 Throttle body ported EGR.

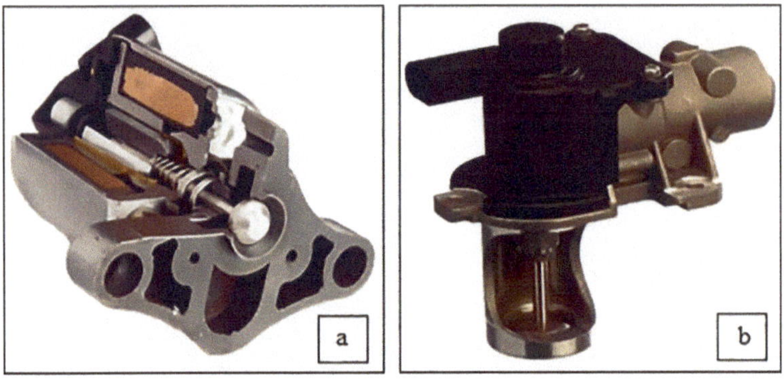

Fig. 7.7 Electrical method of EGR valve actuation: (A) Solenoid, and (B) DC motor.

Table 7.2 Comparison of electrical actuation methods of EGR.

Parameter	Solenoid	DC motor	Torque motor	Stepper motor
Actuation force	*	****	***	***
Response time	****	***	***	**
Compactness	*	**	***	***
Operational reliability	*	**	***	***

*: Low ; ****: High.

added advantage. The characteristics of different electrical means of controlling an EGR valve are compared and presented in Table 7.2.

7.5.3 Electronic/microcomputer control

In the electronic mode of EGR control, the information collected from the engine coolant temperature sensor, the throttle position sensor, and the manifold absolute pressure sensor is utilized by the powertrain control module (PCM) to operate the EGR valve. During electronic EGR operation, the EGR vent solenoid (EGRV), EGR control solenoid (EGRC), and EGR vacuum regulator solenoid (EVR) should switch on and off a number of times for proper recirculation. A failure in any one of the systems may lead to detonation and power loss [25]. A layout depicting the electronic EGR working is shown in Fig. 7.8. In recent times, the electronic control unit (ECU) collectively integrates and controls the common rail direct injection (CRDI), EGR, and after-treatment devices to keep diesel engine emissions to the lowest possible amount. The decision mapping of the EGR control with sensor inputs accomplished by the microprocessor is presented in Table 7.3.

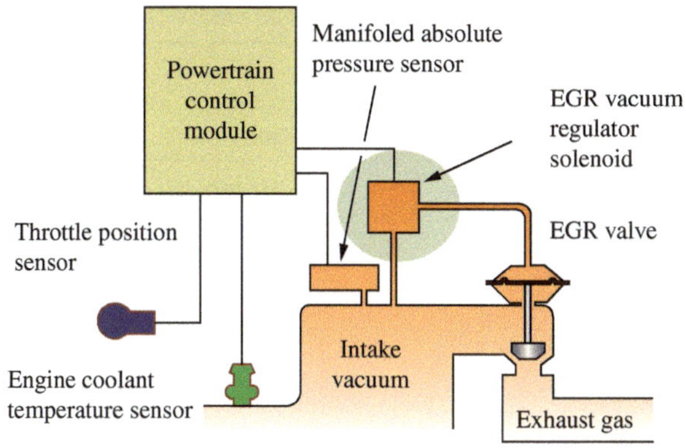

Fig. 7.8 Electronic EGR operation.

Table 7.3 Decision mapping in electronic/microprocessor-based EGR control.

Sensor	Indicator	Condition	Decision for EGR control
Throttle position sensor	Idle	Combustion temperature is low and NO_x emissions are relatively low	EGR cut-off
	Low load	Under low load engine operation, the injected fuel quantity is low and combustion is not so high to establish stable operation	Reduce EGR rate
	High load	This condition necessitates a high power output	Reduce EGR rate
Coolant temperature sensor	Low temp.	Likelihood of combustion instability due to nonuniform distribution of air-fuel mixture in the cylinder	EGR cut-off
	High temp.	No danger of instable combustion and the probability of NO_x generation is high	Increase EGR rate
	$T<35°C$ or $T>90°C$	Danger of combustion instability	EGR cut-off in timely manner
	Starting	The stability of the starting generator and smooth starting operation must be ensured	EGR cut-off

Continued

Table 7.3 Decision mapping in electronic/microprocessor-based EGR control—cont'd

Sensor	Indicator	Condition	Decision for EGR control
Manifold absolute pressure sensor	Intake air condition	If the air temperature is low, there is the possibility of improper combustion and less NO_x generation	EGR cut-off
		If the air temperature is optimum, good combustion and thus more NO_x generation	Increase EGR rate

7.6 EGR implementation in conventional CI engines

The nature of engine operation dictates EGR adoption in CI engines. It is decided as to whether the engine is operated under the steady state or transient state. The following section discusses the different aspects of EGR implementation under those operating states.

7.6.1 Under steady state

When a high speed DI diesel engine is operating under a steady-state low load condition with a high air-fuel ratio, the EGR can be increased to the extent of 50%. This is attributed to the combustion tolerance of the engine to high EGR rates, which is necessary for tremendous NO_x suppression. An EGR rate of more than 50% is also possible with the engine employing a variable geometry turbocharger [1]. To meet the European steady-state cycle (ESC) in the EURO III emission standards, the cooled EGR is sufficient to meet the NO_x/PM trade-off without impairing the fuel consumption and without any after-treatment device [26]. This is shown in Fig. 7.9.

The steady-state EGR experiments conducted on a medium-duty, four-cylinder diesel engine exhibited some interesting results [27]. In-cylinder CO_2 at the start of combustion and exhaust port NO were measured using NDIR and chemiluminescence analyzers, respectively. The variation of in-cylinder CO_2 was found to be increasing in proportion to load and speed. The variation was eminent when the engine was operating under the highest

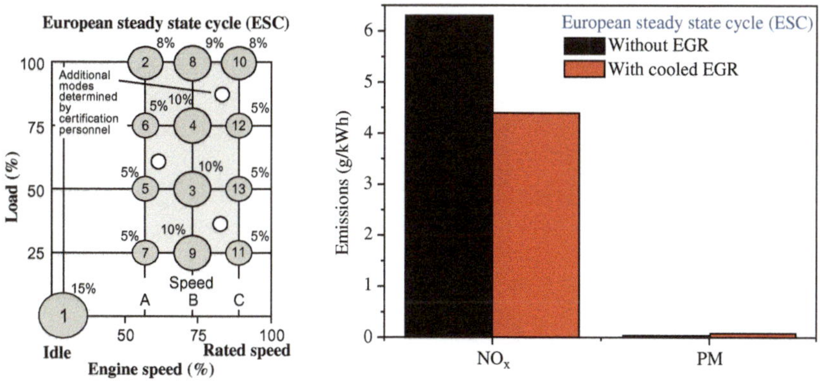

Fig. 7.9 European steady-state cycle and achievable NOx/PM trade-off with cooled EGR.

load and speed. The cyclic variability of CO_2 and NO in a multicylinder diesel engine operated under steady-state conditions is presented in Fig. 7.10. The experiments revealed that the dependence of NO emissions on EGR is nonlinear and this behavior was dominant as the mixture became richer.

7.6.2 Under transient state

The new European transient cycle (ETC) necessitates the use of a high level of cooled EGR and particulate trap to meet the NO_x/PM trade-off as dictated by Euro IV emission standards [28]. The EGR valve in a transient cycle

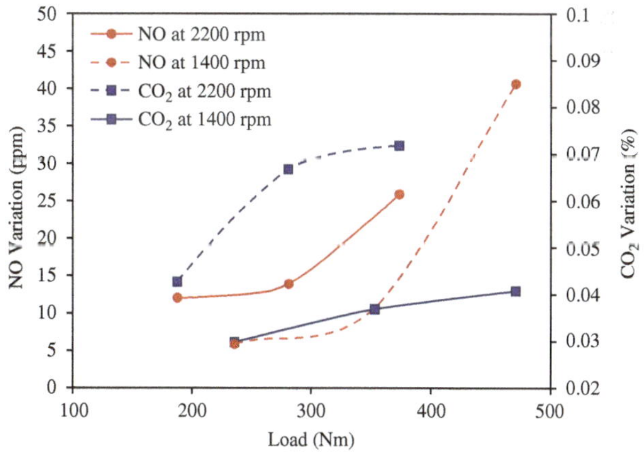

Fig. 7.10 Cyclic variability of CO2 and NO in a diesel engine under steady-state conditions.

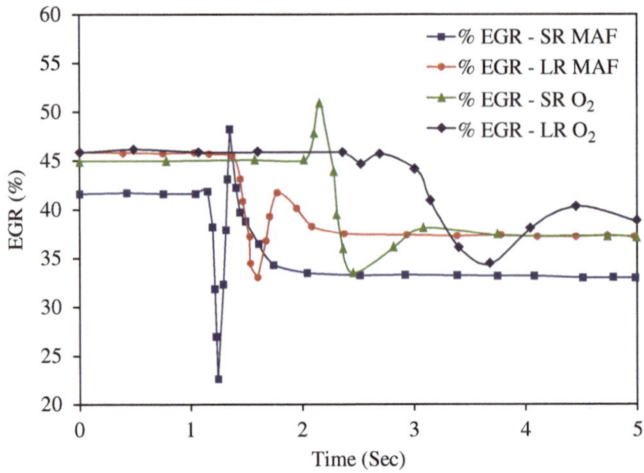

Fig. 7.11 Effectiveness of MAF and O2 concentration methods on EGR control under transient conditions.

can be controlled either by mass airflow (MAF)-based logic or by an intake manifold O_2 concentration-based technique. Out of two strategies mentioned for EGR valve control under transient conditions, the intake manifold O_2 concentration-based technique was found to be more effective, which is evident from Fig. 7.11. A considerable delay exhibited by the long route EGR in controlling the recirculation rate during a transient state produced unfavorable NO_x emissions. While the short route is adopted, the temperature of the recirculated gases will be substantially higher, which mandates an investigation of the effect of hot EGR on all engine-borne emissions. The experiments carried out on six-cylinder, direct injection diesel engines revealed that NO_x emissions are comparable for hot and cooled EGR, but CO and HC emissions are positively influenced by hot EGR [29].

In addition to the above control methods, a neural network-based method was also being considered for controlling EGR and boost pressure [30]. Some researchers have also adopted a lambda-based EGR transition, which is found to be effective for controlling the transient emissions of diesel engines [31]. An appropriate amendment of EGR in advanced combustion concepts such as low-temperature combustion of an IC engine satisfies the stringent emission norms better than the conventional diesel engines. The following section sheds light on the EGR implementation of those advanced combustion diesel engines.

7.7 EGR implementation in advanced combustion CI engines

Most of the modern diesel engines are adopting different versions of low-temperature combustion (LTC) due to their strong reduction potential of NO_x and PM emissions. The highly diluted homogenous mixture utilized in a homogeneous charge combustion ignition (HCCI) engine avoids the local fuel-rich zones and thus NO_x emissions. The homogeneity of the charge also aids in PM reduction [32]. The lean homogenous charge could be prepared outside the engine using manifold induction, fumigation, wide-open throttle, and port fuel injection. Alternatively, it could be produced within the cylinder by adopting early direct injection or late direct injection strategies. To abate the drawbacks associated with HCCI such as limited operating range, knocking, and combustion control, the premixed charge compression ignition (PCCI) combustion concept was devised, which could control the NO_x and PM emissions without impairing HC and CO emissions. Although the combustion phasing in the PCCI engine is controlled by intake charge temperature, injection timing, and EGR, a review of experimental investigations showed EGR to be more effective than other strategies. The desired combustion phasing and magnitude are also achieved by adopting different fuels with dissimilar reactivities; this concept is commonly known as reactivity controlled compression ignition (RCCI). In this mode, the low reactivity fuel is injected in PFI and high reactivity fuel in the cylinder using either single or multiple injection strategies before the onset of premixed charge combustion reactions. EGR in RCCI is varied from low to high depending upon the operating load and speed and meets the stringent emission norms. Generally, all advanced low-temperature combustion concepts adopt EGR to acquire better control over the combustion phasing, knock reduction, operational range extension, and emission advantage. The following section details the application of EGR for LTC in diesel engines.

7.7.1 HCCI

Generally, the heat release in diesel HCCI combustion takes place in two stages. About 7%–10% of the energy is released in the first stage combustion due to low-temperature reactions (LTR) and the remaining energy is released in the second stage combustion due to high-temperature reactions (HTR). The peak heat release rate (HRR) by a single injection diesel engine is about 220 kJ/deg. CA while in HCCI mode, it is only 50 kJ/deg. CA.

This corresponds to about a 77% reduction in HRR. This substantial reduction in HRR reduced NO_x emissions by 98%. Both smoke and NOx could be simultaneously reduced if HCCI combustion is accomplished with 30% EGR. Generally, the maximum NO_x reduction potential is between 76% and 95% under the direct injection mode. But then, the EGR implementation increased the CO and HC emissions by 32% and 135%, respectively. HCCI diesel combustion with multiple injection strategies is capable of reducing both NO_x and smoke emissions when compared to a single injection scheme. EGR manipulation controls the autoignition in HCCI engines. The modulated kinetics (MK) combustion system of Nissan provides good control over mixture homogeneity and HCCI combustion. Retarded injection timing, high swirl rate, and high EGR are prominent characteristics of MK combustion [33]. The fuel conversion of HCCI engines is dominantly influenced by combustion phasing, which in turn is controlled by altering the mixture reactivity, altering the time–temperature history of the mixture, or both. EGR could be utilized in HCCI engines to meet these objectives. Internal EGR aids in achieving mixture homogeneity and smoke reduction while external EGR helps to extend the load range in HCCI combustion [34]. An injection timing of 35–45 degrees bTDC, an excess air factor (λ) of 3–3.7, and an EGR rate of 20%–30% are optimal conditions for establishing efficient HCCI combustion in single-cylinder diesel engines [34–36].

7.7.2 PPCCI and PCCI

To address the issues of limited load range and unacceptable rate of pressure rise, PPCCI and PCCI combustion systems have been devised. In these systems, the fuel is injected at some point during the compression stroke. The injection event and autoignition can be distinguishable by the application of EGR. This system comes under the category of hybrid combustion, as it combines the positive virtues of HCCI and DICI combustion modes, meaning lower emissions and higher efficiency, respectively [37]. A PPCCI combustion engine was running with a double injection strategy, in which 1/6th of the fuel was injected at pilot injection and the remaining during the main injection. It was operated with a 60% EGR rate that resulted in about a 70% reduction in NO_x and 80% in opacity with a 3% drop in power. Typically, the combustion is accomplished at a much faster rate in PCCI combustion than SI and CI [38]. Early in-cylinder fuel injection along with high levels of EGR assist in achieving mixture homogeneity and reduced in-cylinder temperature. A considerable increase in the premixed phase of combustion helps to achieve the simultaneous reduction of NO_x and soot emissions in PCCI engines, although they are slightly higher than those of HCCI engines. However, the PCCI engine produces low HC and CO emissions with higher efficiency than HCCI engines. In this engine, NO_x emissions are dominantly influenced by EGR rate more than the characteristics of fuel such as cetane number and volatility. A maximum of about 50% could be implemented in PCCI engines under light load operation. HC and CO emissions are found to be decreasing while EGR is increased beyond 30% while a substantial change is not observed under low EGR rates. An EGR beyond 20% resulted in an increase in brake-specific fuel consumption and combustion duration. To overcome the hapless combustion stability associated with PCCI engines, several strategies have been attempted such as early injection, double injection, and split injection in combination with varying fuel injection pressures and compression ratios. The knocking and NO_x emissions of PCCI engines are influenced by the start of combustion and combustion phasing. These parameters could be shifted to a favorable region by implementing split fuel injection along with EGR. Generally, advanced main injection timing coupled with high fuel injection pressure and EGR rate amends the PCCI combustion process. The influence ofEGR on the emissions from a PCCI combustion engine is presented in Fig. 7.12.

7.7.3 RCCI

Researchers demonstrated the capacity of RCCI combustion to meet Euro VI norms over a wide range of engine loads. At least two different fuels with different autoignition temperatures are being used to control the combustion phasing and heat release rate [39]. To achieve efficient combustion with ultra-low emissions, it is mandated that a higher portion of energy must be derived from the low reactivity fuel. Better combustion established by RCCI paves the way to attaining the highest thermal efficiency (~60%) in an engine, which is very difficult with HCCI, PPCCI, and PCCI [40]. Numerous experimental/numerical investigations manifested the mastery of RCCI over other techniques in LTC. However, RCCI has its own set of drawbacks such as restrictions in load extension due to high PRR, high HC and CO emissions in crevice volumes, and inferior part load performance. However, these can be mitigated through optimized fuel injection, EGR, and variable valve timing technologies. A schematic of RCCI combustion is shown in Fig. 7.13.

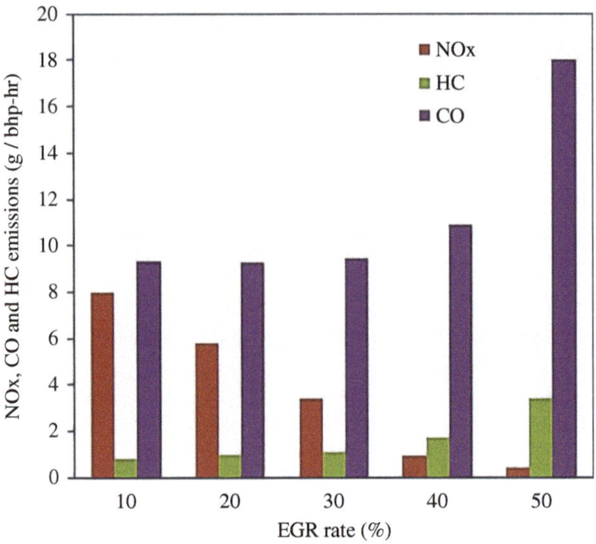

Fig. 7.12 Influence of EGR on the emissions of a PCCI engine.

The impact of different piston bowl geometries (stock, bathtub, and stepped) in addressing high CO and HC emissions in the crevice volumes of the RCCI engine was experimentally investigated using gasoline as a low reactivity fuel and diesel as a high reactivity fuel. The results revealed that both bathtub and stepped piston geometries were able to meet low NO_x emission levels dictated by Euro VI norms as well as better fuel consumption. From the perspective of HC and CO emissions, the stepped piston geometry is not recommended. Another study conducted on a

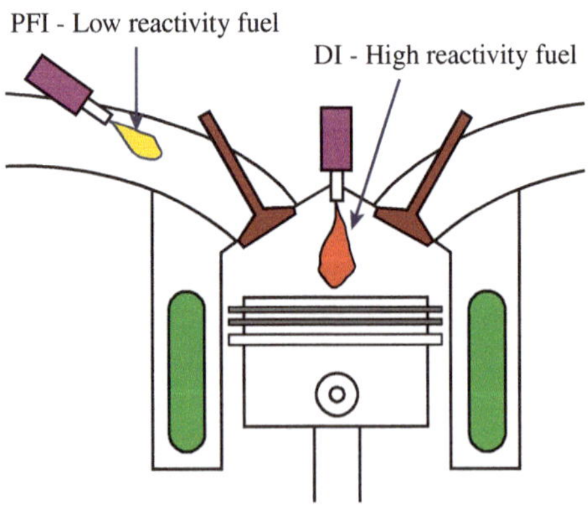

Fig. 7.13 A schematic of RCCI combustion.

water-cooled six-cylinder engine revealed the potential of extending upper and lower limits of RCCI operation. An intake pressure of 0.12 MPa, 30% EGR, 70% gasoline, and advanced injection timing were found to be the optimum operating parameters for a high load regime for RCCI combustion, which also resulted in low levels of NO_x, soot, and HC emissions without compromising the fuel consumption [41]. The low operating range of RCCI combustion could also be smoothed by increasing the diesel fuel proportion and reducing the EGR rate. A dual-mode dual-fuel concept is envisioned to extenuate the high HC and CO emissions at low loads and unacceptable levels of increased pressure rise at high loads [42]. A summary of low-temperature combustion with EGR implementation is presented in Table 7.4.

Table 7.4 Summary of low-temperature combustion with EGR.

Engine type	Concept	Maximum EGR rate	Observation in engine performance and emission characteristics
Single cylinder, four stroke, rated power of 7.4 kW, and premixed ratio of 0%–80%	HCCI-DI	30%	Mean effective pressure was found to be increasing with an increase in premix ratio. Maximum reductions of about 76% and 58% were achieved in NO_x and smoke while HC and CO emissions were increasing with the increased premix ratio.
Single cylinder, rated power of 4.4 kW, common rail	HCCI with external mixture formation	30%	Considerable drop in brake thermal efficiency, pressure rise, and heat release rate were observed with vaporizer and increasing EGR. But NO_x was reduced to the maximum of about 95% and smoke by 83%.

Continued

Table 7.4 Summary of low-temperature combustion with EGR—cont'd

Engine type	Concept	Maximum EGR rate	Observation in engine performance and emission characteristics
Single-cylinder, water-cooled, naturally aspirated DI diesel engine	HCCI with a combination of internal and external EGR	50%	Internal EGR helped in homogeneous mixture formation and smoke reduction while external cooled EGR helped in achieving low NO_x even at high loads without impacting the smoke emissions.
Multicylinder engine with twin stage turbocharger, common rail	PCCI	50%	A substantial reduction of cylinder pressure and heat release rate was observed with 50% EGR. NO_X and comparable HC and CO emissions were reported. A sharp increase in smoke emission was reported with maximum EGR rate.
Single cylinder, rated power of 6.25 kW, common rail	PCCI	30%	Thermal efficiency dropped with increasing EGR; NO_x emissions were reduced below 5 ppm and smoke opacity increased by 6% with increasing EGR.
Four-cylinder engine	PCCI with emission control devices	54%	NO_x, PM, and HC emissions were reduced with an increase in CO emissions.
Single cylinder, four stroke	PCCI	45.4%	A considerable drop in pressure rise rate, NO_x, and smoke emissions.

Table 7.4 Summary of low-temperature combustion with EGR—cont'd

Engine type	Concept	Maximum EGR rate	Observation in engine performance and emission characteristics
Single cylinder (modified from multicylinder), four stroke, port and common rail injection system	RCCI (HRF: diesel; LRF: ethanol)	29%	NO_x emissions reduced by 79% and smoke emissions by 56%. Shorter combustion duration was reported under optimized conditions.
Single cylinder (modified from multicylinder), four stroke, port and common rail injection	RCCI (HRF: diesel; LRF: gasoline)	45%	Increased cylinder pressure and heat release were reported. NO_x emissions were reduced to the extent of meeting the Euro VI norms.
Single cylinder (modified from multicylinder), four stroke, port and common rail injection	RCCI (HRF: diesel; LRF: gasoline)	70%	Indicated mean effective pressure increased with EGR rate while reduced cylinder pressure and heat release rate were reported when EGR varied from 50 to 70%. Reduced emissions (HC, CO, NO_x, and PM) with advanced injection timing and EGR.
Three-cylinder, turbocharged 1.5 L common rail direct injection diesel engine	RCCI (HRF: diesel; LRF: methanol)	40%	The influence of hot and cooled EGR on RCCI combustion revealed that 26% cooled EGR was producing better combustion, performance, and emission characteristics.

Continued

Table 7.4 Summary of low-temperature combustion with EGR—cont'd

Engine type	Concept	Maximum EGR rate	Observation in engine performance and emission characteristics
Single-cylinder (modified), four-stroke, Euro VI diesel engine	RCCI (HRF: diesel; LRF: gasoline)	60%	When EGR rate increased beyond 40%, NO_x emissions were reduced marginally with an abrupt rise in CO emissions.

7.8 EGR implementation for alternate fueled engines

Diesel engines are best known for their multifuel capabilities. A wide range of alternate liquid and gaseous fuels has been tested in diesel engines to study their impact on performance and emission characteristics. The experimental results stand as evidence for the ability of EGR to mitigate emissions from the alternate fueled engines. A selected range of alternate fuels such as fuels synthesized from fossil or biogenic gas (GTL, BTL), fatty methyl esters from different vegetable oils, neat/straight vegetable oils, and recycled waste oil with their results are presented in the following section. With fossil fuel shortages and stringent emission norms, many countries have already started using straight vegetable oils in diesel engines. If the physical and chemical properties are comparable to mineral diesel oil, these straight vegetable oils could be burned as such or with minor modifications. As the properties of neat rapeseed oil and soybean oil meet the DIN V51605 standards, they are recommended for diesel engine applications [43]. Experiments were carried out on a multicylinder turbocharged diesel engine with 100% rapeseed oil with different EGR rates to analyze its influence on NO_x reduction characteristics. During the operation, the EGR was varied between 0 and 20%. When compared to the neat diesel engine, the neat rapeseed oil fueling was found to reduce NO_x emissions by about 43% and 60% with 10% and 20% EGR, respectively. Although the NO_x reduction might also be attributed to the fuel composition change, the contribution was found to be only 27%, so EGR was concluded as the most influential factor [44]. Because of the NO_x-PM trade-off, experiments were carried out on 2.2 L, common rail direct injection with two-stage turbocharging with Jatropha and soybean as neat oils. Irrespective of NO_x levels, Jatropha presented the lowest particulate number and d50. At high EGR, soybean oil exhibited lower d50 and PN levels than diesel while similar trends were observed for all test fuels without EGR [45].

As Jatropha-based biodiesel is free from sulfur, has excellent lubricity, and improved wear resistance, it has become the primary choice for engine researchers. The high bulk modulus, high boiling point, and inherent oxygen of Japtropha biodiesel lead to increased NO_x emissions. The experiment carried out on a single-cylinder, water-cooled diesel engine revealed that 15% EGR was found to be effective in reducing NO emissions lower than diesel-fueled operation under all loads without a compromise in brake thermal efficiency or HC and CO emissions [46]. A twin-cylinder water-cooled direct injection diesel engine running on sunflower biodiesel blends revealed that about a 25% reduction in NO_x emissions could be achieved with a 20% blend and 15% EGR. Under the same operating conditions, the HC and CO emissions were reduced by 5% and 10%, respectively [47]. The combined influence of EGR and nanoparticle-impregnated palm biodiesel blends on the performance and emission characteristics of a single-cylinder air-cooled direct injection diesel engine were studied. With EGR implementation, throughout the operating range lower NO_x emissions were reported for nanoparticle-impregnated palm biodiesel blends over pure biodiesel blends. The additional NO_x reduction was attributed to a lower oxygen concentration and reduced flame temperature [48]. The effectiveness comparison of HPL and LPL EGR was carried out on the NO_x reduction characteristics of a turbocharged diesel engine fueled with rapeseed biodiesel. The experimentation revealed that the EGR rate could be increased to an extent of 45% in LPL; higher effectiveness was reported for HPL EGR without the problem of increased PM emissions. The high oxygen content of biodiesel suppressed soot formation [49].

The experiment carried out with GTL fuel and reformed exhaust gas recirculation (REGR) on a single-cylinder air-cooled direct injection diesel engine yielded some interesting results. Compared to ultralow sulfur diesel fuel, GTL fuel was able to shift the NO_x-PM trade-off to lower values, which was attributed to its spectacular combustion characteristics. REGR was also found to produce more favorable emission characteristics over conventional EGR, irrespective of fuel type and load condition. With 30% REGR, the NO_x and smoke emissions were respectively reduced to the maximum of about 75% and 60% compared to diesel fuel at low loads. Almost 40% reduced NO_x and smoke emissions were reported with 10%

REGR at high load [50]. BTL fuels exhibited low emission characteristics, which included NO_x, soot, HC, and CO emissions. The medium load operation point revealed that the precise adaption of the EGR rate during the combustion of BTL blends led to reduced NO_x with consistent PM emissions [51].

An environmental concern favors the choice of waste oil use in internal combustion engines. A single-cylinder diesel engine running on 100% waste plastic oil reported higher NO_x emissions than diesel fuel. This was reduced by recirculating 20% of flue gases without impacting the smoke, HC, and CO emissions [52]. A 10% EGR rate was found to be optimum for a tire pyrolysis-diesel blend without compromising the engine torque, fuel economy, or HC and CO emissions. A still higher EGR rate could be employed with ester addition [53]. A medium-duty truck engine equipped with a catalytic converter and a cooled EGR system displayed lower NO_x, CO, HC, and PM emissions with natural gas fueling. Besides, D13 driving conditions reported 15% lower CO_2 emissions than conventional diesel operation [54]. The performance of a single-cylinder diesel engine with manifold induction of propane gas and EGR was evaluated at low and high air-fuel ratios. The EGR rate based on the CO_2 approach was varied from 20 to 45%. The dual-fuel operation with EGR caused a considerable reduction in NO_x and PM with a drop in mean effective pressure as compared to neat diesel fueling. The greater sensitivity of the engine toward the EGR rate was attributed to the lower cetane number of propane gas [55]. Another study carried out on an engine that operated with gaseous hydrogen and cooled EGR exhibited about a 6% improvement in brake thermal efficiency due to hydrogen enrichment. In this study, a constant flow rate (20 L/min) of hydrogen was maintained, while the EGR was varied as 15% and 25%. Abrupt drop in NO_x emission was observed with corresponding increase in other regulated emissions (HC, CO, and smoke) [56]. A summary of NO_x control with EGR for different alternate fueled diesel engines is presented in Table 7.5.

7.9 Effect of EGR on oil contamination, engine wear, and soot

Despite its success for NO_x mitigation in light-duty vehicles, EGR implementation has led to an acute increase of soot formation in the cylinder. This, in turn, increases the chance of soot accumulation in the engine oil, changing the chemical properties and contaminating it. The abrasive,

Table 7.5 Summary of NOx control with EGR for alternate fueled engines.

Engine	Name of alternate fuel	EGR	NOx reduction potential	EGR influence
Liquid fuels				
Single-cylinder, air-cooled DI diesel engine	Fish biodiesel–diesel blends (20%–80%)	0%–30%	20% reduced NO_x was reported for 20% biodiesel blend with 30% EGR.	Optimum NO_x–soot trade-off was observed for 20% EGR.
Naturally aspirated, twin-cylinder DI diesel engine	Rice bran methyl ester–diesel blends (0%–50%)	15%	35% reduction in NO_x for 10% blends with 15% EGR.	20% biodiesel blend with 15% EGR was found to produce favorable emission characteristics.
Single-cylinder, water-cooled DI diesel engine	Waste cooking oil biodiesel–diesel blend (20% by vol.)	0%–20%	75% reduced NO_x emissions.	Low EGR (~10%) was recommended for biodiesel blend without deteriorating the combustion quality.
Single-cylinder, naturally aspirated four-stroke, water-cooled diesel engine	Tamarind seed biodiesel (20% by vol.)	0%–30%	58% reduction in NO_x with 30% EGR and 20% blend.	20% EGR was recommended for 20% Tamarind seed biodiesel due to minimum compromise on other emissions.
Single-cylinder, air-cooled, naturally aspirated DI diesel engine	Dimethyl carbonate–diesel blends (15% by vol.)	0%–30%	35% reduced NO_x over diesel operation.	About 70% reduced smoke, 33% reduced CO, and 48% increased HC emissions were reported with 30% EGR.
Gaseous fuels				
Single-cylinder, four-stroke DI diesel engine	Hydrogen in dual fuel mode	10%	45% reduction in NO_x with 10% energy supplied by hydrogen and 10% EGR.	NO_x emissions were also found to shoot up as energy supplied by hydrogen increased.

Continued

Table 7.5 Summary of NOx control with EGR for alternate fueled engines—cont'd

Engine	Name of alternate fuel	EGR	NOx reduction potential	EGR influence
Inline six-cylinder, four-stroke, turbocharged, intercooled DI diesel engine	Natural gas (NG) as neat fuel	30%	NO_x was reduced by about 65% over NG operation without EGR.	Slow combustion at lower engine rpm, was responsible for NO_x reduction. At high rpm, EGR deteriorated the engine performance. Further reduction possible with spark timing adjustment.
Four-cylinder turbocharged DI diesel engine	Liquefied petroleum gas (LPG)	24%	26% low NO_x when 10% diesel was replaced by LPG.	Lowest NO_x emissions were reported when 4%–45% of diesel fuel replaced with LPG.
Single-cylinder, four-stroke, direct-injection, water-cooled CI engine	Syngas	0%–10% (by wt.)	82% reduction in NO_x with multiple injection of syngas with 10% EGR over diesel usage.	Multiple injection with high EGR was recommended under all injection pressures for syngas over single injection strategy.

adhesive, and scuffing wear mechanisms in an IC engine are closely associated with mechanical surface damage, while the corrosion and additive depletion wear mechanisms demand a serious of chemical reactions before resulting wear. The lubricant oil film instituted on the cylinder surface gets disrupted under the different engine operating conditions such as cold start, idling, and rapid acceleration as well as establishing contact between the cylinder surface and accumulated soot in the engine oil. When the accumulated soot particle diameter lies in the range of 0.01–0.8 μm, which is higher than the boundary layer thickness, the abrasive wear prevention is inevitable. The acidic products available in the exhaust will increase the total acid number (TAN) of the lubricant during their recirculation. This, in turn, accelerates the corrosive wear of the lubricated surfaces [57].

The effectiveness of lubricating oils in overcoming the soot-related wear mechanisms under high loads was investigated by conducting API CI-4 CUMMINS-M11 EGR engine tests. Elevated soot loading in lubricant oil ($\geq$ 12%) as well as about a 250% increase in TAN with a drop of 50% in total base number (TBN) were reported. The test results also concluded that abrasion was the dominant wear mechanism, which was aided by corrosion. Another study conducted with simulated carbon black in oil that contains phosphorous-based antiwear additives exhibited antagonistic behavior, which accelerated the engine wear [58]. High levels of soot adversely affect the oil viscosity. The viscosity change was found to be linear up to 1% of soot loading, beyond which the change was abrupt. It is established that friction is proportional to the square root of the lubricant's dynamic viscosity [59]. The friction is upstretched due to a viscosity increase, which in turn increases the fuel consumption and tailpipe emissions. The engine oils formulated with different intensities of phosphorous, dispersant, and sulfonates were examined to ascertain their wear resistance using a three-body wear machine [60]. The scanning electron microscopic (SEM) images taken over the worn-out surfaces revealed that engine wear could be reduced to a greater extent by increasing the phosphorous content.

The literature disclosed that engine wear was principally influenced by soot reactivity rather than the contact surface's mechanical properties. The soot morphology, surface chemistry, and the reactivity are noteworthy concerning engine wear. To abbreviate the negative effect of EGR on oil contamination, a metallic/disposable paper filter can be employed in the recirculating path. By this method, the oil contamination could be reduced to a considerable extent. A considerable increase in total particulate matter (TPM) was observed when a heavy-duty engine was running with 16% EGR. The total particulate matter is a mixture of soluble organic fraction (SOF), solids (SOL), and sulfonate ($SO_4{}^{2-}$). The solids and sulfonate were found to be increasing with increased EGR [61]. Another study related to soot particle morphology revealed the consequence of EGR concentration on particle size distribution. The diameter of the soot particles was respectively influenced by the dilution and thermal effect of EGR at high and low concentration levels. The particle size distribution was analyzed using a multicylinder high-speed direct-injection diesel engine under varying EGR rates (0%–40%). The other engine operating conditions are mentioned in Fig. 7.14. Under 40% EGR, the nucleation mode was found to be negligible and the amount of concentration in this mode was the lowest. The amount of concentration reached its peak in the accumulation mode and large

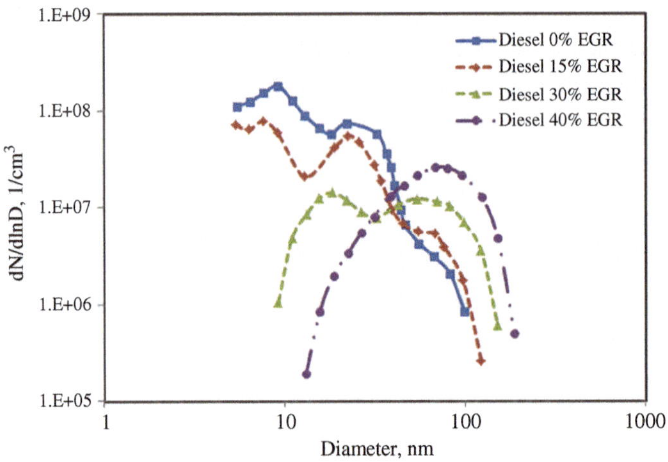

Fig. 7.14 Influence of EGR on exhaust soot particle size distribution.

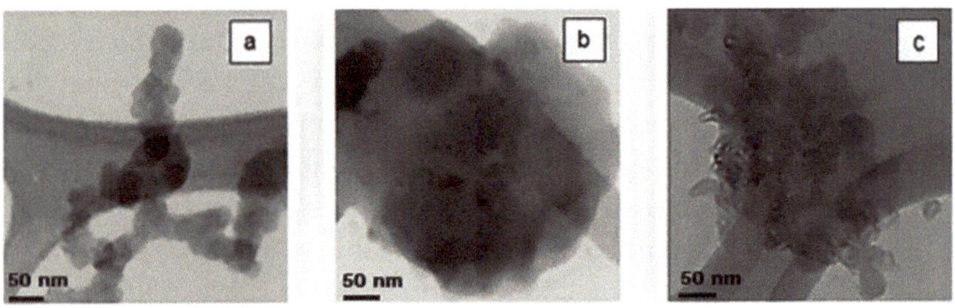

Fig. 7.15 TEM images of soot particles in: (A) conventional diesel combustion; (B) PCCI combustion; and (C) HCCI combustion.

particles were detected for diesel fuel operation, which might be attributed to coagulation, accumulation, condensation of the volatile fraction, and surface growth [62].

A highly ordered graphitic structure and low organic carbon fraction were observed with increasing EGR rate, which might be associated with longer combustion duration and the decrease of the air-fuel ratio. This was also an indicator of reduced soot reactivity under high EGR condition. The transmission electron microscopic (TEM) images of soot particles attained in conventional diesel combustion and PCCI combustion are presented in Fig. 7.15. The average diameter of the primary particles is substantially higher in PCCI combustion (700–1200 nm against 30 nm in conventional diesel combustion). Besides, the morphological investigation revealed a chain-like structure composed of distinct spherical particles in conventional combustion and larger clumps of carbonaceous matter with irregular shapes in PCCI mode combustion. Although the morphology of soot particles in HCCI combustion was comparable to conventional combustion, the tendency to form a chain-like structure was found to be missing.

7.10 EGR in conventional/advanced SI and CI engines–A comparison

Although the major aim of EGR implementation in an SI or CI engine is NO_x suppression, the charge dilution by EGR will allow a larger throttle position that in turn reduces the pumping losses in SI engines. Thus, the efficiency of an SI engine could be increased, which is not possible with CI engines, as they are unthrottled. Dual fuel diesel engines could anticipate an improvement in thermal efficiency, as hot recirculated exhaust gases help in raising the intake temperature and reburning the unburned fuel in recirculated gases [63]. As a high rate of exhaust gas recirculation will result in combustion instability, conventional SI engines normally use a lower EGR rate (~10 to 25%). The relatively high EGR rate adopted in CI engines is accompanied by penalties such as poor specific fuel consumption and increased smoke emissions. However, the desirable outcome of maximum NO_x reduction with reduced HC and smoke emissions could be met by combining the techniques of supercharging with EGR in diesel engines.

EGR implementation for advanced SI and CI engines is not unique and their expected outcomes are also different. As compared to a conventional SI engine, the residual gas recirculation could be increased to the maximum of 50% under the stratified mode of gasoline direct injection (GDI) engines while it is limited to about 25% in port fuel injection (PFI) engines. Generally, cooled EGR is recommended for PFI engines to derive maximum gains. But no conclusive decision is made on a suitable EGR configuration for GDI engines and many research works are still going on. As compared to a conventional CI engine, higher EGR rates are being followed in almost all low-temperature combustion engines. The maximum EGR rate adopted in LTC engines follows the order of RCCI > PCCI > HCCI. Although internal EGR is adopted in some of these engines to achieve mixture homogeneity and smoke reduction, the majority of research works support the recommendation of external cooled EGR for implementation in LTC engines irrespective of their categories for load range extension without impacting smoke emissions.

7.11 Conclusion

This chapter starts with background on EGR, the design variants such as external/internal, the SR/LR/hybrid, and the cooled/hot EGR. This is followed by an elaboration on the different actuation methods such as mechanical, electrical, and electronic/microprocessor control. The impact of EGR on the steady and transient state performance, combustion, and emission characteristics of a diesel engine is elucidated in detail. EGR operating window, their significance and their consequences on advanced diesel combustion especially low temperature combustion such as homogeneous charge compression ignition, partially premixed/premixed charge compression ignition and reactivity controlled compression ignition have expatiated. The positive outcomes of EGR for alternate liquid and gaseous fueled diesel engines are also expounded. The relative merits and demerits of different EGR systems are reviewed and compared based on fuel consumption and various flow rates of exhaust gas recirculation. An EGR rate as high as 70% could be adopted for advanced combustion control. The ignition timing control and homogeneous mixture preparation are positively promoted by EGR implementation. Lower in-cylinder temperature and oxygen displacement by EGR resulted in low NO_x but increased smoke, HC, and CO emissions in alternate fueled diesel engines. EGR-generated soot was found to give rise to lube oil contamination and accelerated engine wear. The increased smoke penalty at high loads due to EGR implementation could be avoided by following a combinational approach such as a retarded injection scheme with low EGR.

The EGR prospects for future diesel engines are identified and presented below. Substantial EGR utilization in low-temperature combustion demands superior quality fuels. Hence, the combustion system of future diesel engines will depend upon tailor-made fuels. The detrimental consequences of long-term EGR must be addressed by developing novel lubricants with suitable additives to overcome the oil contamination and accelerated engine wear at high EGR rates. During the engine transient condition, the precise and metered quantity of exhaust gases must be recirculated depending upon the engine load, which necessitates the development of accurate electronic/microprocessor-based EGR control systems. EGR was found to result in increased particle concentration in accumulation mode and large particles in the diesel exhaust that demand improved diesel particulate filter design.

In general, a high EGR rate with a more efficient after-treatment system that incorporates an oxidation catalyst, a de-NO_x catalyst, an HC and diesel exhaust fluid injector system integrated with several sensors, and a closed-loop control is expected to meet stringent emission norms imposed on future diesel engines with ease. The next chapter talks about NO_x reduction in the tailpipe of an engine via the implementation of an effective catalytic converter. The performance and constructional aspects of different catalytic converters are explained in detail. Modifications imposed on the catalytic converter design by the use of alternative fuels and dual fuel mode operations are also detailed in Chapter 8.

References

[1] Hawley JG, Wallace FJ, Cox A, Horrocks RW, Bird GL. Reduction of steady state NOx levels from an automotive diesel engine using optimised VGT/EGR schedules. SAE Trans 1999;1:1172–84.
[2] Vignesh R, Ashok B. Critical interpretative review on current outlook and prospects of selective catalytic reduction system for De-NOx strategy in compression ignition engine. Fuel 2020;276:117996.
[3] Ladommatos N, Abdelhalim S, Zhao H. The effects of exhaust gas recirculation on diesel combustion and emissions. Int J Engine Res 2000;1(1):107–26.
[4] Manoharan AK, Ashok B, Kumarasamy S. Numerical prediction of NOx in the exhaust of a CI engine fuelled with biodiesel using in-cylinder combustion pressure based variables. SAE Technical Paper; 2016.
[5] Okubo M, Kuwahara T. New technologies for emission control in marine diesel engines. Butterworth-Heinemann; 2019 Aug 29.
[6] Thangaraja J, Kannan C. Effect of exhaust gas recirculation on advanced diesel combustion and alternate fuels—a review. Appl Energy 2016;180:169–84.
[7] Reifarth S. EGR-systems for diesel engines; 2010. Doctoral dissertation.
[8] Zamboni G, Moggia S, Capobianco M. Hybrid EGR and turbocharging systems control for low NOx and fuel consumption in an automotive diesel engine. Appl Energy 2016;165:839–48.
[9] Millo F, Giacominetto PF, Bernardi MG. Analysis of different exhaust gas recirculation architectures for passenger car diesel engines. Appl Energy 2012;98:79–91.
[10] Maiboom A, Tauzia X, Shah SR, Hétet JF. Experimental study of an LP EGR system on an automotive diesel engine, compared to HP EGR with respect to PM and NOx emissions and specific fuel consumption. SAE Int J Engines 2010;2(2):597–610.
[11] d'Ambrosio S, Finesso R, Spessa E. Calculation of mass emissions, oxygen mass fraction and thermal capacity of the inducted charge in SI and diesel engines from exhaust and intake gas analysis. Fuel 2011;90(1):152–66.
[12] Snyder DB, Adi GH, Hall CH, Shaver GM. Control-variable-based accommodation of biodiesel blends. Int J Engine Res 2011;12(6):564–79.
[13] Zu XH, Yang CL, Wang HC, Wang YY. Experimental study on diesel engine exhaust gas recirculation performance and optimum exhaust gas recirculation rate determination method. R Soc Open Sci 2019 Jun 5;6(6):181907.
[14] Park Y, Bae C. Experimental study on the effects of high/low pressure EGR proportion in a passenger car diesel engine. Appl Energy 2014;133:308–16.

[15] Ladommatos N, Balian R, Horrocks R, Cooper L. The effect of exhaust gas recirculation on combustion and NOx emissions in a high-speed direct-injection diesel engine. SAE Technical Paper; 1996 Feb 1.
[16] Baert RS, Beckman DE, Veen A. Efficient EGR technology for future HD diesel engine emission targets. SAE Trans 1999;1:381–93.
[17] Kohketsu S, Mori K, Sakai K, Hakozaki T. EGR technologies for a turbocharged and intercooled heavy-duty diesel engine. SAE Trans 1997;1:482–9.
[18] Raptotasios SI, Sakellaridis NF, Papagiannakis RG, Hountalas DT. Application of a multi-zone combustion model to investigate the NOx reduction potential of two-stroke marine diesel engines using EGR. Appl Energy 2015;157:814–23.
[19] Iden M, Goetzke M, Bandyopadhyay D. Experimental exhaust gas recirculation (EGR) on EMD SD59MX 2.238 megawatt diesel-electric freight locomotives. In: Internal combustion engine division fall technical conference, vol. 55096. American Society of Mechanical Engineers; 2012 Sep 23. p. 583–90.
[20] De Serio D, de Oliveira A, Sodré JR. Effects of EGR rate on performance and emissions of a diesel power generator fueled by B7. J Braz Soc Mech Sci Eng 2017;39(6):1919–27.
[21] Saravanan S. Effect of EGR at advanced injection timing on combustion characteristics of diesel engine. Alex Eng J 2015 Sep 1;54(3):339–42.
[22] Erjavec J, Thompson R. Automotive technology: A systems approach. Cengage learning; 2014 Feb 28.
[23] Berry IJ, Brunt MF. Improved control of EGR during speed/load transients. SAE Technical Paper; 1996 Feb 1.
[24] Gray C. A review of variable engine valve timing. SAE Trans 1988;1:631–41.
[25] Ashok B, Ashok SD, Kumar CR. A review on control system architecture of a SI engine management system. Annu Rev Control 2016;41:94–118.
[26] Zelenka P, Aufinger H, Reczek W, Cartellieri W. Cooled EGR-a key technology for future efficient HD diesels. SAE Technical Paper; 1998 Feb 23.
[27] Kalsi KS, Collings N, Heaton DM, Faulkner SA. Study of steady state and transient EGR behaviour of a medium duty diesel engine. SAE Technical Paper; 2008 Oct 6.
[28] Millo F, Ferraro CV, Bernardi MG, Barbero S, Pasero P. Experimental and computational analysis of different EGR systems for a common rail passenger car diesel engine. SAE Int J Engines 2009;2(1):527–38.
[29] Narusawa K, Odaka M, Koike N, Tsukamoto Y, Yoshida K. An EGR control method for heavy-duty diesel engines under transient operations. SAE Trans 1990;1:991–1004.
[30] Galindo J, Lujan JM, Serrano JR, Hernández L. Combustion simulation of turbocharger HSDI diesel engines during transient operation using neural networks. Appl Therm Eng 2005;25(5–6):877–98.
[31] Liu B, Zhang F, Zhao C, An X, Pei H. A novel lambda-based EGR (exhaust gas recirculation) modulation method for a turbocharged diesel engine under transient operation. Energy 2016;96:521–30.
[32] Assanis DN, Najt PM, Dec JE, Eng JA, Asmus TN, Zhao F. Homogeneous charge compression ignition(HCCI) engines. vol. 646. Warrendale, PA: Society of Automotive Engineers; 2003 Mar.
[33] Kimura S, Aoki O, Kitahara Y, Aiyoshizawa E. Ultra-clean combustion technology combining a low-temperature and premixed combustion concept for meeting future emission standards. SAE Trans 2001;110:239–46.
[34] Chidambaram K, Packirisamy T. Smart ceramic materials for homogeneous combustion in internal combustion engines: a review. Therm Sci 2009;13(3):153–63.
[35] Das P, Subbarao PM, Subrahmanyam JP. Control of combustion process in an HCCI-DI combustion engine using dual injection strategy with EGR. Fuel 2015;159:580–9.

[36] Singh AP, Agarwal AK. Combustion characteristics of diesel HCCI engine: an experimental investigation using external mixture formation technique. Appl Energy 2012;99:116–25.
[37] Dempsey AB, Walker NR, Gingrich E, Reitz RD. Comparison of low temperature combustion strategies for advanced compression ignition engines with a focus on controllability. Combust Sci Technol 2014;186(2):210–41.
[38] Saiteja P, Ashok B. A critical insight review on homogeneous charge compression ignition engine characteristics powered by biofuels. Fuel 2021;285:119202.
[39] Fang W, Kittelson DB, Northrop WF. Optimization of reactivity-controlled compression ignition combustion fueled with diesel and hydrous ethanol using response surface methodology. Fuel 2015;160:446–57.
[40] Splitter D, Wissink M, DelVescovo D, Reitz RD. RCCI engine operation towards 60% thermal efficiency. SAE Technical Paper; 2013 Apr 8.
[41] Wang Y, Yao M, Li T, Zhang W, Zheng Z. A parametric study for enabling reactivity controlled compression ignition (RCCI) operation in diesel engines at various engine loads. Appl Energy 2016;175:389–402.
[42] Benajes J, García A, Monsalve-Serrano J, Boronat V. Gaseous emissions and particle size distribution of dual-mode dual-fuel diesel-gasoline concept from low to full load. Appl Therm Eng 2017;120:138–49.
[43] Bragadeshwaran A, Kasianantham N, Balusamy S, Muniappan S, Reddy DM, Subhash RV, Pravin NA, Subbarao R. Mitigation of NOx and smoke emissions in a diesel engine using novel emulsified lemon peel oil biofuel. Environ Sci Pollut Res Int 2018;25:25098–114.
[44] Labecki L, Ganippa LC. Effects of injection parameters and EGR on combustion and emission characteristics of rapeseed oil and its blends in diesel engines. Fuel 2012;98:15–28.
[45] Koder A, Schwanzer P, Zacherl F, Rabl HP, Mayer W, Gruber G, Dotzer T. Combustion and emission characteristics of a 2.2 L common-rail diesel engine fueled with jatropha oil, soybean oil, and diesel fuel at various EGR-rates. Fuel 2018;228:23–9.
[46] Pradeep V, Sharma RP. Use of HOT EGR for NOx control in a compression ignition engine fuelled with bio diesel from Jatropha oil. Renew Energy 2007;32(7):1136 54.
[47] Rajan K, Senthilkumar KR. Effect of exhaust gas recirculation (EGR) on the performance and emission characteristics of diesel engine with sunflower oil methyl ester. JJMIE 2009;3(4):306–11.
[48] Karthickeyan V, Thiyagarajan S, Geo VE, Ashok B, Nanthagopal K, Chyuan OH, Vignesh R. Simultaneous reduction of NOx and smoke emissions with low viscous biofuel in low heat rejection engine using selective catalytic reduction technique. Fuel 2019 Nov 1;255:115854.
[49] Kawano D, Ishii H, Goto Y, Noda A, Aoyagi Y. Optimization of engine system for application of biodiesel fuel. SAE Trans 2007;116:680–6.
[50] Abu-Jrai A, Rodríguez-Fernández J, Tsolakis A, Megaritis A, Theinnoi K, Cracknell RF, Clark RH. Performance, combustion and emissions of a diesel engine operated with reformed EGR. Comparison of diesel and GTL fuelling. Fuel 2009;88(6): 1031–41.
[51] Moeltner L, Konstantinoff L, Schallhart V. Hydrotreated vegetable oils, biomass to liquid and fatty acid methyl ester as biogen admixtures for diesel engines in passenger cars. SAE Int J Fuels Lubr 2017;10(2):251–61.
[52] Mani M, Nagarajan G, Sampath S. An experimental investigation on a DI diesel engine using waste plastic oil with exhaust gas recirculation. Fuel 2010;89(8):1826–32.
[53] Paul Daniel M, Kumar KV, Durga Prasad B, Puli RK. Performance and emission characteristics of diesel engine operated on plastic pyrolysis oil with exhaust gas recirculation. Int J Ambient Energy 2017;38(3):295–9.

[54] Ishida A, Nishimura A, Uranishi M, Kihara R, Nakamura A, Newman P. The development of the ECOS-DDF natural gas engine for medium-duty trucks: exhaust emission reduction against base diesel engine. JSAE Rev 2001;22(2):237–43.
[55] Lee J, Choi S, Kim H, Kim D, Choi H, Min K. Reduction of emissions with propane addition to a diesel engine. Int J Automot Technol 2013;14(4):551–8.
[56] Saravanan N, Nagarajan G, Kalaiselvan KM, Dhanasekaran C. An experimental investigation on hydrogen as a dual fuel for diesel engine system with exhaust gas recirculation technique. Renew Energy 2008;33(3):422–7.
[57] Aldajah S, Ajayi OO, Fenske GR, Goldblatt IL. Effect of exhaust gas recirculation (EGR) contamination of diesel engine oil on wear. Wear 2007;263(1–6):93–8.
[58] Salehi FM, Khaemba DN, Morina A, Neville A. Corrosive–abrasive wear induced by soot in boundary lubrication regime. Tribol Lett 2016;63(2):19.
[59] Green DA, Lewis R. The effects of soot-contaminated engine oil on wear and friction: a review. Proc Inst Mech Eng D 2008;222(9):1669–89.
[60] Gautam M, Chitoor K, Durbha M, Summers JC. Effect of diesel soot contaminated oil on engine wear—investigation of novel oil formulations. Tribol Int 1999;32(12): 687–99.
[61] Kreso AM, Johnson JH, Gratz LD, Bagley ST, Leddy DG. A study of the effects of exhaust gas recirculation on heavy-duty diesel engine emissions. SAE Trans 1998; 107:665–80.
[62] Labecki L, Lindner A, Winklmayr W, Uitz R, Cracknell R, Ganippa L. Effects of injection parameters and EGR on exhaust soot particle number-size distribution for diesel and RME fuels in HSDI engines. Fuel 2013;112:224–35.
[63] Abd-Alla GH. Using exhaust gas recirculation in internal combustion engines: a review. Energy Convers Manag 2002;43(8):1027–42.

CHAPTER 8

NOx reduction in IC engines through after treatment catalytic converter

G. Sathish Sharma, M. Sugavaneswaran, and R. Prakash
School of Mechanical Engineering, Vellore Institute of Technology, Vellore, India

8.1 Introduction

Engine emission formation and chemical kinetics were discussed briefly in the previous chapter. The factors that influence engine emissions such as engine design, engine operating parameters, and effects of exhaust gas recirculation (EGR) were also explained. This chapter explains the measures taken to reduce NO_x emissions using a three-way catalytic converter. It also discusses the stages in the development of TWC and the chemical kinetics behind NOx reduction through TWC.

Automobiles NO_x release hydrocarbons (HC), carbon monoxide (CO), nitrogen oxides (NO_x), and soot particulates. These poisons have huge unfavorable impacts on human health and will deteriorate the ecological balance. To reduce the pollution emitted by vehicles, all nations worldwide have implemented norms to be followed that restrict the emissions release rate from each vehicle. To cope with these stringent norms, all vehicles must have an after-treatment device. There are various after-treatment methods such as afterburners, exhaust manifold reactors, catalytic converters, and exhaust gas recirculation that are used to treat automotive exhaust gases [1, 2]. The catalytic converter has a greater advantage and a better conversion rate. The catalytic converter has a catalyst coated exhaust system in which poisonous gases are converted to nonpoisonous gases by carrying out oxidation and reducing chemical reactions. The catalyst that is coated over the exhaust system is responsible for the chemical reaction that is carried out to reduce the harmful gases. It is impossible to have complete combustion in internal combustion (IC) engines, due to the lack of time required for complete combustion because it is running at a very high speed [3–5]. As a result of the lack of time required, incomplete combustion happens, forming poisonous gases such as HC, NO_x (NO and NO_2), and CO inside the

NOx Emission Control Technologies in Stationary and Automotive Internal Combustion Engines
https://doi.org/10.1016/B978-0-12-823955-1.00008-5

engine cylinder [3]. An installed catalytic converter will induce the chemical reaction of HC and CO with the oxygen available to produce nonpoisonous gases such as carbon dioxide and water vapor. It will also reduce the soot particles. CO is an outcome of the incomplete combustion of HC present in the fuel. It is consistently present when there is an insufficient supply of oxygen during the combustion process; the oxygen supply always depends upon the engine air/fuel ratio [3]. There are various causes of CO in fumes; the most common is due to an absence of oxygen that occurs when the air/fuel blend is rich [3, 6]. NO_x formation is elevated at high combustion temperatures. During high temperature, oxygen tends to combine with nitrogen to form NO_x [7–12]. NO_x formation also increases during lean-burn conditions; these conditions are maintained to improve the fuel consumption rate. The major difficulty faced during lean-burn conditions is the reduction of NO_x emissions. A separate mechanism or different modification needs to be incorporated in the catalytic converter and injection methods to reduce the NO_x during lean-burn conditions. A modern generation of devices needs to be designed and fabricated to fulfill the latest strict emission norms. To do that, it must be redesigned from numerous points of view. The latest advancements in the modern day's catalytic converter, testing methods and ongoing patterns of the exhaust system are detailed in this section.

The series of processes that reduces the poisonous gases emitted to the atmosphere through the tailpipe is known as the after treatment process. Its purpose is to reduce the emission of harmful pollutants into the environment. The HC, NO_x (NO and NO_2), and CO emissions formed as byproducts of combustion could be reduced by two different methods: in-cylinder techniques and after-treatment techniques. In-cylinder treatments include options such as precise fuel metering, supplying quality air for combustion, optimized air-fuel mixture, using homogeneous mixtures, lowering the in-cylinder temperature during combustion, optimizing the timing of ignition, and ECU to control the functions of the engine based upon the application. These methods are only adequate to meet a certain percentage of the current discharge guidelines to improve the current and to meet future extreme emission standards. A separate system must be installed to reduce the emissions from the engine; this is known as the after-treatment device.

An afterburner is a device where additional air is introduced into the exhaust gases of an IC engine for further reaction with the end products of combustion to promote the reaction that will reduce the emissions. Additional air is supplied to the engine exhaust system by an externally driven pump. The gases present in the combustion chamber are oxidized by adding extra air. The temperature of the mixture will be raised well above the ignition temperature. This chamber is called as direct flame afterburner [13]. Ignition is done in this type of chamber with the help of a glowplug or a sparkplug. This after-treatment process will make only oxidation reactions possible. Both HC and CO emissions could be reduced, whereas NO_x emissions will not be controlled with this kind of set-up. The thermal exhaust manifold reactor is used to diminish carbon base outflows in the fumes of a petrol engine. It acts as a replacement for the exhaust manifold present in the engine where it has a thermally protected chamber. When outside air is included upstream of the reactor, homogeneous responses occur to oxidize the emissions [14]. The thermal exhaust manifold reactor should be designed effectively in such a manner that it promotes self-ignition when engine exhaust mixes with excess secondary air. NO_x cannot be controlled by using this exhaust manifold reactor.

Exhaust systems used to convert harmful fumes and noxious gases into less poisonous gases with help of catalyst-coated exhaust are known as catalytic converters. They are located between the engine manifold and the exhaust tailpipe. The toxic gases streaming out of the engine will pass through the catalyst-coated substrate and react with the catalyst, which is present over the porous wash coat. This exhaust system is comprised of a steel spread plate or steel box, a solid substrate, and a ceramic substrate with a wash coat over which a catalyst such as Pt, Rh, Pd, or TiO_2/CoO is coated [15–17]. The catalytic converters are broadly classified into two types: a two-way catalytic converter and a three-way catalytic (TWC) converter. A two-way catalytic converter contains only an oxidation catalyst which is capable of oxidizing CO and HC emissions without changes in NO_x emissions. Primarily, two-way catalytic converters are present in diesel engines to reduce emissions such as hydrocarbons and carbon monoxide. Some of the two-way catalytic converters were also used in SI engines in the United States during 1981. Even in SI engines, they will reduce only HC and CO emissions. A two-way catalytic converter will perform only an oxidation reaction; hence, only HC and CO will be reduced during the process. The inability of the two-way converter to reduce the NO_x makes the three-way converters the preferred option. A two-way catalytic converter performs two oxidation reactions simultaneously:

- Oxidation of carbon monoxide to carbon dioxide: Allows oxidation of CO to less-harmful CO_2.

$$2CO + O_2 \rightarrow 2CO_2$$

- Oxidation of unburnt and partially burnt hydrocarbons to carbon dioxide and water: Allows oxidation of HC to CO_2 and H_2O.

$$C_xH_{2x+2} + [(3x+1)/2]\,O_2 \rightarrow xCO_2 + (x+1)\,H_2O \text{ (a combustion reaction)}$$

Two-way converters operate relatively efficiently with a lean fuel mixture. The ineffectiveness in controlling NO_x led to the introduction of three-way converters. A TWC converter has both an oxidation and a reduction catalyst, so it can perform both oxidation and reduction reactions [18] that will reduce the HC, CO, and NO_x emissions. The detailed working principles and parts are elaborated upon in this chapter.

8.2 Evolution of catalytic converter

TWC converters were conceptually put forward in 1968 [19]. The first experimental evidence of the selective removal of nitric oxide in the presence of oxygen under conditions close to stoichiometry was observed in 1971 [20]. In 1978, Volvo, using an Engelhard supplied catalyst, was the first automotive company to actually implement a TWC converter in conjunction with electronically controlled fuel injection equipped with closed-loop control [21]. In the intervening seven years, there was an evolution of the catalyst system and, advancements have made to understand the chemical phenomena taking place on the catalyst surface. This understanding has led to the design of much improved catalysts.

8.2.1 First-generation catalytic converter

In the first era, the three-way catalyst innovation was formed, completely actualized between the mid-1970s and the mid-1980s, and continued to be advanced. Its name indicates that the three-way catalyst does the oxidation of CO and HC to carbon dioxide (CO_2) and water vapor (H_2O). Simultaneously, it will also reduce the emissions of nitrogen oxides (NO_x=NO and NO_2) by removing the oxygen molecule from NO_x to form nitrogen (N_2) and oxygen (O_2) [21]. Utilizing oxide-upheld Pt-bunch metal (Pt, Pd, or potentially Rh) catalysts, more than 95% change for every one of these three responses can be figured out. In any case, this degree of reactant execution for the valuable metal-based three-way catalyst is just reachable over a limited range of air-to-fuel ratio for the intake of the IC engine. These extents are basically at the stoichiometric or "equivalence" proportion where there is enough oxygen to complete the combustion of the fuel, ordinarily signified as k = 1. At a lower air-fuel ratio, CO and HC transformations drop while NO_x remains unchanged. At a high air-to-fuel mixture, NO_x transformations drop abruptly. As a result of these imperatives, the accomplishment of the three-way catalyst innovation is reliant on different air-to-fuel ratios. This is rectified by an "oxygen-stockpiling material" as a feature of the three-way catalyst, normally made out of metal oxides, ceria-zirconia, etc.

8.2.2 Second-generation catalytic converter

The three-way catalytic converter has been remarkably successful. The necessity for keeping up the stoichiometric air-to-fuel ratio significantly within a certain range is considered a major drawback in the catalytic converter. It also doesn't work for successful NOx transformation under lean-burn conditions. To overcome those disadvantages, a solution was obtained by the introduction of the second era of vehicle emission control [21]. Essentially, the US government–in collaboration with the country's three significant vehicle producers, General Motors, Ford, and Chrysler–built a partnership for a new generation of vehicles (PNGV) in 1993, with the objective to create exceptionally ecofriendly vehicles while proposing vehicle emission guidelines. Because of their greater innate ecofriendliness, diesel engines became an essential focal point of the PNGV program with a critical "lean-NO_x" issue. A couple of years later [22], there were energizing reports of the amazing reactivity of the Cu/ZSM-5 catalyst for NO_x reduction ($2NO \rightarrow N_2 + O_2$). It also reduces hydrocarbons under a various range of lean-burn conditions that could accommodate fundamentally higher vehicle fuel efficiencies. Previously metal oxides were predominantly used to control emission but after the introduction of ammonia in SCR it has been used widely. NH_3-SCR possess lesser thermal degradation when compared to metal oxide catalyst. Consequently, the researchers concentrated on the NH_3-SCR system for NO_x reduction that provides significantly improved thermal stability.

8.2.3 Modern catalytic converter

Current TWC converters are equipped to have almost 100% conversion efficiency when the catalyst is appropriately warmed and the air-fuel proportion is controlled in a limited band around the stoichiometric ratio with the help of ECU-controlled engines [23].

8.2.3.1 Three-way catalytic converter for SI engines

TWC has the additional ability to reduce NO_x emissions along with HC and CO emissions. Since 1986, automobiles have been equipped with TWC to reduce the toxic gas emission rate due to regulations adopted by the United States, Europe, and many other countries. Fig. 8.1 shows the various components of the modern TWC converter. A detailed description and functions of various parts are listed below:

Substrate: It is made upon either a metal or ceramic material that has a structure that possesses numerous parallel channels through which exhaust gases flow. This provides a large surface area for the engine exhaust to flow. The purpose of the substrate is to promote uniform flow without any back pressure, along with providing a larger surface area.

Washcoat: Mostly alumina is used as a washcoat due to its porous structure that helps to hold the catalyst; it also has high mechanical and thermal stability and increases the surface area of the substrate.

Catalyst: Precious noble metals are used as catalysts in TWC; this is the heart of the catalytic converter. It helps to perform both oxidation and reduction reactions.

Mat: It acts like an insulation material that restricts the heat transfer from the substrate to the casing. It also acts as a protective barrier that nullifies the effect of mechanical vibration.

Case: It is a metal package provided to enclose the other parts.

Heat Shields: They provide thermal stability to various parts of the catalytic converter and also provide resistance to thermal shocks [24]. The cordierite/ceramic substrate has several advantages such as lighter weight, high surface area, reduced backpressure, and cost efficiency. The metallic

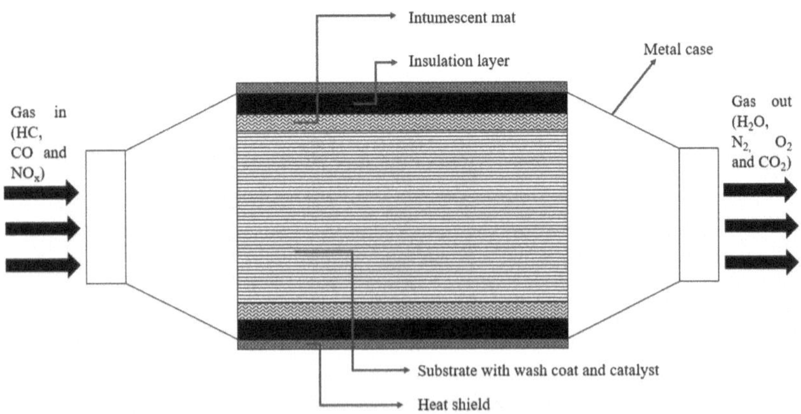

Fig. 8.1 Schematic diagram of a modern TWC set-up for an SI engine.

honeycomb-shaped monolith and corrugated-shaped monolith are the major types of metal substrates most commonly used for auto-exhaust catalytic converters [25]. The advantages are thermal shock fracture resistance, high mechanical strength, low pressure drop due to a thinner cell wall, and better thermal conductivity [26, 27]. The selection of the substrate material is based on the applications used. Ceramic catalytic converters are used when lower thermal stability is needed, but mostly metallic substrates are used in automobiles. Undesirable responses happen in the TWC converter such as the development of hydrogen sulfide and alkali, which could be rectified by using a modified washcoat and different catalysts. Even then, it is hard to wipe out every one of these unwanted reactions altogether. Hence, proper optimization in all selection processes must be carried out to produce the TWC with desired output based on different applications.

8.2.3.2 Three-way catalytic converter for CI engines

Challenges in implementing three-way catalytic converters in CI engines

The NO_x emissions of CI engine cannot be controlled by TWC alone. It must require a separate control device along with TWC to reduce NO_x in CI engine. It will need greater effort to reduce the emissions from the CI engine and it will come at a greater cost. Subsequently, commercially used CI engines work with lean burn ignition and contain high centralizations of oxygen in their fumes gases at all working conditions. Consequently, TWC is not utilized for NO_x control on diesel applications. Diesel emissions can be controlled with high productivity by oxidation impetus innovations. Impetus frameworks that have been created for decreasing NO_x from diesel motors can work through lean NO_x stockpiling followed by a rich decrease (NO_x absorber catalysts) or through a particular synergist decrease utilizing alkali (SCR catalysts), as summed up in Table 8.1.

Table 8.1 Different after-treatment techniques used in diesel engines to reduce emissions.

After-treatment methods	Working principle	Emissions reduced
DOC	Oxidation reaction	CO, HC, PM
Lean NO_x reduction	Selective catalytic reduction by hydrocarbons (HC-SCR)	NO_x, CO, HC
NO_x absorber	Trapping of NO_x from lean exhaust and release under rich conditions	NO_x, CO, HC
SCR	Selective catalytic reduction by NH_3	NO_x

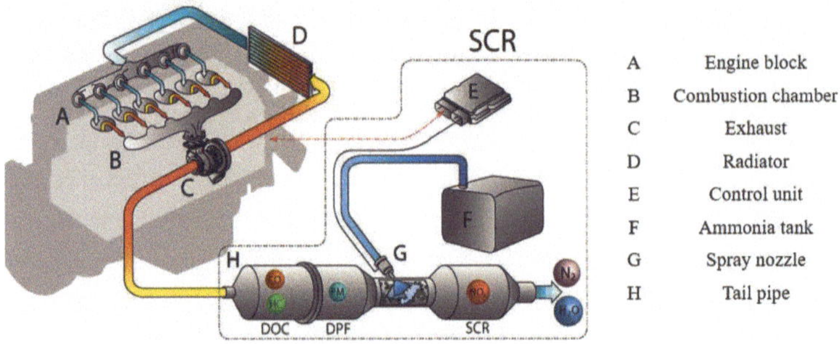

Fig. 8.2 Schematic diagram of after-treatment device set-up for CI engines.

In compressed ignition (CI) engines, the most commonly used after-treatment devices are diesel oxidation catalysts (DOC), selective catalytic reduction (SCR), NO_x traps (or NO_x absorbers), and nonselective catalytic reduction (NSCR). Fig. 8.2 shows a schematic diagram of the after-treatment set-up for CI engines.

Primarily, DOC is made of a stainless steel substrate with many parallel passages through exhaust gas flows. It contains no moving parts, and it has a large amount of interior surface area. The inside surfaces are covered with valuable metals, for example, platinum or palladium as oxidation catalysts. It is called an oxidation catalyst in light of the fact that this after-treatment device converts exhaust gas poisons into innocuous gases by oxidation reactions [28]. Fig. 8.3 shows the various parts in DOC and its purpose.

On account of the CI engine, this device oxidizes CO and HC while the unburnt hydrocarbons are adsorbed on carbon particles. In the field of discharge control, hydrocarbons are adsorbed on the carbon particles present in the exhaust; this emission is known as a soluble organic fraction (SOF). Diesel oxidation catalysts are proficient at changing over the SOF of diesel

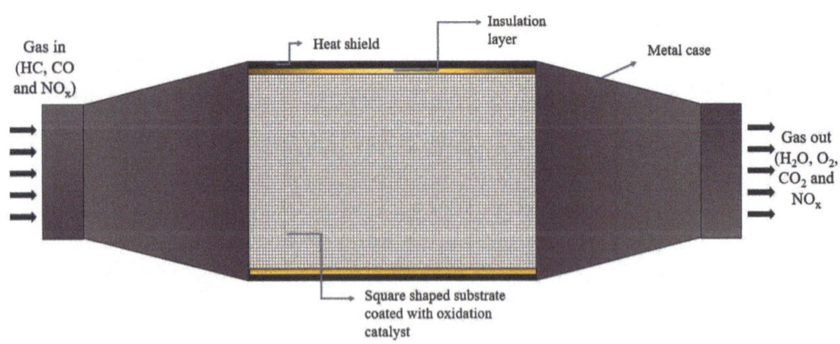

Fig. 8.3 Schematic diagram of DOC.

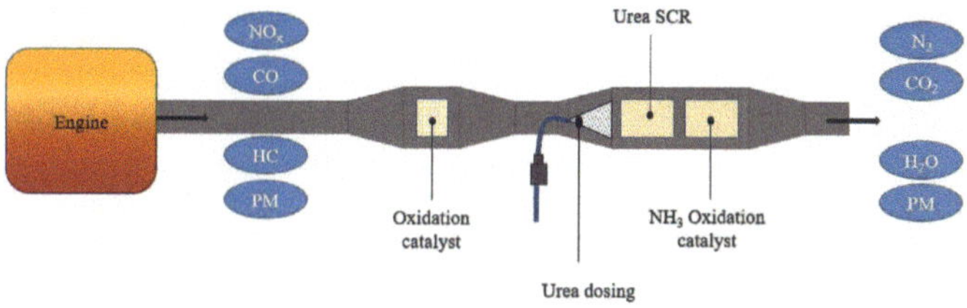

Fig. 8.4 Schematic diagram of SCR.

particulate into CO_2 and H_2O. Under certain working conditions, DOCs have accomplished SOF conversion efficiencies of 80–90%. The conversion efficiency of the DOCs is decreased due to various factors such as engine type, size, life span, obligation cycle, atmosphere condition, maintenance procedure, pattern discharges, test technique, manufacturing process, and the sulfur content present in the fuel. Because DOCs can perform only oxidation reactions, it does not reduce NOx formations. In order to reduce the NO_x content in the CI engine, two primary procedures are used: selective catalytic reduction (SCR) and $NOem_x$ traps (or NO_x absorbers) [29]. A detailed schematic diagram in Fig. 8.4 shows the various components inside the SCR and its need. SCR is an advanced emissions control technology system in which a liquid-reductant agent is injected into the exhaust stream of a CI engine. The primarily used reductant source is urea, also termed diesel exhaust fluid (DEF).

The NO_x traps (or NO_x absorbers) are used to reduce the intermediate compound N_2O which is formed during combustion process. The NO_x formed from this particular procedure is controlled using a catalytic decomposition technique. The nonselective catalytic reduction (NSCR) is an after-treatment device specifically designed to reduce the NO_x emissions. When atmospheric air is introduced into the exhaust system, it will react with O_2 and NO_x to form N_2, CO_2, and H_2O [30]. NSCR will reduce the NO_x emissions even in lean-burn conditions.

8.3 Design and fabrication of three-way catalytic converters

To accomplish the optimum design for the TWC converter there are various factors to be considered. Based upon different applications and working conditions, the design of the TWC converter will be varied. The following parameters play vital roles in designing the optimum TWC converter [31].

Table 8.2 Properties of two different ceramic and two different metal substrates.

Substrate material	Thermal capacity $J/m^2/K$	Geometric surface area m^2	Thickness Mm
Ceramic with 350 cpsi	158	2.79	0.14
Ceramic with 470 cpsi	149	3.03	0.13
Metal with 400 cpsi	82	3.69	0.04
Metal with 500 cpsi	81	4.04	0.04

8.3.1 Heat capacity–catalytic surface area, cell density, wall thickness

The heat capacity of an exhaust system is basically controlled by the material used, the cell density, and the wall thickness. To achieve light-off temperature quickly, the design must be developed in such a way that catalyst heating with maximum effectiveness should be promoted to achieve the optimal use of exhaust thermal energy. The least thermal capacity is required to produce a greater active catalytic surface. The design should also promote the maximum heat transfer from the exhaust gas to the catalytic structure. The accompanying Table 8.2 shows an examination between two different TWCs with different materials used.

It is evident from the table that the metal substrate provides more surface area with less thermal capacity. At the low-temperature working conditions, the reduced cell density shows progressive results. At the high-temperature working conditions, the higher cell density possesses an advantage; hence, based upon the working temperature, the cell density should be finalized.

8.3.1.1 Significance

The lower heat capacity of the substrate increases the conversion efficiency of the TWC converter.

To reduce cold start emissions, a high surface area with a lower heat capacity must be selected.

8.3.2 Catalyst diameter

The light-off temperature is greatly influenced by the TWC converter's diameter. Decreasing the diameter from the inlet to the outlet will lower the heat capacity of the substrate and increase the heat capacity of the catalyst. Hence, an increase in conversion efficiency could be achieved. Cold start emissions will be reduced with a smaller diameter.

8.3.2.1 *Significance*

- Smaller diameters promote an earlier light-off temperature.
- Smaller diameters also provide better flow distribution.

8.3.3 Flow distribution

The thermal aging and light-off temperature will be influenced by the flow distribution of the exhaust gas within the TWC converter. The nonuniform flow will have the advantage of a small diameter but at the same time will burn out the catalyst quicker than anticipated. The nonuniform flow will develop higher pressure loss and reduce the conversion efficiency of the TWC converter. A design must be developed in such a way that the flow distribution will satisfy all needs.

8.3.3.1 *Significance*

Uniform flow distribution will improve the conversion efficiency of the TWC converter, reduce the backpressure, improve the thermal aging resistance behavior, and also make the light-off temperature range lesser.

8.3.4 Coating

A better coating will provide good heating behavior of the catalyst by increasing the heat capacity of the catalyst. Various washcoats are used in automobiles; these are given in Table 8.3. From these, it is evident that alumina possesses a greater advantage.

Table 8.3 Properties of different washcoats used in TWC converters.

Washcoat	Specific Washcoat Mass [g/l]	Heat capacity of substrate after coating with wash oat [J/l/K]
TiO_2	156	523
SiO_2	162	529
Zeolite	213	588
Al_2O_3	233	596

A good coating will improve the thermal stability of the TWC converter and also reduce dust accumulation.

8.3.4.1 *Significance*

The washcoat mass has a negative influence on catalyst heating-up time.

The washcoat must have a higher surface area with high mechanical and thermal stability.

8.3.5 Catalyst length

The exhaust system length has an immediate impact on flow distribution. Uniform flow will be achieved with a longer TWC converter. The longer TWC converter also reduces the pressure loss and improves the light-off temperature. The longer the length of the TWC converter, cold start emissions and lesser thermal aging are achieved.

8.3.5.1 *Significance*

- Catalyst length influences the flow distribution positively.

8.3.6 Fabrication of the three-way catalytic converter

Modern TWC converters are fabricated after undergoing a series of processes such as designing, optimization based upon CFD analysis, and material characterization; this is explained in Fig. 8.5.

The next stage is the fabrication of a catalytic converter, which undergoes process such as manufacturing the substrate and metal casing, sol–gel preparation to coat the catalyst over the substrate, insulating the

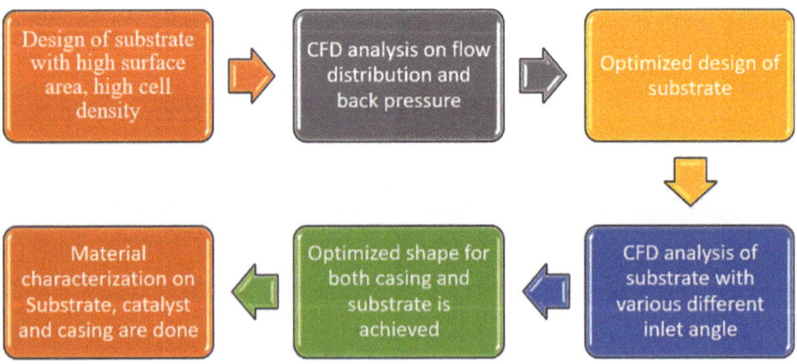

Fig. 8.5 Flow chart for designing a TWC converter.

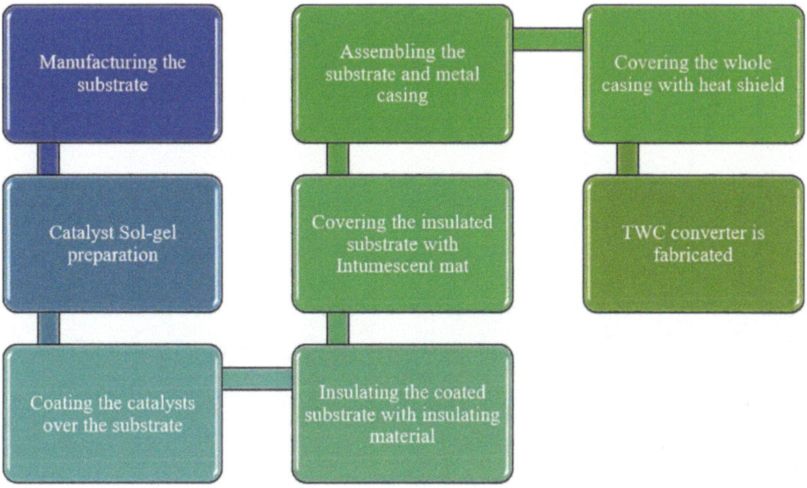

Fig. 8.6 Flow chart for fabrication of the optimized design of a TWC converter.

coated substrate to protect it from heat dissipation, and finally enclosing it with a metal casing to protect from any other physical damage; this is explained in Fig. 8.6.

8.4 Catalysts for NOx control

As in the exhaust system, the catalyst is used to decrease the activation energy and engine temperature during the reduction of NO_x, increase the N_2 selectivity in the reduction of NO_x, and avoid other side reactions to improve the conversion efficiency for TWC. The choice of catalyst is significant. The catalyst must have certain properties such as high deNO_x action, solid antipoisoning capacity, high mechanical quality, and high thermal stability. Numerous catalysts have been demonstrated to be dynamic for reduction reactions to reduce the NO_x level in exhaust emissions. Basically, TiO_2, Al_2O_3, SiO_2, zeolite, and carbon are used as washcoats to carry the catalysts. These catalysts have diverse reduction properties, preferences, and drawbacks, which are summed up in Tables 8.4 and 8.5.

Table 8.5 it shows the advantages, disadvantages, and properties of the catalyst used for nonmetal substrates.

Table 8.4 Description of different catalysts used in metal-based substrates.

Catalyst used	Coating and preparation method	Conversion efficiency of NOx (%)	Light-off temperature (°C)	Catalyst poisoning effect on SO_2, H_2O, alkali, and NH_3	References
$V_2O_5/WO_3/CeO_2$-TiO_2	Incipient wetness impregnation	100	250–475	Good resistance for SO_2	[32]
$V_2O_5/WO_3/Ce/TiO_2$	Wet impregnation	90	200–475	Good resistances for SO_2 and H_2O over 200 °C	[33]
Promoter (W/Ce)/V/TiO_2	Step-by-step impregnation and coimpregnation	>90	225–375	–	[34]
Zr,Ce	Coprecipitation and step-by-step impregnation	92	225–375	Poor resistance for H_2O but good resistance on SO_2.	[33, 35, 36]
Ce, Carbon nanotubes	Sol–gel	98	175–300	Poisoning effects of SO_2 are heavy.	[37]
Sb	Impregnation	100	225–400	Provides good resistance for SO_2	[38]
Sb, Ce	Deposition-precipitation and incipient wetness coimpregnation	>90	220–450	Provides better resistance for SO_2	[39]
Pr	Sol–gel and impregnation	98	220–400	Provides very high resistance for SO_2 and H_2O	[40]
VFe/TiO_2	Sol–gel	<60	280	Good H_2O resistance	[41]

CuO	Wet impregnation	90	230–320	Provides very high resistance for SO_2 and H_2O	[42]
Mo, Ru	Wet impregnation	86	240	Provides resistance on both SO_2 and NH_3	[43]
$CaCl_2$	Impregnation and sol–gel	88.93	210	Good H_2O resistance	[44, 45]
Pd	Wet impregnation	90	200	Good heat resistance	[46]
Rh	Sol–gel and impregnation	95	215	Provides resistance on both SO_2 and NH_3	[47]
Cobalt oxide	Step-by-step impregnation	85	250	Good H_2O resistance	[48]
NiO_2	Wet impregnation	80	260	Strong SO_2 and H_2O Resistances	[49]

Table 8.5 Description of different catalysts used in nonmetal-based substrates.

Catalyst used	Coating and preparation method	Conversion efficiency of NOx (%)	Light-off temperature (°C)	Catalyst poisoning effect on SO_2, H_2O, alkali, and NH_3	References
$F/V_2O_5/TiO_2$	Sol–gel and impregnation	73	280–320	Poisoning effects of SO_2 are heavy.	[50]
$S/\ V_2O_5/TiO_2$	Sol–gel and impregnation	87	210–240	Strong SO_2 and H_2O resistances	[51]
Boron mono fluoride mono sulfide	Successive impregnation	80	–	Excellent SO_2 and H_2O resistances	[52]
Hydroxypropyl acrylate	Wet impregnation	100	250–300	Very good resistance to alkali	[53]

8.5 NOx reaction mechanism and chemical kinetics in three-way catalytic converter

In the three-manner catalysis framework, various responses happen all the while between various reactants (NO_x, CO, O_2, C_xH_y, H_2O, H_2). The chemical kinetics behind the reduction reactions of NO_x are briefly discussed below.

Reduction reactions mechanism:

As CO is one of the reduction agents in exhaust fumes, NO_x decreases when it reacts with CO because CO provides a prevailing response for NO_x reduction from the exhaust emissions. Various writings about the energy and system of this response have been recorded and the most significant works are summed up in Tables 8.6 and 8.7. The general mechanism of this reaction is listed below [60]:

$$NO + CO \rightarrow 1/2N_2 + CO_2$$
$$2NO + CO \rightarrow N_2O + CO_2$$
$$N_2O + CO \rightarrow N_2 + CO_2.$$

At a high temperature, N_2 and CO_2 are formed as a result of the reduction reaction while at average engine working temperatures, N_2O development is supported. The definite system of this response happened on Pd as a catalyst; the majority of the researchers [54, 56, 61] embraced the Langmuir-Hinshelwood method. NO_x and CO initially go after the adsorption on Pd and afterward the adsorbed NO_x species separate on the closest vacant site (G). The dissociative N particle responds with the adsorbed NO_x species to deliver N_2O or consolidate to N_2; the adsorbed O will respond with adsorbed CO to produce CO_2.

$$NO + G \leftrightarrow NO_b K_1$$
$$CO + G \leftrightarrow CO_b K_2$$
$$NO_b + G \rightarrow N_b + O_b K_3$$
$$NO_b + N_b \rightarrow N_2O + 2G\,K_4$$
$$N_b + N_b \rightarrow N_2 + 2G\,K_5$$
$$CO_b + O_b \rightarrow CO_2 + 2G\,K_6$$
$$r_{NO} = K_1 P_{NO} / K_2 P_{CO}$$

where r_{NO} represents the rate of reduction for NO; K_1, K_2, K_3, K_4, K_5, and K_6 are the constants of NO, CO, NO_b, (NO_b, N_b), (N_b, N_b), and (CO_b, O_b) adsorption, respectively; and P_{NO} and P_{CO} represent both NO and CO partial pressures. The equation will indicate that CO is a major inhibitor in the system that indicates the rate of reduction of NO.

Table 8.6 Chemical mechanism and kinetics research results of the CO+NO reaction on the Pd catalyst.

Conditions		Mechanism and rate equations	Conclusive remarks	Reference
Catalysts % loading Light-off Temperature Feedstock	Pd/Al_2O_3 0.006 wt% 350°C CO-NO mixture in He	$NO + G \leftarrow \rightarrow NO_b$ $CO + G \leftarrow \rightarrow CO_b$ $NO_b + G \rightarrow N_b + O$ $NO_b + N_b \rightarrow N_2O + 2G$ $N_b + N_b \rightarrow N_2 + 2G$ $CO_b + O_b \rightarrow CO_2 + 2G$ $r_{NO} = K_1 P_{NO}/K_2 P_{CO}$	1. The adsorption of NO on Pd depends upon the rate of reactions. 2. NO has a reduction rate of +1 and −1 for NO and CO. 3. The absorption rate will decide the reaction rate of NO, whereas CO acts as an inhibitor	[54]
Catalysts % loading Light-off Temperature Feedstock	Pd/Al_2O_3 1 wt% 300°C CO-NO mixture in He	$NO + G \leftarrow \rightarrow NO_b$ $CO + G \leftarrow \rightarrow CO_b$ $NO_b + G \rightarrow N_b + O$ $NO_b + N_b \rightarrow N_2O + 2G$ $NO_b + N_b \rightarrow N_2 + O_b + G$ $CO_b + O_b \rightarrow CO_2 + 2G$ $r_{NO} = 2K_3$ $\lambda_{NO} P_{NO}/(1 + \lambda_{NO} P_{NO} + \lambda_{CO} P_{CO})2$	1. N atoms dominate the formation of N2 via adsorbed NO. 2. The rate of NO transformation is acquired by steady-state approximation to adsorbed N atoms.	[55]
Catalysts % loading Light off Temperature Feedstock	$Pd/CeO_2/Al_2O_3$ 4 wt% 300°C CO-NO mixture in He	$NO + G \leftarrow \rightarrow NO_b$ $CO + G \leftarrow \rightarrow CO_b$ $NO_b + G \rightarrow N_b + O$ $NO_b + N_b \rightarrow N_2O + 2G$ $N_b + N_b \rightarrow N_2 + 2G$ $CO_b + O_b \rightarrow CO_2 + 2G$	1. The rate of NO transformation is acquired by steady-state approximation to adsorbed N atoms. 2. The order of the reaction for CO is negative. 3. The order of reaction can be positive, negative, or zero, which depends upon the partial pressure of CO.	[56]

Catalysts % loading Light-off Temperature Feedstock	Pd/Al_2O_3 1.8 wt% – CO-NO mixture in He	$Pd^0 + NO \leftarrow\rightarrow Pd^0 - NO$ $Pd^0 + CO \leftarrow\rightarrow Pd^0 - CO$ $Pd^0 + NO \leftarrow\rightarrow Pd^0 - NO^-$ $2Pd^0 + CO \leftarrow\rightarrow (Pd^0)_2 - CO$ $(Pd^0)_2 - CO \rightarrow Pd^0 - CO + Pd^0$ Surface reaction $Pd^0 - NO + Pd^0 \leftarrow\rightarrow Pd^0 - N + Pd^0 - O$ $Pd0 + Pd0 - O \leftarrow\rightarrow (Pd^+)_2O^{2-}$ $Pd + + NO \leftarrow\rightarrow Pd^- NO^+$ $Pd^0 - N^+ Pd^{0-} N \leftarrow\rightarrow 2Pd^0 + N^2$ $Pd^0 - NO + Pd^{0-} N \leftarrow\rightarrow 2Pd^0 + N_2O$ $Pd^0 - CO + Pd - O \leftarrow\rightarrow 2Pd^0 + CO_2$	1. The rate of reaction is controlled by adsorbed O. 2. The reaction rate of oxidization of Pd^0 to Pd^+ by NO is faster than the reduction of Pd^+ to Pd^0 by CO.	[57]

Table 8.7 Chemical mechanisms and kinetics research results of the H2 + NO reaction on the Pd catalyst.

Conditions		Mechanism and rate equations	Conclusive remarks	Reference
Catalysts % loading Light-off Temperature Feedstock	Pd/Al_2O_3 1 wt% 70°C H_2-NO mixture $NO_a + H_a \rightarrow N_a + OH_a$	$NO + G \leftarrow \rightarrow NO_b$ $H_2 + 2G \leftarrow \rightarrow 2H_b$ $NO_b + N_b \rightarrow N_2 + O_b + G$ $NO_b + N_b \rightarrow N_2O + 2G$ $N_b + H_b \rightarrow NH_b + G$ $NH_b + H_b \rightarrow NH_{2,b} + G$ $NH_{2,b} + H_b \rightarrow NH_3 + 2G$ $O_b + H_b \rightarrow OH_b + G$ $OH_b + H_b \rightarrow H_2O + 2G$	1.NO dissociation is assisted with the help of chemisorbed H atoms. 2. Reduction rate for NO is −0.39 and +0.67 for NO and H_2, respectively, and their activation energy is 68 kJ/mol.	[58]
Catalysts % loading Light-off Temperature Feedstock	Pd/L_aCoO_3 1 wt% 110°C H_2-NO mixture	$NO + G \leftarrow \rightarrow NO_b$ $H_2 + G \leftarrow \rightarrow H_{2,b}$ $O + H_{2,b} \rightarrow V + H_2O + G$ $NO_b + V \rightarrow O + N_b$ $NO_b + H_{2,b} \rightarrow NH_b + OH_b$ $NO_b + N_b \rightarrow N_2 + O_b + G$ $NO_b + N_b \rightarrow N_2O + 2G$ $N_b + H_{2,b} \rightarrow NH_b + H_b$ $NH_b + H_{2,b\rightarrow} NH_{2,b} + H_b$ $NH_{2,b} + H_{2,\ b} \rightarrow NH_3 + H_b + G$ $O_b + H_{2,b} \rightarrow H_2O + G$ $O_b + H_b \rightarrow OH_b + G$ $OH_b + H_b \rightarrow H_2O + 2G$ "O" is oxygen species on support; 'V': anionic vacancies from support	1. Instead of dissociative atoms, the H_2 absorbs on the catalyst surface as the molecular precursor. 2.The reduction rate of NO is −0.7 and +1 for NO and H_2, respectively, and its activation energy is 79 kJ/mol. 3. Applying the steady-state approximation, the reaction rate was obtained for the anionic vacancies from the support "V."	[59]

Dhainaut et al. [58] assumed that the separation of NO is helped by chemisorbed H molecules, and utilized the Langmuir-Hinshelwood system to portray the NO + H_2 response on the Pd/Al_2O_3 catalyst kinetics:

$$NO + G \leftrightarrow NO_b K_1$$

$$H_2 + 2G \leftrightarrow 2H_b K_2$$

$$NO_b + H_b \rightarrow Nb + OH_b K_3$$

$$NO_b + N_b \rightarrow N_2 + O_b + S\,K_4$$

$$NO_b + N_b \rightarrow N_2O + 2G\,K_5$$

$$N_b + H_b \rightarrow NH_b + G\,K_6$$

$$NH_b + H_b \rightarrow NH_{2,b} + G$$

$$NH_{2,b} + H_b \rightarrow NH_3 + 2\,G$$

$$O_b + H_b \rightarrow OH_b + G$$

$$OH_b + H_b \rightarrow H_2O + 2\,G$$

$$r_{NO} = \left(2K_3 \lambda_{NO} P_{NO} \sqrt{(\lambda_{H2} P_{H2})}\right) \Big/ \left(1 + \lambda_{NO} P_{NO} + \sqrt{(\lambda_{H2} P_{H2})}\right)^2$$

N_2O is formed as a product in reactions with NO and H_2 at a temperature below 200 °C while N_2O is found above 250 °C during reactions between NO and CO. The NH_3 formation was observed between 150 and 225oC during the reactions between NO and H_2. In Table 8.6, different chemical kinetics and their advantages proposed by other researchers are explained. Table 8.7 presents the chemical mechanisms and kinetics research results of the H2 + NO reaction on the Pd catalyst.

8.6 Factors affecting performance of three-way catalytic converters

8.6.1 Thermal stability

The metal in the catalyst is inclined to deactivation by sintering due to exhaust streams with temperatures up to 1000 °C. The exhaust systems become decayed by thermal loading. Thermal disintegration happens because of the introduction of the catalyst to higher-temperature conditions. Prolonged exposure of the TWC converter with high temperature will lead to the thermal degradation of the catalyst, which will reduce the surface area and change the phase of the washcoat and catalysts [62].

8.6.2 Backpressure

Backpressure on the exhaust system will influence the conversion efficiency of the TWC converter. Lesser backpressure will have a higher conversion efficiency due to uniform flow distribution inside the substrate. Lower back pressure will have lesser mechanical loading over the substrate, hence it can withstand a longer period [63]. Also, lesser pressure drop decreases in the light-off period were noticed [64].

8.6.3 Flow distribution

Flow distribution inside the converter is influenced by the shape of the substrate and the inlet angle of the converter. A substrate with a high surface area and high cell density will lead to uniform flow distribution. A uniform flow inside the catalytic converter will increase the conversion efficiency of the catalytic converter. There are many benefits in having uniform flow in catalytic converters such as less back pressure, high structural strength, attain light off temperature in quick span of time, provides uniform heat transfer and lesser thermal degradation.

8.6.4 Conversion efficiency

Conversion efficiency is a term used to denote the amount of reduction that happened in the converter as well as to quantify the performance of the TWC converter. Conversion efficiency depends upon various parameters such as uniform flow distribution, high surface area, high cell density, light-off temperature, thermal aging of the catalyst, and lesser backpressure. The factors influencing the conversion efficiency of the catalytic converter are shown in Fig. 8.7.

8.6.5 Catalyst light-off temperature

The catalyst in the TWC converter will get activated only after reaching a certain range of temperature; this is known as the light-off temperature. Beyond that temperature, only the converter will begin to reduce the emissions. Mostly, all catalysts used in converters will have the light-off temperature within the range of 300–350°C. A typical meaning of a light-off temperature is the point at which the conversion efficiency arrives at 50% [65]. The TWC converter overall has low thermal inertia, and will be built of materials that have a low specific heat with high thermal conductivity that will get to the working temperature quickly. The emission norms at all part of the globe are getting more and more stringent. In order to cope up with the current regulations, advanced emission control strategies need to be implemented. The TWC converter can be heated before turning over the engine or quickly once it has begun.

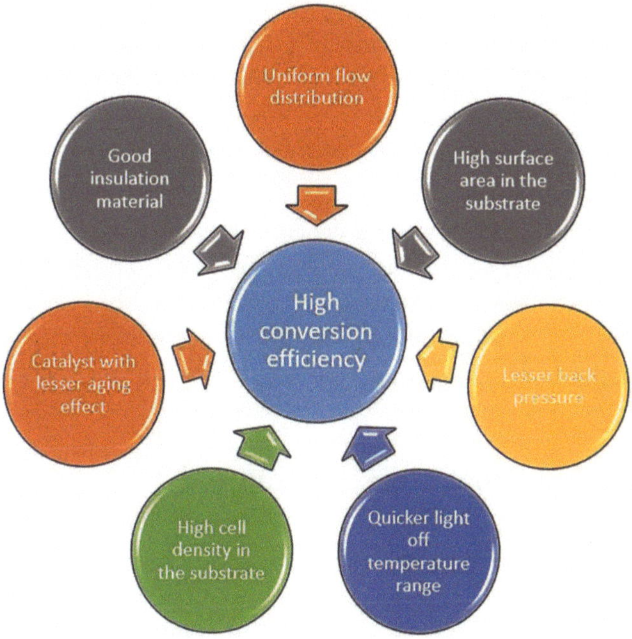

Fig. 8.7 Factors influencing conversion efficiency. *No Permission Required.*

8.6.6 Cold start emission

Cold start emissions are an important factor that influences the performance of the catalytic converter. When the ambient temperature is around −7°C to 15°C during ignition, the automobiles will release a huge amount of emissions. During cold start conditions, the engine will consume more fuel, so it will emit more particulate matter that will settle inside the converter. This will reduce the performance of the TWC converter. In cold start conditions, due to the lower ambient temperature the fuel will also be at a lower temperature, which will affect its chemical properties. This will also lead to improper combustion and emissions of more HC, CO, and NO_x. Cold start emission conditions will also make the converters take more time to reach the light-off temperature, so there will be more emissions [66].

8.6.7 Lean burn emission

The lean burn is the combustion of lean air and fuel mixture. Engine combustion is viewed as lean when an overabundance of air is brought into the engine alongside the fuel. It happens at high air/fuel proportions, which is useful for good fuel economy [67]. The disadvantage of a lean burn is the huge measure of NO_x being created (more oxygen and nitrogen will combine to form NO_x at the high temperature inside the combustion chamber) and a complex TWC converter is required to reduce the emissions [68, 69].

8.6.8 Durability analysis of catalytic converters

Catalyst aging is defined as the loss of catalytic action after some time that has a significant impact on controlling the emission rate in the TWC converter. The TWC converter will lose its catalytic ability due to catalyst deactivation, which happens when the structure and conditions of the catalyst get changed. Catalyst deactivation normally consumes longer time exposure to the exhaust gases and due to exposure to high temperature [70]. Fig. 8.8 shows the factors that influence TWC performance as well as remedies.

8.6.9 Control of engine air-fuel ratio with ECU

The air/fuel combination of an engine is generally controlled by changing the timing of opening every one of the fuel injector valves. The change of the proportion will be made exactly to empower the exhaust system to work viably; hence CO, HC, and NO_x are converted into innocuous substances. One of the methods adopted to control the air/fuel mixture is the two-sensor method. This system is introduced in the fumes line of the engine upstream of the exhaust system, which is shown in Fig. 8.9. The sensor reacts to changes in exhaust gas emission stoichiometric conditions. Due to this, the sensor will deliver an unexpected voltage bounce in its yield signal, a high voltage relating to a rich admission combination. The voltage signal from the k-sensor is communicated to a corresponding basic PC, which sends signal to control the opening and closing of injections valves. The

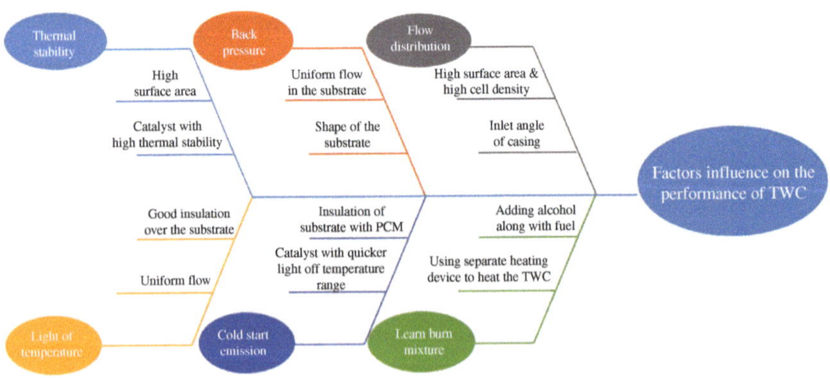

Fig. 8.8 Factors influencing TWC performance and remedies.

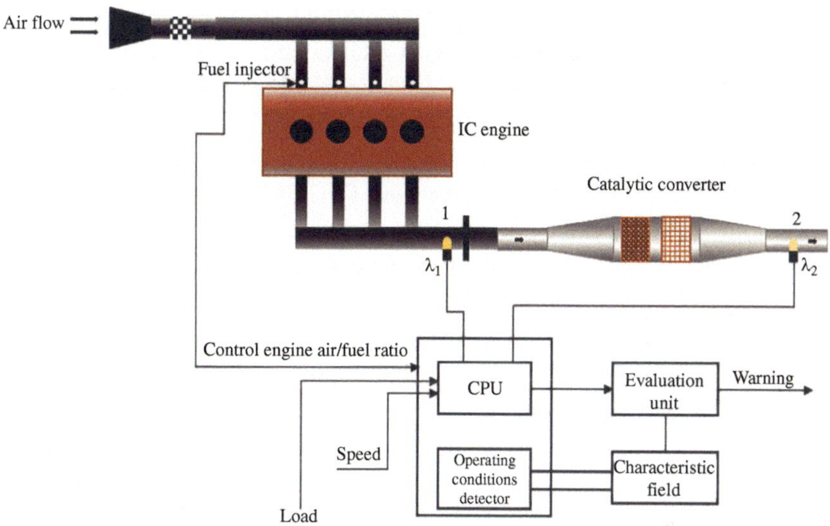

Fig. 8.9 Modern SI engine set-up with sensors for controlling air/fuel combination.

injections valve control helps to achieve the required air-fuel ratio as per the demand. When the air/fuel blend gets rich because the oxygen that has been ingested and put away in the exhaust system is delivered, HC and CO are oxidized. As an outcome, NO_x, HC, and CO can be filtered simultaneously.

With continued use, the catalytic converter becomes degraded by warm maturing, sintering, and crystallization of the dynamic surface layer. It is harmed by substances such as S, Pb, P, Zn, and Mg, with the result that the oxygen putting away capacity of the converter falls. The reduction in the performance of the exhaust system to control could be controlled by monitoring the free oxygen entry into the catalyst system. Consequently, the downstream sensor mirrors the abatement of transformation movement with a more rich-to-lean variance of oxygen, because of the expansion in untreated fumes gases going through the exhaust system. The temperature of the exhaust system is likewise a significant factor that can impact the right assurance of the crumbling of an exhaust system by methods for oxygen or air/fuel sensors. The oxygen stockpiling limit of the exhaust systems, all in all, differs with an adjustment in the temperature, which causes a variation in the sign of the sensor put downstream of the exhaust system. An average motor design for observing the productivity of an exhaust system is shown in Fig. 8.9. It shows an exhaust system having an oxygen test 1 organized upstream of the converter and an oxygen test 2 orchestrated downstream of the converter as well as a worth deciding unit, an assessment unit, and a trademark field. The control unit with a CPU and an operating condition detector. Which can fill in the required air/fuel blend as per requirement. The calculation unit gets an upstream test signal from upstream test 1 and a downstream test signal 2 from downstream test 2. From these signs, the calculation unit assesses the emission loss of the exhaust system. The calculation unit plays out this assessment during, for example, a consistent state activity of the motor to which the exhaust system is associated. The working state is declared by an identifier when the motor is worked in a consistent state. The worth controlled by the calculation unit is contrasted with a trademark field an incentive in the assessment unit with the trademark field esteem being provided from the trademark field. On the off chance that the examination shows that the exhaust system has an more misfortune which is impermissibly high, at that point the assessment unit emanates a shortcoming sign to the driver notice implies for example a Malfunction Indication Lamp (MIL). Estimations of factors are provided to the working state identifier and the trademark field and show the working condition of the inward ignition motor. Such factors are motor speed, vehicle speed, water coolant temperature, load, and the temperature of the fume gases.

The control unit measures all the information and directs the air/fuel combination provided to the motor.

8.7 Recent developments in catalytic converters

It is evident that most problems occurred at the substrate region, so various attempts have been made to develop an optimum design and material to counter the problems. This leads to the high use of metal as a substrate instead of ceramic. Metal substrates are designed in such a way that they possess both high surface area and greater porosity [71]. Due to the higher cost and rare availability of noble metal catalysts, various other catalysts such as metal oxides, zeolite, and pyrokites are used in TWC converter. To overcome cold start emissions, electric heating coils are wound over the substrate of the TWC converter to achieve the light-off temperature at a quicker rate [72]. Some manufactures also use a small TWC fitted before the main TWC converter to preheat the main TWC; this type of small TWC is called a precat. Even with phase change material (PCM) insulation over the substrate, cold start emissions could be reduced significantly. Along with gasoline, a small proportion of other additives such as ethanol, methanol, butanol, and other alcohol-based fuels are used for combustion in order to reduce

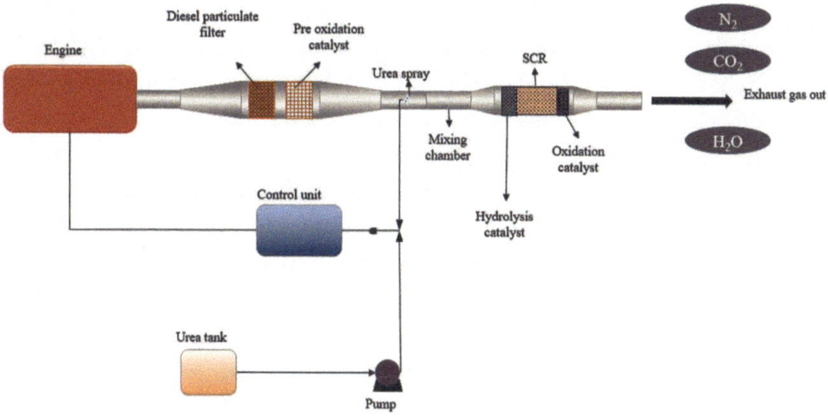

Fig. 8.10 Schematic diagram of the modern emission control system in CI engines.

the regulated emissions such as HC, CO, and NO_x. Due to more oxygen content in alcohol-based fuels, HC and CO get reduced. Alcohol fuels pose a high latent heat of vaporization, which reduces the engine in-cylinder temperature; this will lower the possibilities of NO_x formation. Water is also added as an emulsion to reduce the in-cylinder temperature, which will also reduce NO_x formation. Fig. 8.10 shows a recent after-treatment device used in CI engines to reduce HC, CO, and NO_x emissions.

8.8 Conclusion

Creating an optimum design with more surface area, new materials with better physical and chemical properties, and better catalysts for improved conversion efficiency of the TWC converter are continuous processes. The corrugated structured substrate has a greater surface area for a given specific length compared with other structures. Alumina is most commonly used as a washcoat due to its high surface area and porous nature. The thermal stability of the metallic and ceramic substrate for the TWC converter shows that the metallic substrate has an edge over the ceramic substrate due to the turbulent flow of exhaust gas, which induced in a metallic substrate ensures mixing and sufficient contact of the gases with the catalysts. A TWC converter that provides a smaller pressure drop, a catalyst with high heat capacity, a substrate with low heat capacity, and uniform flow will have a higher conversion efficiency. The use of noble metals as catalysts has increased the overall cost of the TWC converter; this could be reduced by replacing appropriate catalysts depending upon the application. The metal substrate will help to achieve the light-off temperature earlier to achieve a higher conversion efficiency, thereby reducing cold start emissions. PCM insulation over the substrate will also reduce cold start emissions.

In this chapter, the control measures taken to reduce NO_x formation using the TWC were explained briefly, along with the chemical kinetics of reducing the emissions. The TWC will alone not be able to reduce NO_x and other emissions to meet the present and upcoming government norms regarding emissions. In order to cope with the government norms, new techniques must be developed to reduce engine emissions. The latest techniques used are NO_x adsorbents, selective catalytic reduction, fuel reformulation with the inclusion of alcohol content in the fuel used, and advanced combustion strategies (HCCI, PCCI, and RCCI) to reduce engine emissions. These will be explained in detail in the forthcoming chapters. The next chapter deals with the function and mechanism of adsorbing techniques used to reduce NO_x.

References

[1] Ganesan V. Internal combustion engines. McGraw Hill Education (India) Pvt Ltd; 2012.

[2] Jaichandar S, Annamalai K. Jatropha oil methyl ester as diesel engine fuel—an experimental investigation. Int J Automot Mech Eng 2016;13:3248–61.

[3] Heywood JB. Internal combustion engine fundamentals. New York: Mcgraw-hill; 1988.

[4] Hasan MM, Rahman MM. Homogeneous charge compression ignition combustion: advantages over compression ignition combustion, challenges and solutions. Renew Sust Energ Rev 2016;57:282–91.

[5] Yasin MHM, Mamat R, Aziz A, Yusop AF, Ali MH. Investigation on combustion parameters of palm biodiesel operating with a diesel engine. J Mech Eng Sci 2015;9:1714–26.

[6] Mathur M, Sharma R. Internal combustion engines. Dhanpat Rai Publ; 2005.

[7] Sher E. Handbook of air pollution from internal combustion engines: Pollutant formation and control. Academic Press; 1998.

[8] Muharam Y, Mahendra M, Gayatri D, Kartohardjono S. Simulation of ignition delay time of compressed natural gas combustion. Int J Automot Mech Eng 2015;12: 3125–40.

[9] Noor MM, Wandel AP, Yusaf T. The simulation of biogas combustion in a mild burner. J Mech Eng Sci 2014;6:995–1013.

[10] Noor MM, Wandel AP, Yusaf T. Effect of air-fuel ratio on temperature distribution and pollutants for biogas MILD combustion. Int J Automot Mech Eng 2014;10:1980–92.

[11] Kumaran P, Gopinathan M, Kantharrajan S. Combustion characteristics of improved biodiesel in diffusion burner. Int J Automot Mech Eng 2014;10:2112–21.

[12] Hamada KI, Rahman MM, Abdullah MA, Bakar RA, A. Aziz AR.. Effect of mixture strength and injection timing on combustion characteristics of a direct injection hydrogen-fueled engine. Int J Hydrog Energy 2013;38:3793–801.

[13] Vallero DA. Chapter 13—air pollution control technologies. In: Air pollution calculations quantifying pollutant formation, transport, transformation, fate and risks. Elsevier; 2019. p. 377–428.
[14] Gei Hold BW, Mellor AM. Operation of a homogeneous exhaust manifold reactor. Symp Combust 1975;15(1):1225–32.
[15] Dowden DA. Catalytic hand book. Verlag New York, Inc; 1970.
[16] Yao HC, Stepien HK, Gandhi HS. J Catal 1980;61:54750.
[17] Kalam MA, Masjuki HH, Redzuan M, Mahlia TMI, Fuad MA, Mohibah M, Halim KH, Ishak A, Khair M, Shahrir A, Yusoff A. Development and test of a new catalytic converter for natural gas fuelled engine. Sadhana 2009;34(Part 3):467–81.
[18] de Vries JE, Yao HC, Baird RJ, Gandhi HS. J Catal 1983;84:8–14.
[19] Gross GP, Biller WF, Greene DF, Kearby KK. U.S. Patent 3,370,914; 1968.
[20] Jones JH, Kummer JT, Otto K, Shelef M, Weaver EE. Environ Sci Technol 1971;2:790–8.
[21] Peden CHF. Cu/Chabazite catalysts for 'lean-burn' vehicle emission control. J Catal 2019;373:384–9.
[22] Iwamoto M, Furukawa H, Mine Y, Uemura F, Mikuriya SI, Kagawa S. J Chem Soc Chem Commun 1986;(16):1272.
[23] Venkatesh L, Logeshkumar R, Jayaprakash G, Dinesh N. Control and reduction of emissions using catalytic converter. Int Res J Eng Technol 2017;04(11):788–94 e-ISSN: 2395–0056, p-ISSN: 2395–0072.
[24] Kulkarni SP, Kharat SB, Kulkarni AK, Joshi VM, Deore H, Pandhare A. Modification of catalytic converters thereby bettering their performance and reducing the cost. National Conference on Emerging Trends in Science and Technology; 2018. ISSN [ONLINE]: 2395-1052.
[25] Govender S, Friedrich HB. Monoliths: a review of the basics, preparation methods and their relevance to oxidation. Catalysts 2017;7:62. https://doi.org/10.3390/catal7020062.
[26] Avila P, Montes M, Miro EE. Monolithic reactors for environmental applications: A review on preparation technologies. Chem Eng J 2005;109:11–36.
[27] Farrauto RJ, Heck RM. Catalytic converters: state of the art and perspectives. Catal Today 1999;51:351–60.
[28] Zhang J. Catalytic converter—part I of automotive after-treatment system. Bowmannz; 2008.
[29] Cho S. Properly apply selective catalytic reduction for NOx removal. Chemical Engineering Progress; January 199439–45.
[30] Heck RM. Catalytic abatement of nitrogen oxides–stationary applications. Catal Today 1999;53:519–23.
[31] Worsztynowicz B, Uhryński A, Borkowski B, Pluta M. The analysis of thermal state of catalytic converter depending on fuel supply and engine load using thermo-vision. Combust Engines 2017;170(3):30–6. https://doi.org/10.19206/CE-2017-305.
[32] Cheng K, Zhang T, Li JM, Wei YC, Jiang GY, Duan AJ, Liu J, Zhao Z. J Environ Sci 2014;26:2106–13.
[33] Chen L, Li JH, Ge MF. J Phys Chem C 2009;113:21177–84.
[34] Youn S, Song I, Kim DH. J Nanosci Nanotechnol 2016;16:4350–6.
[35] Zhang YP, Zhu XQ, Shen K, Xu HT, Sun KQ, Zhou CC. J Colloid Interface Sci 2012;376:233–8.
[36] Zhao X, Huang L, Namuangruk S, Hu H, Hu XN, Shi LY, Zhang DS. Cat Sci Technol 2016;6:5543–53.
[37] Li Q, Hou XX, Yang HS, Ma ZX, Zheng JW, Liu F, Zhang XB, Yuan ZY. J Mol Catal A 2012;356:121–7.
[38] Phil HH, Reddy MP, Kumar PA, Ju LK, Hyo JS. Appl Catal B 2008;78:301–8.

[39] Lee KJ, Kumar PA, Maqbool MS, Rao KN, Song KH, Ha HP. Appl Catal B 2013;142–143:705–17.
[40] Chao JQ, He H, Song LY, Fang YJ, Liang QM, Zhang GZ, Qiu WG, Ran Z. Chem J Chin Univ 2015;36:523–30.
[41] Schill L, Putluru SSR, Jensen AD, Fehrmann R. Catal Commun 2014;56:110–4.
[42] Yu LM, Zhong Q, Zhang SL, Zhao W, Zhang R. Combust Sci Technol 2015;187:925–36.
[43] Chen WM, Ma YP, Yan NQ, Qu Z, Yang SJ, Xie JK, Guo YF, Hu LG, Jia JP. Fuel 2014;133:263–9.
[44] Zhao LK, Li CT, Zhang J, Zhang XN, Zhan FM, Ma JF, Xie YE, Zeng GM. Fuel 2015;153:361–9.
[45] Zhang MZ, Wang P, Dong Y, Sui H, Xiao DD. Chem Eng J 2014;253:243–50.
[46] Biffis A, Centomo P, Del Zotto A, Zecca M. Pd metal catalysts for cross-couplings and related reactions in the 21st century: a critical review. Chem Rev 2018;118:2249–95.
[47] Paranjpe R, Suresh AK, Aghalayam P. Understanding Pt–Rh synergy in a three-way catalytic converter. Int J Chem React Eng 2013;11(1):1–8.
[48] Deng X, Tuysuzz H. Cobalt-oxide-based materials as water oxidation catalyst: recent progress and challenges. ACS Catal 2014;4(10):3701–14.
[49] Bhalothia D, Chen P-C, Yan C, Wang K-W, Chen T-Y. Heterogeneous NiO2-to-Pd epitaxial structure performs outstanding oxygen reduction reaction activity. J Phys Chem C 2020;124(4):2295–306.
[50] Zhao W, Zhong Q, Zhang TJ, Pan YX. RSC Adv 2012;2:7906–14.
[51] Zhao W, Zhong Q, Pan YX, Zhang R. Chem Eng J 2013;228:815–23.
[52] Tran T, Yu J, Gan LN, Guo F, Phan D, Xu GW. Catalysts 2016;6:56/1–56/14.
[53] Putluru SSR, Schill L, Godiksen A, Poreddy R, Mossin S, Jensen AD, Fehrmann R. Appl Catal B 2016;183:282–90.
[54] Muraki H, Shinjoh H, Fujitani Y. Reduction of NO by CO over alumina supported palladium catalyst. Ind Eng Chem Prod Res Dev 1986;25:419–24.
[55] Granger P, Dhainaut F, Pietrz S, Malfoy P, Mamede AS, Leclercq L, Leclercq G. An overview: comparative kinetic behaviour of Pt, Rh and Pd in the NO + CO and NO + H2 reactions. Top Catal 2006;39:65–76.
[56] Holles JH, Switzer MA, Davis RJ. Influence of ceria and lanthana promoters on the kinetics of NO and N2O reduction by CO over alumina-supported palladium and rhodium. J Catal 2000;190:247–60.
[57] Almusaiteer K, Chuang SSC. Dynamic behavior of adsorbed NO and CO under transient conditions on Pd/Al2O3. J Catal 1999;184:189–201.
[58] Dhainaut F, Pietrzyk S, Granger P. Kinetics of the NO + H2 reaction over supported noble metal based catalysts: support effect on their adsorption properties. Appl Catal B Environ 2007;70:100–10.
[59] Dhainaut F, Pietrzyk S, Granger P. Kinetic investigation of the NO reduction by H2 over noble metal based catalysts. Catal Today 2007;119:94–9.
[60] Shelef M, Graham GW. Why rhodium in automotive three-way catalysts. Catal Rev Sci Eng 1994;36:433–57.
[61] Mamede AS, Leclercq G, Payen E, Granger P, Grimblot J. In situ Raman characterisation of surface modifications during NO transformation over automotive Pd-based exhaust catalysts. J Mol Struct 2003;651–653:353–64.
[62] Vinod M. Janardhanan & Olaf Deutschmann, Computational fluid dynamics of catalytic reactors, Indian Institute of Technology Hyderabad
[63] Windmann J, Zacke P. Impact of the inlet flow distribution on the light off behaviour of three-way catalytic converter. Society of Automotive Engineers; 2002.
[64] Milton BE. Control technologies in spark-ignition engines. Handbook of air pollution from internal combustion engines. Academic Press; 1988189–258.

[65] Leman AM, Rahman F, Jajuli A, Zakaria S, Feriyanto D. Emission treatment towards cold start and back pressure in internal combustion engine against performance of catalytic converter: a review, In: MATEC web of conferences, vol. 87; 2017. p. 02021. https://doi.org/10.1051/matecconf /20178702021.
[66] Ye Z, Li Z, Mohamadian H. Research on NOx emission aftertreatment of lean burn gasoline engine using adsorber reduction catalyst, In: ASME/IEEE 2007 joint rail conference and internal combustion engine division spring technical conference; 2007.
[67] Ando H, Kuwahara K. A keynote on future combustion engines. SAE Paper 2001-01-0248.
[68] Ye Z. GDI engine exhaust after treatment system analysis and oxygen sensor based identification, modeling & control of lean NOx trap, In: Proceedings of the ASME ICE division 2003 spring technical conference; 2003. p. 713–9 May 11-14. Austria.
[69] Samimi O, Abdolmaleki S, Rudaki J. Comparison of aged and fresh automotive three-way catalyst in real driving. In: Sixth international conference on internal combustion engines; 2009.
[70] Yu D, Salmi MT. Catalytic kinetics 2nd edition. Springer; 2016345–446.
[71] Rayn McKay October 2006 Beat the law import tuner (Enthusiast Network) 122-12
[72] Paia D, Prabhub MK. Recent advances in substrate materials and thermal analysis of catalytic converters. Mater Today Proc 2018;5:24221–30.

CHAPTER 9

NO_x reduction in IC engines through adsorbing technique

S. Sathishkumar and M. Mohamed Ibrahim
Automotive Research Centre, School of Mechanical Engineering, Vellore Institute of Technology, Vellore, Tamil Nadu, India

9.1 Introduction

Due to increasing air pollution and growing concern about public health in the world, stringent emission regulations are being implemented in various sectors. These regulations are used to reduce the emissions of nitrogen oxides (NO_x), carbon monoxide (CO), hydrocarbons (HCs), and particulate matter (PM) from different sectors [1–3]. NO_x is an extreme polluting gas that is mostly produced through burning of biomass and fossil fuels. NO_x includes nitric oxide (NO) and nitrogen dioxide (NO_2), but NO is the predominant form in nearly 95% of the combustion products [4–6]. To reduce NO_x emissions from the major source (vehicles), the three-way catalyst (TWC) was introduced in gasoline engines to meet the emission standards and to reduce NO_x as well as HC and CO with the help of a precious metal-based catalyst. However, leaded gasoline fuel poisoned the TWC catalyst. Therefore, unleaded gasoline is used as fuel if TWC is incorporated in vehicles. For proper functioning of TWC in the gasoline engine the air-fuel ratio needs to be maintained at the stoichiometric point. When the air-fuel ratio is increased for a typical engine, the overall energy consumption of the engine was reduced nearly 30% with a decrease in CO_2 formation. This operation can be attained by maintaining a lean air-fuel mixture within the cylinder for both spark ignition (SI) and compression ignition (CI) engines. However, the TWC ineffectively reduces NO_x to N_2 for lean-burn engines due to difficulty in operation. This is the situation where the lean NO_x trap (LNT) was developed due to the ineffective reduction of NO_x by TWC at the excessive oxygen in the exhaust. The LNT technology was developed by Toyota in the 1990s [7,8]. In LNT, basic elements adsorb the NO_x during the lean operation of the engine. The stored NO_x is then desorbed and reduced with the help of available reductants such as HC and CO during engine operation under fuel-rich conditions.

NOx Emission Control Technologies in Stationary and Automotive Internal Combustion Engines
https://doi.org/10.1016/B978-0-12-823955-1.00009-7

Therefore, the engine needs to be operated in lean and rich conditions. The engine operates in the lean mixture for a longer period to store the NO_x on the catalyst surface. The engine runs in the rich mixture for a shorter period to regenerate and reduce NO_x emission. The LNT does not demand external reductants such as ammonia or urea like selective catalytic reduction (SCR). A typical LNT catalyst is a ternary level that consists of active elements (Ba and Pt) and alumina support in the base. For instance, Pt-BaO/Al_2O_3 is considered for a typical LNT system. In an excessive oxygen condition, the available NO_x is oxidized and stored on Ba. Once its reached its saturation point, the stored NO_x is reduced to N_2 in the oxygen-less atmosphere on Pt. Ba is considered the most effective storing material in the LNT system among the alkali and alkaline earth metals [7]. To reduce sulfur oxide (SO_x) adsorption, titanium oxide (TiO_2) is generally added to the catalyst. This is because sulfur poising is one of the main drawbacks of LNT catalysts and surface sulfates are more stable in LNT compared to nitrates. To improve hydrogen formation, rhodium (Rh) and zirconium oxide (ZrO_2) are added to the ternary catalyst [8,9]. Due to the kinetic limitations offered by LNT in cold start conditions, NO_x may not be adsorbed in the adsorbent surface at low-temperature conditions (less than 200°C). To address this issue, a passive NO_x adsorber is used to store the NO_x at low temperature, and the NO_x is released once the exhaust gas temperature increases above 200°C. This desorbed NO_x will be reduced through the downstream NH_3-SCR system once it is active because the minimum temperature to decompose the urea was reached and the effective reduction of NO_x can be obtained [10]. This chapter discusses both active NO_x adsorbing and passive NO_x absorbing techniques as well as their material properties, working characteristics, adsorption temperature, influence of exhaust gas species, influence of space velocity, and interaction of sulfur on the adsorbing material. In addition, selective NO_x recirculation (SNR) is also discussed in this chapter.

9.2 Active NO_x adsorption or lean NO_x trap (LNT)

The lean-burn engine offers a high fuel conversion efficiency and reduces CO_2 emissions when compared to stoichiometric engines. However, NO_x reduction from the exhaust of the lean-burn engine is a difficult task for the automotive industry. To solve this issue, a promising technology in the after-treatment systems is called lean NO_x trap (LNT). It is also called NO_x storage and Reduction (NSR), NO_x adsorber catalyst (NAC) and deNO_x. This is particularly suitable for gasoline lean-burn and light duty diesel engines. The LNT performance can be enhanced when the downstream of LNT is connected with selective catalytic reduction (SCR) [11–14]. The LNT catalyst includes a storage element (BaO); precious metals such as Pt, Pd, and Rh; and support materials such as Al_2O_3, CeO_2, and ZrO_2. It is operated in two phases: (1) lean and (2) rich conditions. During lean operation, the precious metal is used to oxidize NO into NO_2 while the BaO is used to store the NO_x in the form of nitrate. In the second phase during the rich operation, the adsorbed NO_x is released from the storing element and reduced to N_2 using reductants (HC, CO, H_2), which are produced from the incomplete fuel combustion. Here, the NO_x absorbing performance depends on two important properties of a precious material: dispersion and particle size [15,16]. The LNT catalyst is deactivated when sulfur comes into contact with the catalyst surface because sulfur acts as a poisoning agent. During processing of the NO_x emission, the only unwanted pollution from LNT is ammonia. The level of ammonia generation is varied depending on the catalyst in LNT. For example, Pt/Al_2O_3 at high temperature produces a high level of NH_3, whereas Pd/Al_2O_3 is less effective with respect to temperature change [12,17].

9.2.1 LNT working characteristics

Under lean operating conditions, NO presented in the exhaust is oxidized to NO_2 on the platinum surface. This NO_2 is stored on barium carbonate ($BaCO_3$) in the form of barium nitrate ($Ba(NO_3)_2$). The typical storing element includes calcium, barium, zirconium, lanthanum, and potassium. The stored barium nitrate is loosed from the trap in the form of barium oxide (BaO) under rich mixture condition. NO_x is reduced to N_2 with the presence of HC, CO, and H_2 on the Pt catalyst. Eventually, the barium carbonate is restored. The steps involved in the adsorption and desorption process are as follows [18]:

$$NO + \frac{1}{2}O_2 \leftrightarrow NO_2\,(\text{step}:1) \tag{9.1}$$

$$BaCO_3 + 2NO_2 \leftrightarrow Ba(NO_3)_2\,(\text{Step}:2) \tag{9.2}$$

$$Ba(NO_3)_2 \leftrightarrow BaO + 2NO_2\,(\text{Step}:3) \tag{9.3}$$

$$2NO_2 + 2CO/HC \xrightarrow{\text{Pt Catalyst}} N_2 + 2CO_2\,(\text{step}:4) \tag{9.4}$$

$$BaO + CO_2 \rightarrow BaCO_3\,(\text{Step}:5) \tag{9.5}$$

The adsorbed NO_x is converted to nitrogen by the NO_x regeneration cycle. For a short while during this cycle, the engine should run in rich condition to reduce the NO_x into nitrogen. A graphical representation of the LNT operation is depicted in Fig. 9.1 [19]. The NO_x regeneration can be achieved in four different ways: fuel injection in the exhaust, post injection during combustion, exhaust gas recirculation, and intake air throttling. The LNT is most suitable for small diesel engines such as passenger cars, which are more cost effective than SCR. In LNT, the NO_x storage element is normally an alkaline metal oxide or alkali metal such as barium oxide, coupled with Pt and Rd. catalysts. NO_x is adsorbed into the storing element during lean diesel operation; however for nearly 60–120 s, the engine is running with more fuel, so hydrocarbon and carbon monoxide regenerate the barium nitrate into harmless nitrogen [20].

In one of the NSR cases at low-temperature conditions, $Pt/Ba/Al_2O_3$ was used as the catalyst; Pt indicates a noble metal, Ba indicates the adsorbing element, and Al_2O_3 is the supporting element. In this case, hydrogen was used as a reductant to reduce nitrate into nitrogen. This catalyst showed sufficient NO_x storage capacity; however, the rate of NO_x release is slow and regeneration is insufficient. Moreover, a huge amount of NH_3 was produced with a high level of H_2 injection [21,22]. When comparing $Pt/Ba/Al_2O_3$ and $Pt/K/Al_2O_3$ catalysts in NSR for a complete cycle, the Ba-included catalyst showed superior performance to the other one, but there was no ammonia generated for the $Pt/K/Al_2O_3$ catalysts [23]. Sulfur poisoning is the major problem in NSR catalysts. The sulfur-containing fuel led to sulfur oxides (SO_x) emission in the engine exhaust. While SO_2 come in contact with the catalyst surface, SO_2 interacts with alumina to generate aluminum sulfate that obstructs the NO_x reaction with BaO. Further, SO_2 reacts with BaO to form barium sulfate

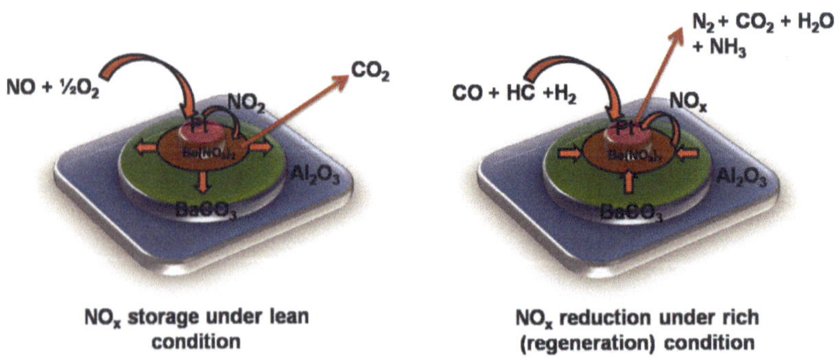

Fig. 9.1 Graphical illustration of NO_x storage and reduction for LNT.

($BaSO_4$). However, the use of titanium dioxide (TiO_2) put down the sulfate formation without affecting the efficiency of the NO_x storage material [13,24].

9.3 Influences of exhaust gas species, temperature, and hydrogen in LNT

This section describes the effect of exhaust gas species and temperature on NO_x adsorption and reduction. In addition, hydrogen addition on NO_x reduction is studied in LNT.

9.3.1 Influences of CO_2 and H_2O on NO_x adsorption

In LNT, the rate of reaction of NO_x storage is greatly influenced by the exhaust gas species such as H_2O, CO_2, and a combination of both. Table 9.1 lists the influence of H_2O, CO_2, and H_2O+CO_2 along with the base case for different temperature levels in the Pt-Ba/γ -Al_2O_3 (1/20/100 w/w) catalyst [25]. Here, it has been performed at two different conditions: (1) at NO_x breakthrough and (2) at the saturation point. The NO_x breakthrough point indicates that NO_x storage is enhanced. As seen from the table, the highest level of NO_x stored at the NO_x breakthrough point is 3.5×10^{-4} Mol_{NOx}/g_{cat} at a temperature of 350°C. In most cases at the NO_x breakthrough point, the level of NO_x adsorption increased with rising temperature for the base case, H_2O, CO_2, and H_2O+CO_2 conditions; this effect is excluded at 400°C for the base case and H_2O. The presence of CO_2 in the base case decreased the NO_x storage significantly compared with the H_2O case. The combined effect of CO_2 and H_2O increased the NO_x storage because the presence of H_2O decreased the negative effect of CO_2 in the base case. The addition of H_2O in the base case increased the NO_x storage capacity at low temperature and decreased the storage capacity at high temperature. This is applicable for NO_x breakthrough and saturation conditions. This is because water enhances NO_x storage at low temperature and weakens NO_x storage at high temperature. With water in the adsorption surface, NO_x adsorption occurs mainly on $Ba(OH)_2$ sites. As a result, water is released at an adsorber outlet.

9.3.2 Influence of temperature on NO_x reduction

Fig. 9.2 represents the temperature influence on the typical LNT catalyst (Pt-Ba/Al_2O_3) during NO_x reduction with the use of hydrogen (2000 ppm) when CO_2 and H_2O are absent [25]. This case was studied with

Table 9.1 Influence of H_2O, CO_2, and H_2O+CO_2 in LNT [25].

Temperature (°C)	At NO_x breakthrough (Mol NO_x/gcat)				At saturation point (Mol NO_x/gcat)			
	Base case (*10^{-4})	Presence of H_2O (*10^{-4})	Presence of CO_2 (*10^{-4})	Presence of H_2O+ CO_2 (*10^{-4})	Base case (*10^{-4})	Presence of H_2O (*10^{-4})	Presence of CO_2 (*10^{-4})	Presence of H_2O $+CO_2$ (*10^{-4})
150	0.0165	0.2367	0.0018	0.0018	1.3621	2.1014	0.4521	0.9355
200	1.0037	2.1046	0.0055	0.5266	1.783	3.745	0.8446	0.9868
250	1.8074	2.2844	0.0091	0.9633	3.7962	3.2844	3.1991	1.6636
300	3.345	2.3725	0.2991	1.1798	5.8095	4.5868	4.4446	3.5346
350	3.5982	2.7725	0.9376	1.0661	4.0692	3.4721	3.3868	3.927
400	3.1725	2.4386	1.4111	1.8881	3.2389	2.4996	2.0446	2.329

Base case: NO (1000 ppm) and O_2 (3%,); Presence of H_2O: NO (1000 ppm), O_2 (3%), and H_2O (1%); Presence of CO_2: NO (1000 ppm), O_2 (3%), and CO_2 (3000 ppm); Presence of H_2O+CO_2: NO (1000 ppm), O_2 (3%), CO_2 (3000 ppm), and H_2O (1%).

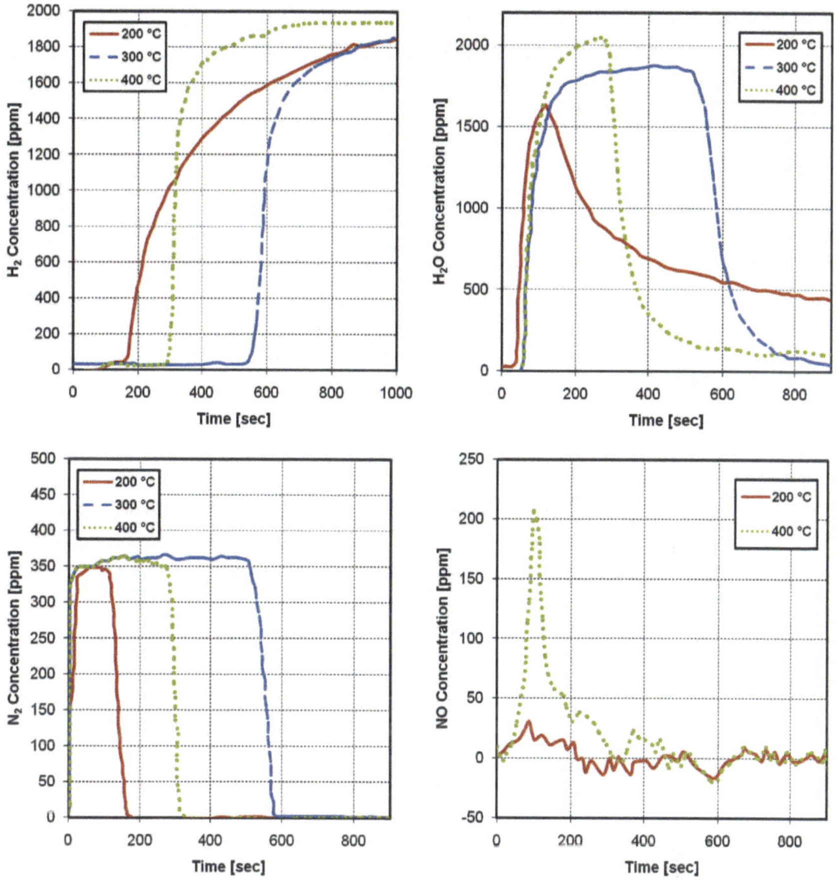

Fig. 9.2 Influence of temperature on adsorbed NO_x reduction with the use of hydrogen [25].

three different temperature conditions at 200°C, 300°C, and 400°C. As seen from the figure, the hydrogen was fully consumed during the NO_x reduction for 200°C. As a result, the H_2 concentration is not detected in the outlet of the adsorber, whereas the adsorbed NO_x is reduced to N_2 by the reaction of H_2. Therefore, an N_2 trace has been detected in the outlet and reached 360 ppm until the reduction of the adsorbed NO_x. This is similar to 300°C and 400°C, but the period of the N_2 trace is depends on the stored NO_x because the level of NO_x adsorption is different at various temperatures.

A small level of NO of about 200 ppm at 400°C was noticed in the outlet during the beginning of the reduction process. Ammonia was not noticed in the outlet of the adsorber; it was found that a small difference in the level of Pt (50% to 65%) in the catalyst avoided ammonia formation. It has been reported that based on the level of NO_x adsorbed and the temperature in the catalyst, ammonia formation is affected. While NO_x reduction occurred, H_2O was also produced and stored in the form of $Ba(OH)_2$ on the Ba sites. It is evident from the figure that even after the N_2 concentration was reduced to zero in the outlet, H_2 and H_2O traces were observed, especially at 200°C. This was due to H_2O release from the initially stored $Ba(OH)_2$ and the reduction of oxygen species in the catalyst. For the temperatures of 300°C and 400°C, H_2O traces were observed until N_2 was released fully in the outlet.

9.3.3 Influence of hydrogen on NO_x reduction

The reduction of stored NO_x on the Pt-Ba/ γ -Al_2O_3 (1/20/100 w/w) catalyst up to the saturation point with use of hydrogen at a constant temperature of 350°C is shown in Fig. 9.3 [25]. Here, the hydrogen addition is carried out at nearly 2000 ppm in the reactor. As a result, the adsorbed NO_x was reduced to N_2 at a level of 360 ppm by completely consuming H_2; as seen from the figure, it remained constant until 400 s. This NO_x reduction reaction was purely dependent and limited on the

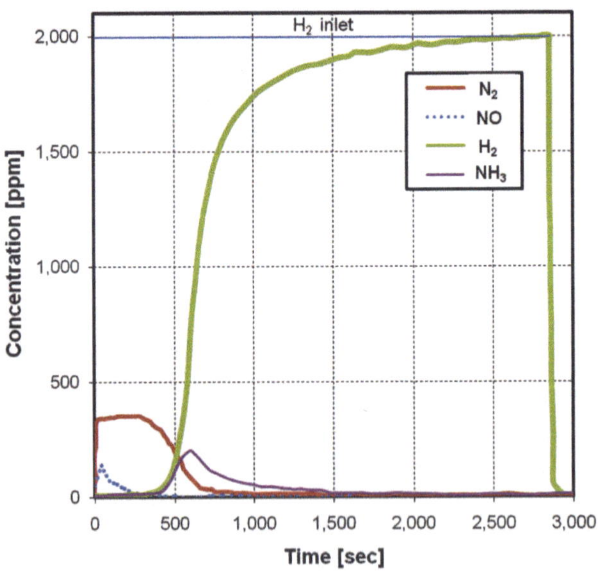

Fig. 9.3 Influence of hydrogen on NO_x reduction at constant temperature [25].

concentration of hydrogen. Further, a small level of NO was noticed once the hydrogen concentration was initiated on the catalyst. Then, the nitrogen level started to decline after 400 s and the hydrogen concentration and ammonia trace were observed at the same time. As mentioned in an earlier section, the byproduct of ammonia is unavoidable on LNT while reducing the stored NO_x species. While the NO_x reduction led to the formation of water due to the interaction of NO_x and H_2, this formed water will not be released immediately. There is a delay in water release of nearly 50 s because it is adsorbed on the Ba sites in the form of $Ba(OH)_2$. The addition of hydrogen on the catalyst increased the temperature of the catalyst by 3–5°C because of the exothermic reaction during the NO_x reduction. This indicates that the operation was performed with less thermal effects. After the reduction process, a small level of nitrates exist on the catalyst surface, even though most of the NO_x was removed from the catalyst. While performing the NO_x reduction by hydrogen, the important reactions involved in the catalyst are as follows:

$$Ba\left(NO_2\right)_2 + 3\,H_2 \rightarrow BaO + N_2 + 3\,H_2O \; Ba\left(NO_3\right)_2 + 5\,H_2 \rightarrow BaO + N_2 + 5\,H_2O \tag{9.6}$$

9.4 Selective NO_x recirculation (SNR)

The SNR approach was initially developed in 1994 and is related to exhaust gas treatment for lean combustion in a diesel engine [26]. The SNR technique is similar to LNT except for the NO_x reduction process. Here, NO_x reduction occurs by the recirculation of NO_x back to the engine cylinder. The typical SNR technique is illustrated in Fig. 9.4. As seen from the

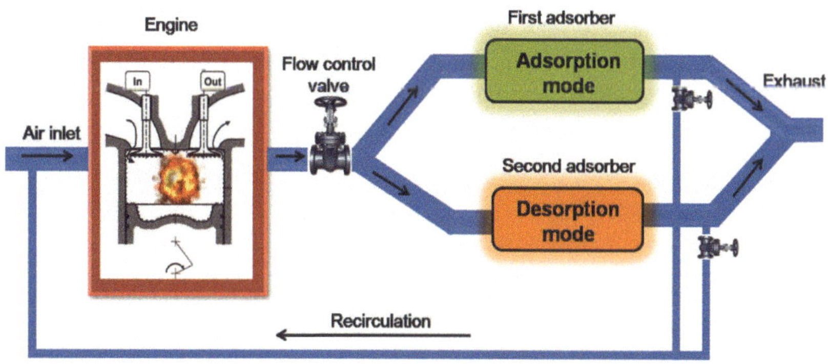

Fig. 9.4 Schematic representation of the SNR technique.

figure, two adsorbers are connected parallel to the exhaust; one is operating in adsorption mode and the other one is operating in desorption mode alternatively. In the intake of the adsorbers, the flow control valve is kept to split the exhaust gas coming from the engine. In the outlet of the adsorbers, return valves are provided to send the desorbed NO_x to the engine. The intake and outlet valves communicate for better operation of the adsorption and desorption modes [27,28]. Initially, the control valve is switched to direct the exhaust gas to one of the adsorbers, which is working in adsorption mode. The NO_x is adsorbed into the adsorber of the surface. The adsorption performance is influenced by the type of material and the exhaust gas temperature. Here, the adsorption and desorption modes are performed alternatively by the two adsorbers. Once the first adsorbent reaches the saturation level, the control valve changes the direction of the exhaust gas toward the second adsorber. The saturated first adsorber releases the NO_x, which is sent to the engine intake manifold through the return line. Then, it is mixed with the intake air and passed to the combustion chamber, where it is thermally decomposed [27,29]. One of the SNR studies on natural gas, gasoline, and diesel engines showed that the level of intake NO concentration on the efficiency was not significantly affected for a gasoline engine as compared with the rest of the engine types. The rates of NO_x decomposition for natural gas, diesel, and gasoline engines were 22%, 57%, and 87%, respectively. Here, the highest NO_x decomposition rate was achieved for the gasoline engines due to the operation of the stoichiometric ratio with huge EGR for a long period of time during the NO_x reduction stage.

9.4.1 NO_x adsorbing catalyst materials

There are different type of adsorbing materials used to store the NO_x from the exhaust during the lean operation, including carbonaceous materials, metal oxides, spinels, zeolites, perovskites, and heteropolyacids. Based on the property of the material, the adsorber can store NO or NO_2 only, but some cases can store both of them. The NO_x adsorption mechanisms are categorized into four types: (1) In a first mechanism, with increasing temperature, the adsorbed species (NO_2, NO^+, NO, N_2O_3) are desorbed from the catalyst. With a further rise in temperature, the adsorbed species decomposed into N_2 and oxygen (in the case of perovskites). When the surface area of the adsorbent increases, the adsorption capacity of the material is also increased. (2) Only at elevated temperature is the nitrate or nitrate salt produced at a sufficient level; this is not bounded by the surface area of the adsorbent. These salts are reduced with their oxides with the increased generation of NO_x. (3) Intercalation is the third mechanism, which is related to layered oxides (e.g., hydrotalcites, titanates, cuprates). The NO_x molecules are reversibly included within the layers without any adjustment in the crystal structure. (4) The carbon material adsorption type operates in a small and low-temperature range, therefore it is less important for lean exhaust gas applications [29].

9.5 Passive NO_x adsorber or low-temperature NO_x adsorber (LTNA)

Selective catalytic reduction (SCR) with a urea injection system is used to reduce NO_x emissions effectively from diesel engines when the catalyst temperature reaches 200°C. Once the exhaust gas temperature moves to 180°C, urea is decomposed [$CO(NH_2)_2$] into two ammonia (NH_3) molecules. During lean operations, this NH_3 reacts with NO_x over the catalyst of SCR to decompose NO_x into N_2 and H_2O. But, the SCR catalyst takes 3min to reach 200°C due to the slow warm-up rate of the diesel engine, though an exhaust heating system coupled with the SCR catalyst can be used to warm up the exhaust gas. As a result, NO_x reduction using urea is suspended during the warm-up period in the SCR system [30–33]. Due to stringent emission standards, NO_x emissions during a cold start also have to be controlled. During a cold start, the passive NO_x adsorber stores the NO when the exhaust gas comes into contact with the adsorber surface. Then, the stored NO is released under a lean operation as the exhaust temperature reaches more than 200°C. This desorbed NO moves to the SCR catalyst for reducing NO to N_2 with the help of ammonia (NH_3), as seen in Fig. 9.5. Because the temperature of the exhaust gas is very low during a cold

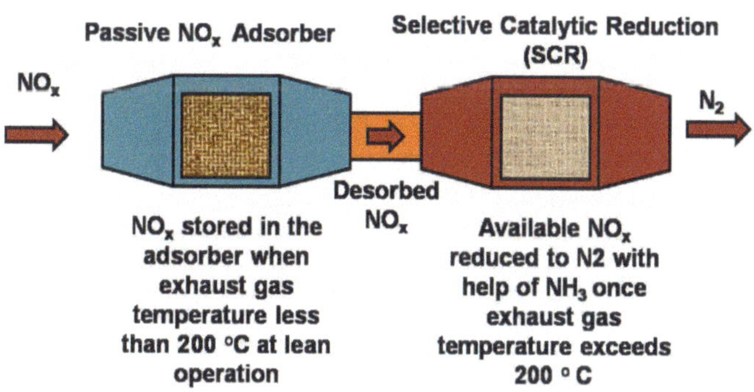

Fig. 9.5 Passive NO_x adsorber system.

start, the oxidation of NO to NO_2 is very difficult. Hence, it would be worthwhile to use a passive NO_x adsorber to adsorb and release the NO without the requirement to oxidize the NO to NO_2 [33–35].

9.5.1 Metal oxides and zeolite for passive NO_x adsorption

In passive NO_x adsorption, three different materials have been studied extensively: Pt/Pd/Al_2O_3, Pt/Pd/ceria, and Pd/zeolites. A few studies used Pd/cerium oxide (CeO_x) as a catalyst to adsorb NO when the exhaust gas temperature is below 200°C. When the temperature exceeds 200°C, the adsorbed NO_x is released as NO, then it is decomposed in the SCR catalyst. The authors noticed that a major problem with the ceria-based catalyst for NO storage is sulfur poisoning [36,37]. The diesel cold start concept (dCSC) also used Pd/ceria as a catalyst. Similar to earlier work, the system stored the high levels of HC and NO_x with enhanced efficiency; some of the stored NO_x and HC was decomposed directly without being desorbed [38]. Another study used monolithic catalysts that consist of Pt or Pd with 0.92 g/liter on a washcoat of Al_2O_3 or ceria/zirconia (CZO) to examine the NO_x storage performance of the precious metal and carrier. The results suggested that the Pd on CZO combination showed the best NO_x storage performance at above 100°C, when ethylene (C_2H_4) and H_2O are presented. Most importantly, total NO_x was released when the temperature is below 400°C, also small N_2O and NO_2 generations were noticed. As the catalyst was oxidized, the Pd/CZO catalyst provided better NO_x storage performance than the Pd/Al_2O_3 catalyst, which says that the supporting element CZO played a vital role in the performance of NO_x storage [39]. A study examined high Pd (2.76 g/liter) and low Pd (0.92 g/liter) in a Pd/CZO catalyst with different feed gas compositions. When the temperature is below 100°C, C_2H_4 and H_2O resist the performance of NO_x storage for both cases of Pd. However, when the temperature was above 100°C, C_2H_2 enhanced the performance of NO_x storage with the presence of H_2O. When the temperature was below 100°C, the storage performance was improved by CO with H_2O. The presence of C_2H_4 and CO enhanced the performance of the catalyst; this depends on the interaction of NO and reductants to create intermediate chemicals on the catalyst. As LTNA contacted CO and NO at 100°C, cyanate (NCO) was formed in the Pd/ceria powder catalyst [40].

The combined loading of both Pd (50 g/ft^3) and Pt (50 g/ft^3) in the Pt/Pd/CeO_2 catalyst showed the notable NO_x storage capacity at the low-temperature stage and the cold start catalyst concept [38]. Some important works in the literature [41–44] also analyzed the Pt/Pd-related catalyst washcoat on ceria and alumina. The type of noble metal presence in the support material also signifies the results of the passive NO_x adsorber. For instance, alumina and ceria are used as support materials in cold start applications. The results revealed that the Pd/CeO_2-based catalyst showed better NO_x storage performance than the Pt/CeO_2 catalyst. The oxidative environment or oxygen helps Pd to store NO_x, whereas the NO_x storage capacity of the CeO_2 is kept constantly. On the other hand, this material group shows a loss of NO_x storage capacity when it is exposed to sulfur [45,46]. In dCSC, the use of Pd/ceria as a catalyst in the passive NO_x adsorber is eliminated. Instead, Pd is seated on three different types of zeolite [beta (BEA), zeolite socony mobil–5 (ZSM-5), and chabazite (CHA)] to extend the NO_x storage temperature range. With the H_2O presence at 100°C, Pd/CHA stores the NO_x at the exchange sites of Pd. Moreover, unlike the Pd/ceria catalyst, none of the three Pd/zeolite catalysts were affected by sulfur poisoning. This Pd/zeolite-based catalyst showed similar adsorption performance on subsequent desorption cycles; however, based on the zeolite structure, the response to aging and the nature of Pd were varied [47,48]. On the whole, it is clear from the literature that Pd is most commonly used as a precious metal for low-temperature NO_x storage. However, depending on the loading of the support and the Pd, the amount of Pd is changed; in some cases, Pt and Pd are used simultaneously. Hence, the NO_x storage capacity is proportioned to platinum group metals (PGM) as seen from Fig. 9.6. It is evident from the figure that a high NO_x to PGM value is noticed for zeolite Pd catalysts when the weight percentage of Pd is less than or nearly 1%. In most cases, the NO_x to PGM ratio is decreased as the Pd amount is doubled. Alumina and ceria Pd also showed the same trend, but these materials store the NO_x at a slightly lesser level and lower rate than the zeolite-based catalyst.8

9.6 Operating conditions for NO_x adsorption

This section discusses the influences of adsorption temperature, space velocity, exhaust gas species, ethene (C_2H_4), and sulfur poisoning on NO_x adsorption to understand the required operating conditions for NO_x adsorption.

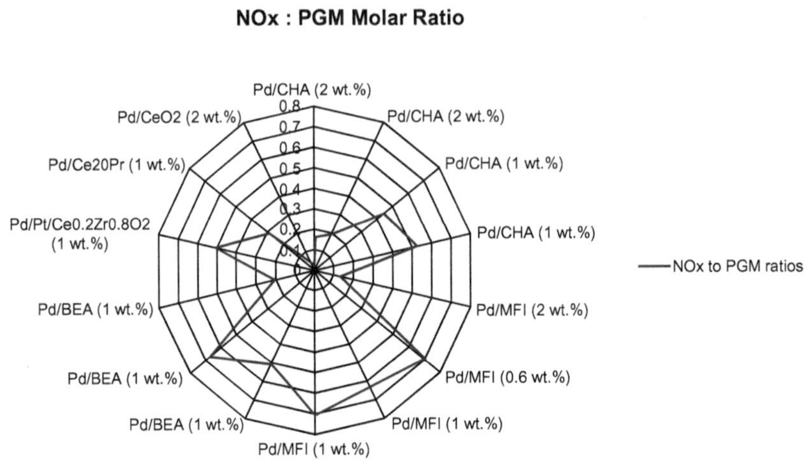

Fig. 9.6 NOx:PGM ratios for different passive NO_x adsorber catalysts [10].

9.6.1 Influence of adsorption temperature

Because of condensable exhaust gas components, it is difficult to conduct low-temperature passive NO_x adsorption experiments [49]; the specific problem is water condensation. In addition, the working temperature determines the adsorption rate and water impact. Therefore, the influence of temperature on NO_x storage has to be examined to gain insights into the efficiency of NO_x storage in a more practical approach. The NO_x storage efficiency is a ratio of adsorbed NO_x and inlet NO_x concentration, which is calculated as follows in Eq. (9.7) [44].

$$NO_x\,\text{storage efficiency} = \frac{\int_0^t \left([NO_x]_{inlet} - [NO_x]_{outlet}\right)^*\text{flow rate}\,dt}{\int_0^t \left([NO_x]_{inlet}\right)^*\text{flow rate}\,dt} \times 100(\%) \tag{9.7}$$

With increasing adsorption temperature, the NO_x storage efficiency is also increased for both Pd/CeO_2 and Pt/CeO_2, as illustrated in Fig. 9.7. This is because rising oxygen movement on the surface of the CeO_2 improved the NO_x storage capacity. In addition, it is evident from the figure that at different adsorption temperatures (80°C, 120°C, 160°C) the NO_x storage efficiency is higher for Pt/CeO_2 than Pd/CeO_2. This

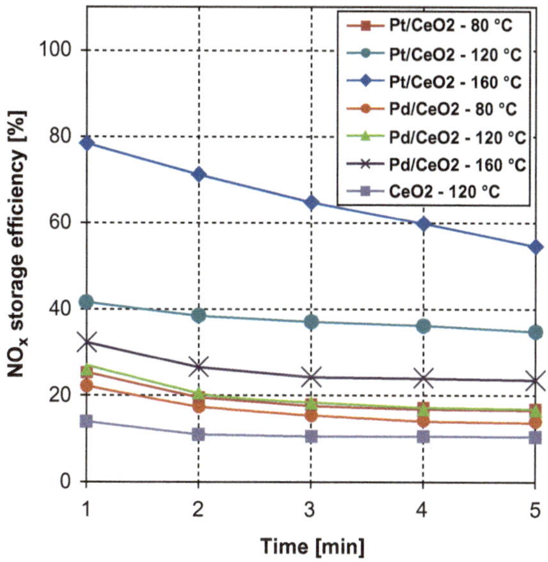

Fig. 9.7 Influence of adsorption temperature on NO_x storage efficiency for the CeO_2 catalyst [44].

was due to the Pt-based catalyst being more active than the Pd-based catalyst for oxidizing NO to NO_2, and the increasing adsorption temperature raised the rate of this reaction. Eventually, the NO_2 is stored on Pt/CeO_2 in the form of nitrates, so it leads to higher NO_x storage efficiency. However, in the case of Pd/CeO_2, the stored NO_x was released significantly when the temperature was below 300°C, whereas the Pt/CeO_2 catalyst regenerated the stored NO_x as the temperature went above 300°C. This is because Pd/CeO_2 and Pt/CeO_2 stored the NO_x in the form of nitrites and nitrates, respectively. The overall NO_x desorption efficiency was decreased after reaching 500°C for both Pd/CeO_2 and Pt/CeO_2 catalysts due to the rise in stability of the stored NO_x species. Similar to the above catalysts, the NO_x storage efficiency was increased with rising storage temperature for both La-doped Pt/Al_2O_3 and pure Pt/Al_2O_3 [42,44]. Upon increasing the storage temperature from 80°C to 160°C, the NO_x storage efficiency was improved from 30% to 60% for Pt/Al_2O_3 for 60 s during adsorption. However, the NO_x desorption efficiency was decreased with rising adsorption temperature when the temperature was below 250°C because of the greater production of nitrates.

With zeolite-based catalysts, the NO_x storage capacity declined with rising storage temperature within the range of 80–150°C for the Pd-BEA catalyst, whereas the Pd-CHA catalyst increased the NO_x storage capacity with increasing temperature for the temperature range of 80–70°C. In the case of Pd-MFI (ZSM-5), the NO_x storage efficiency was not affected until 170°C [47,50]. Hence, these temperature differences were related to the type of zeolite structure and the NO binding strength. On the whole, the metal oxide-based catalyst increased the NO_x storage efficiency and reduced the NO_x desorption efficiency with increasing adsorption temperature.

9.6.2 Influence of space velocity

The ratio of the flow rate of the exhaust to the volume of the catalyst is called space velocity. When there is an increase in exhaust flow rate or a decrease in catalyst volume, the space velocity is increased, which leads to the quick saturation of NO_x storage sites with a reduction in NO_x storage efficiency. The NO_x slip time decreased from 340s to 10s on the Pd/CZO-based catalyst due to a rise in space velocity to 7.5K/h from 30K/h [40]. This high space velocity led to a decrease in the NO_x slip time on the Ag/Al_2O_3 catalyst. Particularly, the NO_x storage efficiency decreased from 56% to 34% and 57% to 33% for Ag/Al_2O_3 (1.3wt.%) and Ag/Al_2O_3 (4wt.%) because the space velocity increased from 40K/h to 80K/h and 40K/h to 100 per h for Ag/Al_2O_3 (1.3wt.%) and Ag/Al_2O_3 (4wt.%) [51].

9.6.3 Influence of exhaust gas species

This subsection discusses the effect of NO, NO_2, H_2O, CO_2, and C_2H_4 species on a passive NO_x adsorber for both metal oxide and zeolite catalysts.

9.6.3.1 Influence of NO and NO_2 concentration

As a large concentration of NO comes into contact with the catalyst surface, NO storage sites are saturated quickly. As a consequence of this, NO_x slip can happen as the effect of the space velocity. With a decrease in NO concentration, the storage efficiency increased by 54%, 70%, and 100% for the three NO concentrations (400ppm, 200ppm, 130ppm, respectively) on the Pd/CZO catalyst. However, the NO concentration has a similar amount of stored NO_x at the maximum level, this saying that the storage capacity of the sample limits the NO_x storage efficiency of the catalyst. The passive NO_x adsorber stores NO_2 more effectively than NO. The near 100% NO_2 storage efficiency was noticed while performing a transient test for first

450 s on Pd/CZO catalyst in a monolithic passive NO_x adsorber [40]. Wherein, the feed gas has H_2O (5%) and the temperature of the bed was below 100°C for 120 s. In contrast, when the temperature was below 100°C, the presence of H_2O completely restricted NO adsorption on the Pd/CZO catalyst. As the temperature increased from 100°C, the maximum storage efficiency reached 83% only. However, nitrates are formed due to NO_2 storage. Therefore, by the end of the test at a maximum temperature of about 400°C, the sample was not fully purged. The data showed that in the sample, about 30% of the stored NO_x remained at the completion of the test. The majority of the stored NO_x was discharged when the feed gas has NO instead of NO_2; this NO participation led to the formation of nitrites and was easier to reduce.

9.6.3.2 Influence of H_2O and CO_2 on oxide-based catalyst

The results showed that while H_2O approaches adsorption material, the NO_x adsorption sites were occupied with H_2O molecules, as a result, the overall NO_x adsorption capacity was decreased [42,52]. A monolithic passive NO_x adsorber has either Pt or Pd supported on Al_2O_3 or CZO. In this case, the existence of water on the catalyst decreased the NO_x storage efficiency significantly when the temperature of the catalyst bed was less than 100°C. The reduction in NO_x adsorption at this temperature was caused by condensation of water on the storage sites, thus it inhibited NO_x adsorption [40,53]. The presence of CO_2 and H_2O on the passive NO_x adsorber decreased the NO_x storage efficiency [54]. For example, the existence of CO_2 and H_2O in a Pt or Pd washcoat of CeO_2 catalyst decreased the NO_x storage efficiency by nearly 50% [43,44]. The formation of nitrates improved on the Pt/CeO_2 and Pd/CeO_2 catalyst when CO_2 and H_2O were absent. On the other hand, nitrite formation was promoted on the Pd/CeO_2 or Pt/CeO_2 catalyst because of the presence of CO_2 and H_2O [44].

9.6.3.3 Influence of H_2O and CO_2 on zeolite-based catalyst

The zeolite-based catalyst also showed inhibition of NO_x storage when there was water, similar to the metal oxide-based catalyst but with a lesser effect on zeolite. Particularly, NO penetration was decreased from 38 min to 20 min for a combined CeO and beta zeolite based catalyst when the concentration of water is increased from 0.6% to 7.29% in the feed gas. This indicated that the NO detection rate was decreased at the outlet of the reactor on increasing water concentration [55]. This was because the NO storage sites were filled by water molecules. Especially, molecules of water reacting with CeO during adsorption at 40°C resulted in $Ca(OH)_2$, further it reacts with NO_x and produce $Ca(NO_3)_2$. These reactions inhibit the NO_x adsorption sterically. Moreover, the reaction of H_2O and NO_x at an adsorption temperature of less than 100°C led to the formation of HNO_2 and/or HNO_3, which decreased the adsorption capability of the CaO-beta catalyst. This influence of the water reaction on NO_x storage is significantly low for the Pd/zeolite-based catalyst. The effect of water molecules at 100°C on the Pd/chabazite-based material is shown in Fig. 9.8 [47].

Under a dry condition in a NO-DRIFTS analysis, three different bands from FTIR were noticed on a Pd/chabazite catalyst such a way that 1806, 1863 and 2151 cm^{-1}, the band of 1806 and 1863 cm^{-1} are related to linear nitrosyl species on the ions of Pd^{2+}, The band of 2151 cm^{-1} is concerned with the NO storage on Brönsted acid sites. In wet conditions, the peak at 2151 cm^{-1} disappeared, suggesting that molecules of water replaced the stored NO^+ on Brönsted acid sites. However, water molecules did not replace the stored NO_x on Pd sites. There was only a single peak at 1798 cm^{-1} instead of two peaks at 1806 and 1863 cm^{-1} for the dry case but the magnitude was maintained. This indicates that a Pd support on zeolite based catalyst could adsorb NO at low temperatures even though water is presented on the catalyst. Similarly, a study [56] also showed that there is a decline in NO_x storage capacity when H_2O reacted with the

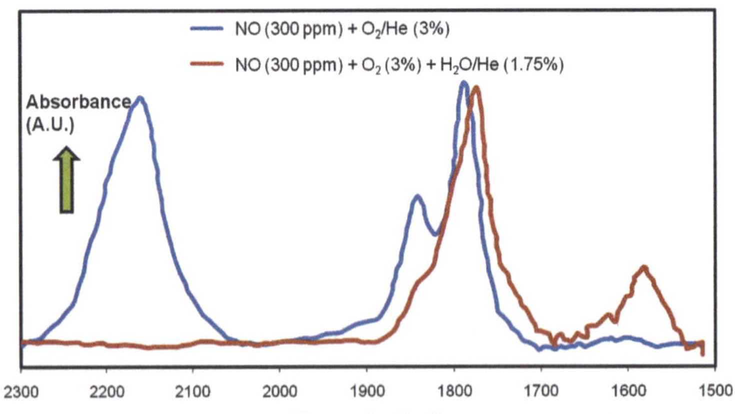

Fig. 9.8 Exposure of NO (300 ppm) and O_2 (3%) with and without H_2O (1.75%) on the Pd/CHA catalyst from NO-DRIFTS spectra recorded at 100°C [47].

Pd surface. The NO_x storage efficiency was degraded when CO_2 contacted the LNT or active NO_x adsorption catalyst, due to the competing interest between NO_x and CO_2 on the barium sites [57]. The presence of CO_2 on the Pd/ZSM-5 catalyst showed that there is no significant influence on the adsorption and desorption of NO_x as the CO_2 concentration was changed from 0% to 13%. This was due to NO_x was stored on Pd sites by generating nitrosyls, thus CO_2 was not adsorbed on the storage sites. Hence, there was no competition between CO_2 and NO_x on the adsorption sites in a Pd/ZSM-5 catalyst [58].

9.6.4 Influence of ethene (C_2H_4)on NO_x adsorption

The presence of HC in the inlet of the adsorption system influenced the characteristics of NO_x adsorption. In this concern, some of the literature [39,40,42] reported the influence of ethene (C_2H_4) on Pd and Pt catalyst washcoat in oxide-based material. Specifically, NO_x storage efficiency was improved when C_2H_4 reacted with the Pd supported in ceria/zirconia (CZO) oxide; the experiment was carried out in both dry and wet operations [39]. The NO_x storage efficiency increased to 49% from 32% with the presence of C_2H_4, as the temperature of the bed was less than 200°C during the transient test with the absence of H_2O. This rise in storage efficiency was because of the generation of alkyl nitrites or alkyl nitrates due to the reaction between NO and C_2H_4 [40,42]. Nevertheless, due to the interaction with carbonate species, alkyl species were not formed, which was confirmed through a DRIFTS investigation. With the presence of C_2H_4, different NO_x storage efficiencies were noticed for the Pt/CZO and Pt/Al_2O_3 catalysts. With the temperature of the bed less than 100°C, the NO_x storage efficiency was decreased to 37% from 72% on Pt/CZO with the presence of C_2H_4 in the feed gas. At this same operating condition, NO_x storage efficiency was also decreased to 49% from 80% on the Pt/Al_2O_3 catalyst [39]. This reduction in efficiency was due to the reaction of C_2H_4 and NO_x. It reduced the NO_2 to NO on Pt, and this formed NO was not more effectively adsorbed than NO_2. Therefore, NO_x storage efficiency was decreased with the presence of C_2H_4 on the Pt-based catalyst. As a whole, with the presence of C_2H_4, the NO_x storage efficiency was increased for Pd-based material and decreased for the Pt-based passive NO_x adsorber. Hence, the influence of C_2H_4 on the NO_x storage efficiency directly depends on the noble material.

9.6.5 Sulfur poisoning of passive NO_x adsorber

The combustion of fuel containing sulfur in the engine leads to the formation of SO_x in the engine exhaust. The presence of SO_x with a passive NO_x adsorber causes deactivation of the adsorbent with the system [38,59]. A study on the influence of sulfur on an Ag/Al_2O_3 catalyst-based passive NO_x adsorber showed that the addition of SO_2 (20 ppm) into the feed gas decreased the NO_x storage capacity to 36% from 56%. It also confirmed that the Ag/Al_2O_3-based catalyst was very sensitive to sulfur poisoning. After the completion of three successive desulfation procedures under rich, stoichiometric, and lean operations until 630°C, the adsorbed sulfur was removed by only 27% from Ag/Al_2O_3 (1.3 wt.%) because it was more complex to regenerate the sulfur-poisoned catalyst [51]. Improving the material acidity is a proper way to ease the desulfation and enhance the sulfur tolerance in a passive NO_x adsorber because SO_2 species are acidic in nature [60]. For instance, doping TiO_2 on an Ag/Al_2O_3 catalyst raised the catalyst material acidity and enhanced the desulfation efficiency [61]. The presence of sulfur in the feed gas does not affect the NO_x storage efficiency in zeolite material compared with oxide material because the zeolite catalyst has a high resistance to sulfur content [50]. For instance, the NO_x storage capacity was decreased when one mole of Pd faces 21 mol of SO_2 on Pd/CeO_2 catalyst at 100°C. However, the NO_x storage efficiency was not affected when SO_2 is contacted on Pd/CHA, Pd/MFI and Pd/BEA catalysts as seen from Fig. 9.9 [47]. Largely diffused Pd^{2+} ions at exchange sites make zeolites adsorb the NO at a low temperature without any notable effect of SO_2. As a whole,

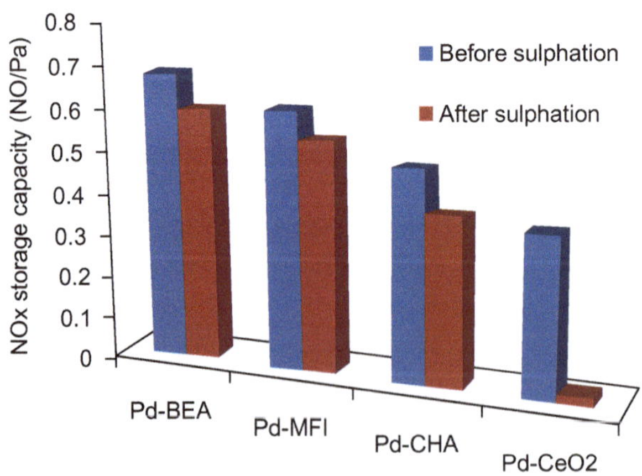

Fig. 9.9 Influence of sulfur on NO_x storage capacity for oxide and zeolite catalysts [47].

the presence of sulfur in oxide-based catalysts affects the performance of the passive NO_x adsorber. With zeolite, the performance characteristics were not affected significantly by the sulfur content in the feed gas.

9.7 NO_x desorption characteristics

In a passive NO_x adsorber, the adsorber stores the NO_x from the feed gas until the saturation point is reached. During the saturation point, the NO_x level in the inlet is the same as the NO_x level in the exhaust of the passive NO_x adsorber. After the adsorber reaches saturation point, this adsorbed NO_x released from the adsorber is called NO_x desorption. Temperature programmed desorption (TPD) is used to monitor NO_x desorption. In real-world conditions, the temperature begins to increase once the adsorption occurs; when the temperature moves to a higher level, desorption starts. NO_x desorption is complex because adsorption sites need to be regenerated for subsequent cold start. Moreover, NO_x desorption has to occur at a relatively high temperature at this point where the SCR catalyst is adequately active. Based on the after-treatment arrangement, this temperature may change. However, 200°C is a suitable NO_x desorption temperature for the after-treatment system that has NH_3-SCR [32,62]. Fig. 9.10 illustrates NO/NH_3 conversion with respect to reaction temperature. NO_x desorption at a low temperature led to ineffective NO_x reduction in the SCR catalyst. On the other hand, NO_x release at a high temperature resulted in the improper recovery of adsorption sites. It may be noted that for NH_3 production, urea decomposition is not considered in Fig. 9.10.

9.7.1 Influence of desorption temperature and exhaust gas species

Due to the notable NO_x adsorption capacity and the ability to withstand temperature and water, metal oxide and zeolite supports with noble metals are used as catalysts for a passive NO_x adsorber. Table 9.2 shows the desorption characteristics for different catalysts in a passive NO_x adsorber. Most catalysts come under the desorption temperature range of 200–400°C. For the same materials, the NO_x release temperature may differ based on the ramp rate, aging conditions, gas composition, and different methods of synthesis. The CO and H_2O during the NO_x release are not influenced more.

As the CO level increases from 0 to 225 ppm in the feed gas, a high level of NO is released from the Pd-BEA catalyst at a high temperature zone;

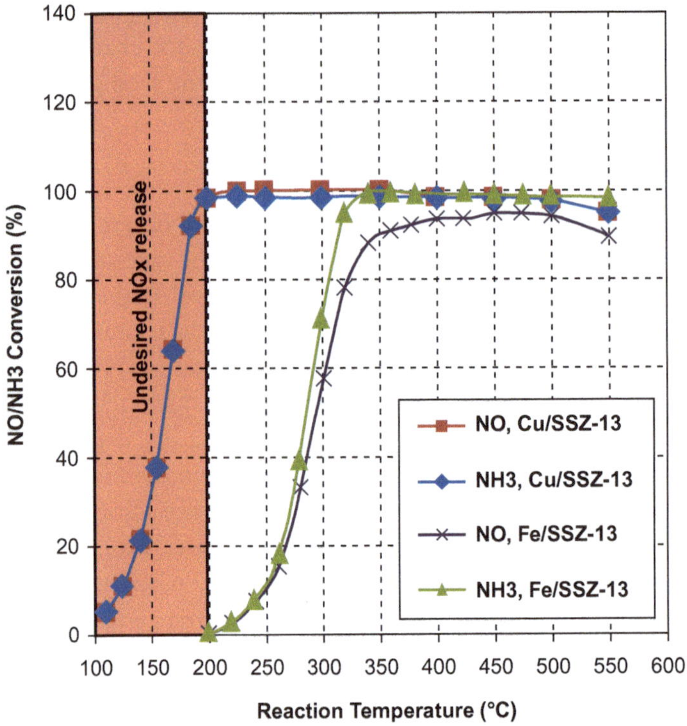

Fig. 9.10 NO/NH_3 conversion with respect to reaction temperature for standard SCR [33].

overall, the NO_x release increased. However, there was no effect on desorption when CO addition occurred at 180°C. NO_x release increased as the temperature moved to 260°C, although the CO level was low [68]. Similar results were also noticed for the unaged catalyzes of Pd-ZSM-5 and Pd-SSZ-13 [48]. Results confirmed through reactor and IR studies that Pd was reduced by CO to a decreased oxidation state. As a consequence of this, the bond between Pd and NO_x was strengthened, leading to an increase in the NO_x release temperature on the Pd-BEA catalyst. Due to the interaction between the NO and CO species, thermally stable surface intermediates (OC-Pd (II)-NO) were produced [56]. A similar enhancement in the NO_x release temperature was also noticed for Pd-CeO_2/ZrO_2, when CO contacted the adsorber [40].

Table 9.2 Desorption characteristics for different catalysts in a passive NO_x adsorber.

Author	Catalyst type	Ramp rate (°C / min)	Gas composition	Hydrothermal aging temperature (°C)	Desorption temperature (°C)
Chen et al. [38]	$Pt/Pd/Al_2O_3$ (140 g/ft^3)	100	H_2O: 5%, CO: 200 ppm and decane: 500 ppm	750	250, 400
Chen et al. [47]	Pd/CHA (1 wt.%)	100	H_2O: 5%, CO: 200 ppm and n-$C_{10}H_{22}$: 50 ppm	750	360
Chen et al. [47]	Pd/MFI (1 wt.%)	100	H_2O: 5%, CO: 200 ppm and n-$C_{10}H_{22}$: 50 ppm	750	275
Chen et al. [47]	Pd/BEA (1 wt.%)	100	H_2O: 5%, CO: 200 ppm and n-$C_{10}H_{22}$: 50 ppm	750	260
Ji et al. [41]	Pt/Pd/Ce0.2Zr0.8O_2 (1 wt. %)	10	H_2O: 3.5% and CO_2: 5%	750	200, 350
Ji et al. [42]	Pt/10La/Al_2O_3 (1 wt.%)	10	H_2O: 3.5% and CO_2: 5%	–	150, 450
Jones et al. [44]	Pt/Ce20Pr (1 wt.%)	10	H_2O: 3.5% and CO_2: 5%	–	150, 400
Lee et al. [63]	Pd/MFI (2 wt.%)	10	H_2O: 5%	–	200, 450
Murata et al. [64]	Pd/MFI (1 wt.%)	20	CO: 1000 ppm	–	300
Ren et al. [51]	Ag/Al_2O_3 (1.3 wt.%)	10	H_2O: 6%, CO_2: 6%, CO: 1000 ppm, C_3H_6: 90 ppm, C_3H_8: 30 ppm, n-dodecane: 40 ppm, m-xylene: 25 ppm, H_2: 500 ppm	650	430
Ryou et al. [65]	Pd/CHA (2 wt.%)	10	H_2O: 5% and CO_2: 5%	750	250, 400
Ryou et al. [66]	Pd/CeO_2 (2 wt.%)	10	H_2O: 5%	750	300, 450
Ryou et al. [67]	Pd/CHA (2 wt.%)	10	H_2O: 5% and CO_2: 5%	750	230, 420
Theis and Lambert [40]	$Pd/CeO_2/ZrO_2$ (1.5 wt.%)	10	H_2O: 5%, CO_2: 5% and C_2H_4: 500 ppm	700	240
Vu et al. [68]	Pd/BEA (1 wt.%)	63	H_2O: 5% and CO: 200 ppm	750	150, 260
Zheng et al. [48]	Pd/CHA (1 wt.%)	10	H_2O: 2.5% and CO: 200 ppm	–	360
Zheng et al. [48]	Pd/MFI (0.6 wt.%)	10	H_2O: 2.5% and CO: 200 ppm	–	250,400
Zheng et al. [48]	Pd/BEA (1 wt.%)	10	H_2O: 2.5% and CO: 200 ppm	–	210

9.7.2 Influence of ramp rate

In the TPD experiments, the temperature at which NO_x desorption occurred was called the NO_x desorption temperature, as discussed earlier. This NO_x desorption temperature may shift either low or high based on the ramp rate. Therefore Table 9.2 also lists the ramp rate as a function of the NO_x desorption temperature. The NO_x desorption peak temperature moves to a high amplitude with a rising heating rate. A different ramp rate for TPD experiments led to larger differences in NO_x desorption temperature, as seen from the table [69]. It is evident from Table 9.2 that most of the experiments have used a slower ramp rate of 10°C/min. This slow ramp rate provided legible resolution between the desorption peaks for kinetic studies. Nevertheless, the mentioned NO_x desorption temperature may change when it goes to real-world conditions because the magnitude of the ramp rate (10°C/min) is lower than the rate of variation of exhaust gas temperature at cold start conditions. The relation between the desorption temperature and ramp rate is not linear. Particularly when a ramp rate was less than 10°C/min, there was a notable variation in NO_x desorption temperature. When the ramp rate was greater than 50°C/min, there was a change in the NO_x desorption temperature as a larger change in the ramp rate [70].

9.8 Conclusions

In a lean diesel engine, NO_x removal is a quite difficult task because of highly oxidizing conditions. The active NO_x adsorber is considered a promising after-treatment technology to remove NO_x from lean-burn engines. It consists of a support material (Al_2O_3, ZrO_2, CeO_2), an adsorbing element (BaO), and a precious metal (Pd, Pt). During lean operation, NO_x from the exhaust is stored by the adsorber. During rich conditions, the stored NO_x decreases with the help of reductants (HC, CO, H_2). An important challenge of an active NO_x adsorber is desulfurization. The unwanted emission from the trap is NH_3, which can be stored separately for the subsequent operation of SCR. The presence of H_2O enhanced the NO_x storage capacity at low temperature and had a negative effect on NO_x storage at high temperature. In many cases at the NO_x breakthrough point, the level of NO_x adsorption increased with rising temperature for the base case, H_2O, CO_2, and $H_2O + CO_2$ conditions. In the case of SNR, the reduction of NO_x occurred by the recirculation of NO_x back to the engine combustion chamber. Therefore, use of reductants can be avoided for SNR.

In a passive NO_x adsorber, initial research was mostly concentrated on metal oxides with precious metals. The NO_x storage efficiency is higher for a Pt/metal oxide catalyst than a Pd/metal oxide catalyst with increasing adsorption temperature. However, an important problem in Pt and metal oxide catalysts is nitrate formation, thus the passive NO_x adsorber is complex to regenerate the nitrates fully and clean up the storage sites for the subsequent cold start. Hence, Pd-based material may be selected as a catalyst for a passive NO_x adsorber because it stores the NO_x in terms of nitrites. Therefore, the regeneration of the adsorber is much easier than a Pt-based catalyst for regular operating conditions. Low adsorption performance, sulfur poisoning, and trap prevention from H_2O at a temperature below 10°C are the notable problems for the oxide based-catalyst. More recently, the use of zeolite materials in passive NO_x adsorbers has emerged. Pd ion-exchanged zeolites can store the NO_x more effectively in the form of nitrosyls on the Pd sites and a high NO_x to PGM ratio is noticed for the Pd/zeolite catalyst. Unlike metal oxides, the zeolite-based passive NO_x adsorber has a high tolerance to H_2O at low temperature as well as sulfur poisoning. As a whole, zeolite-based material better well as a catalyst in a passive NO_x adsorber than metal oxide. When the operating temperature reaches 200°C, the SCR system effectively reduces the desorbed NO_x from the passive NO_x adsorber with the help of ammonia. The operating conditions and performance of the SCR system will be discussed in a detailed manner in the next chapter.

References

[1] Wong AP, Kyriakidou EA, Toops TJ, Regalbuto JR. The catalytic behavior of precisely synthesized Pt-Pd bimetallic catalysts for use as diesel oxidation catalysts. Catal Today 2016;267:145–56. https://doi.org/10.1016/j.cattod.2016.02.011.

[2] Du S, Tang W, Guo Y, Binder A, Kyriakidou EA, Toops TJ, et al. Understanding low temperature oxidation activity of nanoarray-based monolithic catalysts: from performance observation to structural and chemical insights. Emiss Control Sci Technol 2017;3:18–36. https://doi.org/10.1007/s40825-016-0054-y.

[3] Kim MY, Kyriakidou EA, Choi JS, Toops TJ, Binder AJ, Thomas C, et al. Enhancing low-temperature activity and durability of Pd-based diesel oxidation catalysts using ZrO_2 supports. Appl Catal Environ 2016;187:181–94. https://doi.org/10.1016/j.apcatb.2016.01.023.

[4] Anon. Formation and control of nitrogen oxides. Catal Today 1988;2:369–79. https://doi.org/10.1016/0920-5861(88)80002-6.

[5] Fritz A, Pitchon V. The current state of research on automotive lean NOx catalysis. Appl Catal Environ 1997;13:1–25. https://doi.org/10.1016/S0926-3373(96)00102-6.

[6] Janssen F. Emission control from stationary sources. In: Environmental catalysis. London: Imperial College Press; 1999. p. 234–93. https://doi.org/10.1142/9781848160613_0013.

[7] Miyoshi N, Matsumoto S, Katoh K, Tanaka T, Harada J, Takahashi N, et al. Development of new concept three-way catalyst for automotive lean-burn engines. SAE Tech Pap 1995. https://doi.org/10.4271/950809.
[8] Matsumoto S. Catalytic reduction of nitrogen oxides in automotive exhaust containing excess oxygen by NOx storage-reduction catalyst. CATTECH 2000;4:102–9. https://doi.org/10.1023/A:1011951415060.
[9] Engström P, Amberntsson A, Skoglundh M, Fridell E, Smedler G. Sulphur dioxide interaction with NO(x) storage catalysts. Appl Catal Environ 1999;22: L241–8. https://doi.org/10.1016/S0926-3373(99)00060-0.
[10] Gu Y, Epling WS. Passive NOx adsorber: an overview of catalyst performance and reaction chemistry. Appl Catal Gen 2019;570:1–14. https://doi.org/10.1016/j.apcata.2018.10.036.
[11] Wang J, Ji Y, Easterling V, Crocker M, Dearth M, McCabe RW. The effect of regeneration conditions on the selectivity of NOx reduction in a fully formulated lean NOx trap catalyst. Catal Today 2011;175:83–92. https://doi.org/10.1016/j.cattod.2011.02.048.
[12] Takahashia N, Shinjoh H, Iijima T, Suzuki T, Yamazaki K, Yokota K, et al. The new concept 3-way catalyst for automotive lean-burn engine: NOx storage and reduction catalyst. Catal Today 1996;27:63–9. https://doi.org/10.1016/0920-5861(95)00173-5.
[13] Matsumoto S. DeNOx catalyst for automotive lean-burn engine. Catal Today 1996;29:43–5. https://doi.org/10.1016/0920-5861(95)00259-6.
[14] Takeuchi M, Matsumoto S. NOx storage-reduction catalysts for gasoline engines. Top Catal 2004;28:151–8. https://doi.org/10.1023/B:TOCA.0000024344.91688.e4.
[15] Wang XQ, Kim DH, Kwak JH, Wang C, Szanyi J, Peden CHF. Effect of reductive treatments on Pt behavior and NOx storage in lean NOx trap catalysts. Catal Today 2011;175:78–82. https://doi.org/10.1016/j.cattod.2011.03.032.
[16] West B, Huff S, Parks J, Lewis S, Choi JS, Partridge W, et al. Assessing reductant chemistry during in-cylinder regeneration of diesel lean NOx traps. SAE Tech Pap 2004. https://doi.org/10.4271/2004-01-3023.
[17] Doornbos G, Adams E, Carlsson PA, Dahl D, Laurell M, Schyllander H, et al. Comparison of lab versus engine tests in the Development of a highly efficient ammonia formation catalyst for a passive SCR system. SAE Tech Pap 2015. https://doi.org/10.4271/2015-24-2504.
[18] Wang Y, Raman S, Grizzle J. Dynamic modeling of a lean NOx trap for lean-burn engine control. In: Proceedings of American control conference, San Diego, USA; 2003. p. 1208–12. https://doi.org/10.1109/acc.1999.783232.
[19] Vrabie V, Scarpete D, Zbarcea O. The new exhaust after treatment system for reducing NO_X emissions of diesel engines: lean NO_X trap (LNT). A study. Sci Proc XXIV Int Sci Conf 2016;1:110–3.
[20] Casapu M, Grunwaldt JD, Maciejewski M, Wittrock M, Göbel U, Baiker A. Formation and stability of barium aluminate and cerate in NOx storage-reduction catalysts. Appl Catal Environ 2006;63:232–42. https://doi.org/10.1016/j.apcatb.2005.10.003.
[21] Sjöblom J, Papadakis K, Creaser D, Odenbrand CUI. Use of experimental design in development of a catalyst system. Catal Today 2005;100:243–8. https://doi.org/10.1016/j.cattod.2004.09.066.
[22] Sakamoto Y, Motohiro T, Matsunaga S, Okumura K, Kayama T, Yamazaki K, et al. Transient analysis of the release and reduction of NOx using a Pt/Ba/Al2O3 catalyst. Catal Today 2007;121:217–25. https://doi.org/10.1016/j.cattod.2006.05.088.
[23] Malpartida I, Guerrero-Pérez MO, Herrera MC, Larrubia MA, Alemany LJ. MS-FTIR reduction stage study of NSR catalysts. Catal Today 2007;126:162–8. https://doi.org/10.1016/j.cattod.2007.05.014.

[24] Takahashi N, Suda A, Hachisuka I, Sugiura M, Sobukawa H, Shinjoh H. Sulfur durability of NOx storage and reduction catalyst with supports of TiO_2, ZrO_2 and ZrO_2-TiO_2 mixed oxides. Appl Catal Environ 2007;72:187–95. https://doi.org/10.1016/j.apcatb.2006.10.014.

[25] Forzatti P, Castoldi L, Lietti L, Nova I, Tronconi E. Chapter 6 identification of the reaction networks of the NOx storage/reduction in lean NOx trap systems. Stud Surf Sci Catal 2007;171:175–208. https://doi.org/10.1016/S0167-2991(07)80207-2.

[26] Bögner W, Krämer M, Krutzsch B, Pischinger S, Voigtländer D, Wenninger G, et al. Removal of nitrogen oxides from the exhaust of a lean-tune gasoline engine. Appl Catal Environ 1995;7:153–71. https://doi.org/10.1016/0926-3373(95)00035-6.

[27] Krutzsch B, Wenninger G, Weibel M, Stapf P, Funk A, Webster DE, et al. Reduction of NOx in lean exhaust by selective NO x-recirculation (SNR-technique) part I: system and decomposition process. SAE Tech Pap 1998. https://doi.org/10.4271/982592.

[28] Hodjati S, Vaezzadeh K, Petit C, Pitchon V, Kiennemann A. NOx sorption-desorption study: application to diesel and lean-burn exhaust gas (selective NOx recirculation technique). Catal Today 2000;59:323–34. https://doi.org/10.1016/S0920-5861(00)00298-4.

[29] Gómez-García MA, Pitchon V, Kiennemann A. Pollution by nitrogen oxides: an approach to NOx abatement by using sorbing catalytic materials. Environ Int 2005;31:445–67. https://doi.org/10.1016/j.envint.2004.09.006.

[30] Lambert C, Cavataio G. Development of the 2010 Ford diesel truck catalyst system. New York, NY: Springer; 2014. p. 659–89. https://doi.org/10.1007/978-1-4899-8071-7_21.

[31] Kamasamudram K, Currier NW, Chen X, Yezerets A. Overview of the practically important behaviors of zeolite-based urea-SCR catalysts, using compact experimental protocol. Catal Today 2010;151:212–22. https://doi.org/10.1016/j.cattod.2010.03.055.

[32] Forzatti P, Nova I, Tronconi E. New "enhanced NH3-SCR" reaction for NOx emission control. Ind Eng Chem Res 2010;49:10386–91. https://doi.org/10.1021/ie100600v.

[33] Gao F, Wang Y, Kollár M, Washton NM, Szanyi J, Peden CHF. A comparative kinetics study between Cu/SSZ-13 and Fe/SSZ-13 SCR catalysts. Catal Today 2015;258:347–58. https://doi.org/10.1016/j.cattod.2015.01.025.

[34] Theis JR, Ura JA. Assessment of zeolite-based low temperature NOx adsorbers: effect of reductants during multiple sequential cold starts. Catal Today 2020;1–10. https://doi.org/10.1016/j.cattod.2020.01.040.

[35] Chen HY, Liu D, Weigert E, Cumaranatunge L, Camm K, Bannon P, et al. Durability assessment of diesel cold start concept (dCSC™) technologies. SAE Int J Engines 2017;10. https://doi.org/10.4271/2017-01-0955.

[36] Coulson JE, Brisley RJ, Keane O, Phillips PR, Mountstevens EH. Thermally regenerable nitric oxide adsorbent. WO 2008/ 047170; 2008.

[37] Melville JE, Brisley RJ, Keane O, Phillips PR, Mountstevens EH. Thermally regenerable nitric oxide adsorbent. US Patent 8,105,559; 2011.

[38] Chen HY, Mulla S, Weigert E, Camm K, Ballinger T, Cox J, et al. Cold start concept (CSC™): a novel catalyst for cold start emission control. SAE Tech Pap 2013;2:372–81. https://doi.org/10.4271/2013-01-0535.

[39] Theis JR. An assessment of Pt and Pd model catalysts for low temperature NOx adsorption. Catal Today 2016;267:93–109. https://doi.org/10.1016/j.cattod.2016.01.032.

[40] Theis JR, Lambert C. The effects of CO, C_2H_4, and H_2O on the NOx storage performance of low temperature NOx adsorbers for diesel applications. SAE Int J Engines 2017;10:1627–37. https://doi.org/10.4271/2017-01-0942.

[41] Ji Y, Xu D, Bai S, Graham U, Crocker M, Chen B, et al. Pt- and Pd-promoted CeO_2-ZrO_2 for passive NOx adsorber applications. Ind Eng Chem Res 2017;56:111–25. https://doi.org/10.1021/acs.iecr.6b03793.
[42] Ji Y, Bai S, Crocker M. Al_2O_3-based passive NOx adsorbers for low temperature applications. Appl Catal Environ 2015;170–171:283–92. https://doi.org/10.1016/j.apcatb.2015.01.025.
[43] Jones S, Ji Y, Bueno-Lopez A, Song Y, Crocker M. CeO_2-M_2O_3 passive NOx adsorbers for cold start applications. Emiss Control Sci Technol 2017;3:59–72. https://doi.org/10.1007/s40825-016-0058-7.
[44] Jones S, Ji Y, Crocker M. Ceria-based catalysts for low temperature NOx storage and release. Catal Lett 2016;146:909–17. https://doi.org/10.1007/s10562-016-1704-y.
[45] Ryou YS, Lee J, Lee H, Kim CH, Kim DH. Low temperature NO adsorption over hydrothermally aged Pd/CeO_2 for cold start application. Catal Today 2018;307:93–101. https://doi.org/10.1016/j.cattod.2017.02.025.
[46] Ryou YS, Lee J, Lee H, Kim CH, Kim DH. Effect of sulfur aging and regeneration on low temperature NO adsorption over hydrothermally treated Pd/CeO_2 and Pd/$Ce0.58Zr0.42O_2$ catalysts. Catal Today 2017;297:53–9. https://doi.org/10.1016/j.cattod.2017.06.035.
[47] Chen HY, Collier JE, Liu D, Mantarosie L, Durán-Martín D, Novák V, et al. Low temperature NO storage of zeolite supported Pd for low temperature diesel engine emission control. Catal Lett 2016;146:1706–11. https://doi.org/10.1007/s10562-016-1794-6.
[48] Zheng Y, Kovarik L, Engelhard MH, Wang Y, Wang Y, Gao F, et al. Low-temperature Pd/zeolite passive NOx adsorbers: structure, performance, and adsorption chemistry. J Phys Chem C 2017;121:15793–803. https://doi.org/10.1021/acs.jpcc.7b04312.
[49] Garrido Gonzalez N, Baar R, Drueckhammer J, Kaeppner C. The thermodynamics of exhaust gas condensation. SAE Int J Engines 2017;10:1411–21. https://doi.org/10.4271/2017-01-9281.
[50] Rajaram RR, Chen H-Y, Liu D. Passive NOx adsorber. US Patent US10005075B2; 2018.
[51] Ren S, Schmieg SJ, Koch CK, Qi G, Li W. Investigation of Ag-based low temperature NOx adsorbers. Catal Today 2015;258:378–85. https://doi.org/10.1016/j.cattod.2015.02.008.
[52] Li WB, Yang XF, Chen LF, Wang JA. Adsorption/desorption of NOx on MnO_2/ZrO_2 oxides prepared in reverse microemulsions. Catal Today 2010;148:75–80. https://doi.org/10.1016/j.cattod.2009.03.028.
[53] Theis JR, Lambert CK. An assessment of low temperature NOx adsorbers for cold-start NOx control on diesel engines. Catal Today 2015;258:367–77. https://doi.org/10.1016/j.cattod.2015.01.031.
[54] Sarshar Z, Zahedi-Niaki MH, Huang Q, Eić M, Kaliaguine S. MTW zeolites for reducing cold-start emissions of automotive exhaust. Appl Catal Environ 2009;87:37–45. https://doi.org/10.1016/j.apcatb.2008.08.025.
[55] Chang X, Lu G, Guo Y, Wang Y, Guo Y. A high effective adsorbent of NOx: preparation, characterization and performance of ca-beta zeolites. Microporous Mesoporous Mater 2013;165:113–20. https://doi.org/10.1016/j.micromeso.2012.07.040.
[56] Khivantsev K, Gao F, Kovarik L, Wang Y, Szanyi J. Molecular level understanding of how oxygen and carbon monoxide improve NOx storage in palladium/SSZ-13 passive NOx adsorbers: the role of NO+ and Pd(II)(CO)(NO) species. J Phys Chem C 2018;122:10820–7. https://doi.org/10.1021/acs.jpcc.8b01007.
[57] Kobayashi T, Yamada T, Kayano K. Study of NOx trap reaction by thermodynamic calculation. SAE Tech Pap 1997. https://doi.org/10.4271/970745.

[58] Daw S, Pihl J, Choi J, Partridge B, Toops T, Prikhodko V, et al. Joint development and coordination of emissions control data and models (CLEERS analysis and coordination). In: Oral present 2018 annual merit review; 2015.
[59] Corbos EC, Courtois X, Bion N, Marecot P, Duprez D. Impact of the support oxide and Ba loading on the sulfur resistance and regeneration of Pt/Ba/support catalysts. Appl Catal Environ 2008;80:62–71. https://doi.org/10.1016/j.apcatb.2007.10.009.
[60] Matsumoto S, Ikeda Y, Suzuki H, Ogai M, Miyoshi N. NO(x) storage-reduction catalyst for automotive exhaust with improved tolerance against sulfur poisoning. Appl Catal Environ 2000;25:115–24. https://doi.org/10.1016/S0926-3373(99)00124-1.
[61] Tsukamoto Y, Nishioka H, Imai D, Sobue Y, Takagi N, Tanaka T, et al. Development of new concept catalyst for low CO_2 emission diesel engine using NOx adsorption at low temperatures. SAE Tech Pap 2012. https://doi.org/10.4271/2012-01-0370.
[62] Schmeisser V, Weibel M, Sebastian Hernando L, Nova I, Tronconi E, Ruggeri MP. Cold start effect phenomena over zeolite SCR catalysts for exhaust gas after treatment. SAE Int J Commer Veh 2013;6:190–9. https://doi.org/10.4271/2013-01-1064.
[63] Lee J, Ryou YS, Cho SJ, Lee H, Kim CH, Kim DH. Investigation of the active sites and optimum Pd/Al of Pd/ZSM–5 passive NO adsorbers for the cold-start application: evidence of isolated-Pd species obtained after a high-temperature thermal treatment. Appl Catal Environ 2018;226:71–82. https://doi.org/10.1016/j.apcatb.2017.12.031.
[64] Murata Y, Morita T, Wada K, Ohno H. NOx trap three-way catalyst (N-TWC) concept: TWC with NOx adsorption properties at low temperatures for cold-start emission control. SAE Int J Fuels Lubr 2015;8:454–9. https://doi.org/10.4271/2015-01-1002.
[65] Ryou YS, Lee J, Cho SJ, Lee H, Kim CH, Kim DH. Activation of Pd/SSZ-13 catalyst by hydrothermal aging treatment in passive NO adsorption performance at low temperature for cold start application. Appl Catal Environ 2017;212:140–9. https://doi.org/10.1016/j.apcatb.2017.04.077.
[66] Ryou YS, Lee J, Kim Y, Hwang S, Lee H, Kim CH, et al. Effect of reduction treatments (H2 vs. CO) on the NO adsorption ability and the physicochemical properties of Pd/SSZ-13 passive NOx adsorber for cold start application. Appl Catal Gen 2019;569:28–34. https://doi.org/10.1016/j.apcata.2018.10.016.
[67] Ryou YS, Lee J, Lee H, Kim CH, Kim DH. Effect of various activation conditions on the low temperature NO adsorption performance of Pd/SSZ-13 passive NOx adsorber. Catal Today 2019;320:175–80. https://doi.org/10.1016/j.cattod.2017.11.030.
[68] Vu A, Luo J, Li J, Epling WS. Effects of CO on Pd/BEA passive NOx adsorbers. Catal Lett 2017;147:745–50. https://doi.org/10.1007/s10562-017-1976-x.
[69] Falconer JL, Schwarz JA. Temperature-programmed desorption and reaction: applications to supported catalysts. Catal Rev 1983;25:141–227. https://doi.org/10.1080/01614948308079666.
[70] Yakerson VI, Rozanov VV, Rubinshtein AM. Thermodesorption studies on adsorbate-adsorbent interaction. Surf Sci 1968;12:221–46. https://doi.org/10.1016/0039-6028(68)90125-8.

CHAPTER 10

Selective catalytic reduction for NO_x reduction

R. Vignesh and B. Ashok
Engine Testing Laboratory, School of Mechanical Engineering, Vellore Institute of Technology, Vellore, Tamil Nadu, India

10.1 Introduction

At first, the selective catalytic reduction (SCR) system was specially developed for heavy-duty diesel engines by the European motor industry to meet the Euro 5 and Euro 6 emission norms. Interestingly, the installation of a selective catalytic reduction system offers higher-level NO_x reduction compared to other reduction techniques and also indirectly decreases the fuel consumption rate. When the engine is configured with a high combustion temperature, this leads to low fuel consumption, low PM, and higher NO_x production. At this time, the SCR system removes the NO_x from the exhaust gas subsequently. This creates fuel consumption potential along with the legal emission limits. During the initial stage of SCR development, it was not suitable for light-duty vehicles because of the low exhaust gas temperature making it difficult to vaporize the SCR liquid. Recently, SCR technology has evolved to a great extent for the requirement of light-duty vehicles and passenger segment cars. In an SCR system, design is a concern. The use of a reductant agent is responsible for the reduction of NO_x to N_2, and also a catalyst is required to reduce the activation energy (speeding up the reaction) for the NO_x reduction. Typically, an aqueous solution of non-toxic urea is preferred as the reductant. For the catalyst, zeolite (200–400°C) or vanadium (300–450°C) is chosen concerning the exhaust gas temperature range of the engine.

10.2 Overview of SCR system and its components

The SCR system is employed in a modern-day diesel engine to satisfy the stringent emission norms. The SCR system uses a reductant agent along with a catalyst to reduce the NO_x emissions from the exhaust manifold of the engine. This system mainly comprises the sensing unit, controller,

NOx Emission Control Technologies in Stationary and Automotive Internal Combustion Engines
https://doi.org/10.1016/B978-0-12-823955-1.00010-3

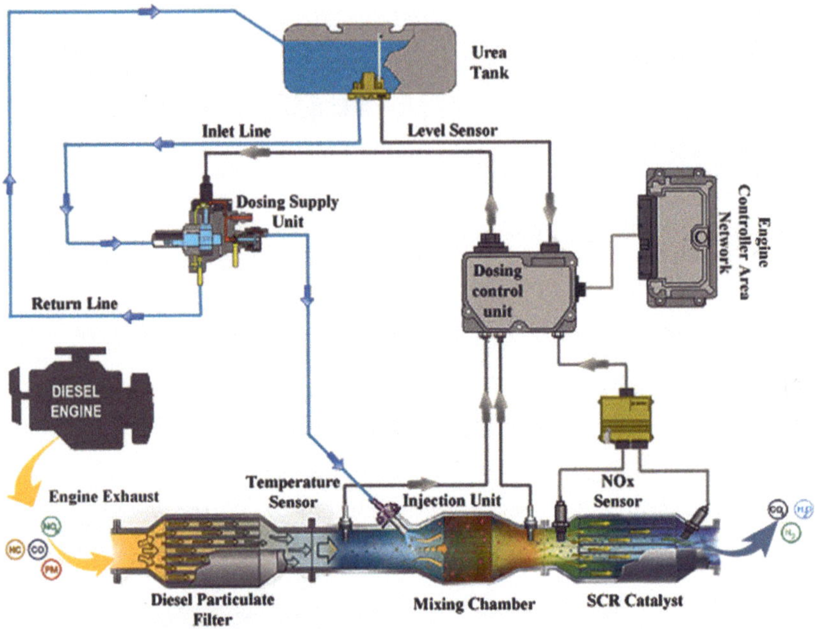

Fig. 10.1 Schematic layout of the NH_3-SCR system integrated with DPF.

reductant system, and dosing system, as shown in Fig. 10.1. The exhaust gas from the engine manifold, which consists of HC, CO, NO_x, PM, etc., after the diesel oxidation catalyst (DOC) enters into the mixing chamber through the diesel particulate filter (DPF) where the trapping of hard particles takes place. In the mixing chamber, the injection of the reductant agent through the dosing pump is actively controlled by the SCR controller. The amount of reductant injection is decided by the inputs from the NO_x and temperature sensor. Once the mixing of the exhaust gas and reductant agent in the mixing chamber takes place, the mixture enters into the SCR catalyst zone. Selective chemical reduction is performed over the surface of the catalyst, where nitrogen oxides (NO_x) are converted into nitrogen (N_2). As a byproduct, water and tiny amounts of carbon dioxide (CO_2) are then expelled through the vehicle tailpipe. The detailed working of the individual components of the SCR system, such as the reductant system, dosing system, sensing system, control unit, and SCR catalyst, are discussed in the following sections.

10.2.1 Reductant system

A reductant system is comprised of the reductant medium and the reductant storage tank. The reductant medium converts nitrogen oxides into harmless water and nitrogen through a reduction reaction with the aid of a catalyst. For wide suitability under different system conditions, the various reductants used in the SCR system include NH_3, CO, CH_4, C_2H_4, C_3H_8, C_3H_6, n-C_4H_{10}, 1-C_4H_8, 1,3-C_4H_6, ethanol, isobutanol, methanol, etc. The reductant medium is stored in the reductant tank. It consists of a temperature control unit that helps to maintain the range of reductant temperature required in the tank; otherwise, excessive high or low temperature will affect the properties of the reductant. It also contains a level sensor, which in turn is connected with the electronic control unit (ECU). For the continuous operation of the SCR emission control system, the reductant agent is essential and must be present in the tank all the time. When the sensor indicates a lower level of reductant agent in the tank, replenishment from an onboard tank is necessary. Without sufficient reductant agent, the controller prevents the vehicle from starting. Overall, the consumption rate of the reductant agent depends on the operation of the vehicle; an aggressive and heavy-load drive will have a higher consumption rate. Usually, the average reductant agent utilization is approximately 2%–4% of fuel consumption.

10.2.2 SCR catalyst

The control of NO_x is carried out using various reductants. However, in the SCR system, the role of a reductant catalyst is essential to enhance the rate of the reaction for the rapid conversion of emissions into elemental gases. Typically, the reductant catalysts are prepared in the form of the full or coated monolith. The preparation of full monolith catalysts could either be an extruded or corrugated catalytic material. The coated monolith is by the washcoating of the active catalyst phase over cordierite, or a metallic substrate. The catalysts favoring the reduction process are silver, aluminum, titanium, cerium, vanadium, tungsten, zeolites, etc. The catalyst performance depends on the right combination of reductant-catalyst pair, the design of the catalyst structure, and their effective activity in both high- and low-temperature conditions. Overall, catalysts are the heart of the SCR system. To better understand the various catalyst performances and their required optimal operating conditions, a detailed discussion about SCR catalysts is presented as a separate section.

10.2.3 Sensing system

The sensing system is comprised of a differential pressure sensor, a temperature sensor, an NO_x sensor, and an NH_3 sensor. The use of a differential pressure sensor is to measure the pressure drop in the DPF zone. The temperature and NO_x sensor data decide the amount of reductant supply required for NO_x reduction. For the precise operation of SCR, two NO_x sensors are essential. One is mounted after the DPF unit and gives a signal to the SCR control unit about the amount of NO_x present in the exhaust gas. The other is after the SCR catalyst zone. This sensor is mounted for diagnostic purposes of ensuring how much of the NO_x is converted to nitrogen. At last, especially in the NH_3-SCR system, the NH_3 sensor fitted above the SCR catalyst region is used to measure the quantity of reductant slippage, which occurs in the tailpipe.

10.2.4 SCR controller

The proper control strategy in controller can able to perform all the SCR components in optimum condition with the support of sensing unit. In the beginning of the SCR process, the control unit receives speed and load combination data from the engine CAN along with temperature data from upstream and downstream of the SCR catalytic converter, which were preprocessed to estimate the NO_x concentration level in the exhaust gas based on the dynamic operating condition of the engine load. The predicted NO_x level is compared with the measured NO_x sensor data in the control unit. To facilitate this feedback, a control strategy is essential for the control unit. So, for a precise operation, the SCR control unit must have combined control strategies (i.e., integration of feed-forward, feed-back, and open-loop). The schematic structure of the combined control strategy for the SCR control unit is depicted in Fig. 10.2.

10.2.5 Dosing system

In SCR, the dosing system consists of a supply unit and an injection unit. The primary function of the supply unit is dosing a precise amount of reductant required for the SCR reaction. The amount of reductant needed depends on the exhaust gas temperature, flow rate, NO_x level, amount of reductant stored in the catalyst, and space velocity. Based on these aspects, the metered amount of reductant quantity is allowed into the injection unit. For the NH3-SCR system, aqueous urea injection at a temperature below 180°C has difficulty in the conversion of aqueous urea to NH_3. If it is

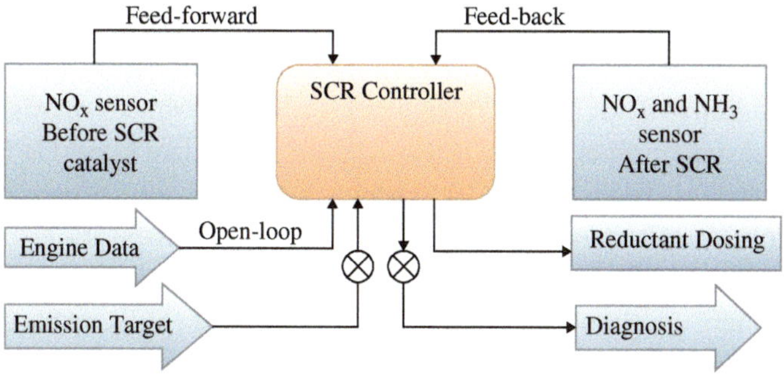

Fig. 10.2 Basic structure of the combined control strategy for the SCR control unit.

insufficient, the fraction of NO_x in the exhaust becomes unconverted, which leads to a drop in NO_x conversion efficiency, or if excessively injected, it will cause an undesirable reductant slippage in the tailpipe. In a urea-SCR system, the NH_3/NO_x ratio in the SCR system is about 1; a ratio higher than 1 leads to ammonia slip. In real time, maintaining the precise ratio is very difficult and also can't achieve the equivalent NO_x. conversion rate. Hence, a ratio between 0.8 and 1 is used to minimize the ammonia slip with satisfactory NO_x conversion. Fig. 10.3 represents the significant effect of the NH_3/NO_x ratio on ammonia slip and NO_x conversion.

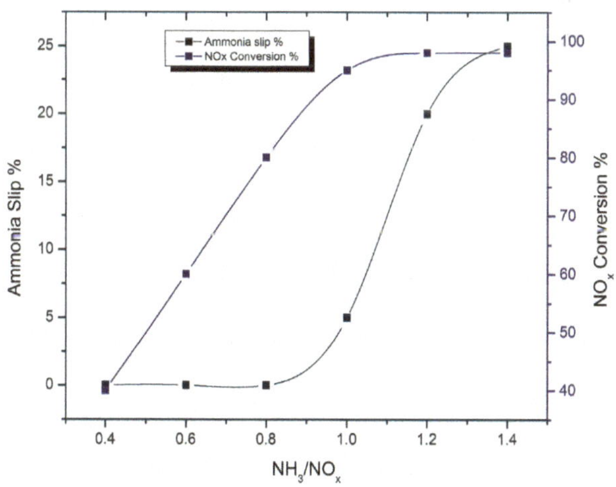

Fig. 10.3 Effect of NH_3/NO_x ratio on ammonia slip and NO_x conversion on a V_2O_5/TiO_2 SCR catalyst.

10.2.6 Emplacement of SCR system

Typically, for the complete removal of hazardous pollutants in the exhaust gas, the engine exhaust manifold has various after-treatment systems, including a diesel oxidation catalyst (DOC), a diesel particulate filter (DPF), and a selective catalytic reduction (SCR) system. The desired function of this system mainly correlates with the serial arrangement of its position. Fig. 10.4 shows various configurations of an SCR system position in the exhaust line. Possibly, there are three different emplacements of the SCR system: (i) The SCR is in place after the DOC and DPF; this configuration offers a few benefits such as enhancing the speed of the SCR reaction, the SCR catalyst remains cleaner, and increasing passive DPF regeneration. However, it has some drawbacks: the SCR efficiency is reduced during cold start, and the SCR catalyst is exposed to high temperatures during active regeneration. (ii) The SCR is placed between the DOC and DPF; here, the SCR catalyst is getaway from the DPF regeneration temperature and also the SCR catalyst nearby to engine increase in SCR catalyst efficiency during cold start because of rapid warm-up. Moreover, the DPF will act in response as an NH_3 slip for the SCR system. As a drawback, the NO_2 level after the SCR system is reduced, which diminishes the passive regeneration of DPF and causes frequent active regeneration. (iii) The SCR is followed by DOC and DPF. The advantage of this configuration is the SCR catalyst is not exposed to high-temperature regeneration and also supports cold start operation. As a downside, an inappropriate NO/NO_2 ratio reduces the speed of the SCR reaction and also reduces the passive regeneration of

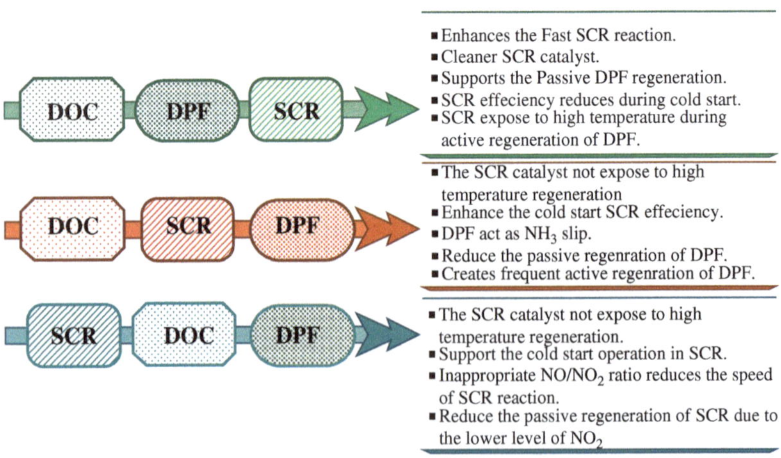

Fig. 10.4 Various configurations of SCR position in the exhaust manifold.

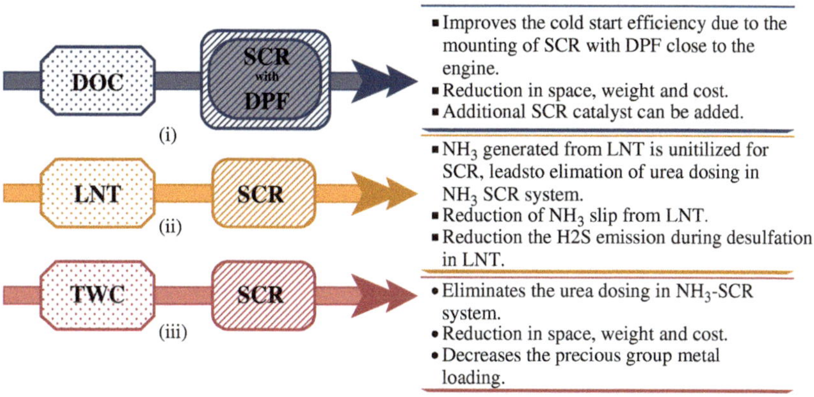

Fig. 10.5 An integrated approach of an SCR system with other after-treatment devices.

SCR due to the lower level of NO_2. Apart from these three configurations, an integrated approach of the SCR catalyst coated over DPF, a lean NO_x trap (LNT) placed before the SCR, and a three-way catalytic converter (TWC) followed by an SCR catalyst are shown in Fig. 10.5. The potential benefit of an integrated approach is that it discards the trade-off between the three configurations while also improving the SCR performance and reducing/removing the consumption of the urea reductant supply in the NH_3-SCR system.

10.3 De-NO_x chemistry in SCR

The SCR system comprises various chemical reactions. To design an efficient SCR system, it is essential to have proper knowledge about the NO_x reduction reaction. Also, one must understand the formation of desirable and undesirable chemical reactions in the whole process. The SCR process has the potential to convert NO_x to N_2 with the help of a reductant agent along with selective catalysts. The reduction technique is the basic process of SCR. Typically, the compounds containing —N are used as reductant agents, including aqueous ammonia or urea, anhydrous ammonia, and cyanuric acid. Moreover, —N free compounds such as hydrocarbons, carbon monoxide, hydrogen, and alcohol can also be used as effective reductant agents. The urea reduction reactions that have taken place are as follows Eqs. (10.1)–(10.3).

$$(NH_2)_2CO\,(aqueous) \xrightarrow[\text{Evaporation}]{} (NH_2)_2CO\,(molten) + H_2O\,(gas) \tag{10.1}$$

$$\text{[Water vaporization heat=2270 kJ/kg]}$$

$$(NH_2)_2CO(molten) \xrightarrow[\text{Thermolysis}]{} NH_3(gas) + HNCO(gas) \tag{10.2}$$

$$[\Delta H298 = +186\ KJ]$$

$$HNCO(gas) + H_2O \xrightarrow[\text{Hydrolysis}]{} NH_3(gas) + CO_2(gas) \tag{10.3}$$

$$[\Delta H298 = -96\ KJ]$$

While spraying atomized urea solution downstream of the exhaust gas, reaction (10.1) takes place. The dispersed solid urea particles in the vaporized water droplets begin to melt above 133°C. Further, by thermolysis, the thermal decomposition of molten $(NH_2)_2CO$ into NH_3 and isocyanic acid is seen in reaction (10.2). The decomposition efficiency depends on the exhaust gas temperature. The majority of the studies reported that thermolysis begins at temperatures of 143°C, 152°C, and 160°C. Next, in the SCR catalyst zone, the hydrolysis (10.3) of isocyanic acid in reaction with water yields NH_3 and CO_2. The endothermic reactions (10.1) and (10.2) occur in the decomposition tube and the exothermic reaction (10.3) in the SCR catalyst zone. As a whole, 1 mol of urea would decompose into 2 mol of NH_3 (stoichiometric molar ratio, about one mole of NH_3 is required to remove one mole of NO_x). It is noted that no complete conversion of the urea solution into NH_3 takes place before the entrance of the SCR catalyst zone. The formation of isocyanic acid as an intermediate delays the liberation of NH_3 due to the short decomposition space, and time makes a mixture of ammonia, isocyanic acid, and urea go into the SCR catalyst zone.

Further, the NH_3, thus formed through decomposition, gets stored in the SCR catalyst. The reaction on the SCR catalyst is shown in Eqs. (10.4)–(10.8), where the reduction of NO_x to nitrogen occurs. The SCR reaction rate is based on the availability of NH_3, NO, and NO_2 in the presence and absence of oxygen. Reaction (10.4) represents "the standard SCR reaction" mechanism, which contains 1:1 stochiometric NH_3: NO with the presence of oxygen, whereas reaction (10.5) happens in the absence of oxygen (neglected at lean combustion).

$$4NO + 4NH_3 + O_2 \rightarrow 4N_2 + 6H_2O\ [\text{standard SCR reaction}] \tag{10.4}$$

$$6NO + 4NH_3 \rightarrow 5N_2 + 6H_2O \tag{10.5}$$

During NO formation, the NO_2 also will form as side reaction. The presence of NO and NO_2 in the exhaust gas reacts with NH_3 with an equimolar quantity of NO and NO_2, the faster SCR reaction (10.6) takes place kinetically 10 times faster than the standard SRC reaction (10.4). Thus, to facilitate "the fast SCR reaction" if the NO_2 concentration is lower, the use of a Pt-based solid oxidation catalyst provides the required NO_2. If the NO_2: NO exceeds a 1:1 fraction, it triggers the slower SCR reaction (10.7). Finally, the rate of SCR reaction is in the order of (10.6) > (10.4) > (10.7).

$$NO + NO_2 + 2NH_3 \rightarrow 2N_2 + 3H_2O \text{ [Fast SCR reaction]} \tag{10.6}$$

$$6NO_2 + 8NH_3 \rightarrow 7N_2 + 12H_2O \text{ [Slow SCR reaction]} \tag{10.7}$$

$$2NH_3 + 2NO_2 \rightarrow NH_4NO_3 + N_2 + H_2O \tag{10.8}$$

$$NH_4NO_3 \rightarrow N_2O + 2H_2O \tag{10.9}$$

$$2NH_3 + 2O_2 \rightarrow N_2O + 3H_2O \tag{10.10}$$

During the SCR reaction, the reasons for the formation of undesirable side reactions (10.8)–(10.10) are: (i) at temperatures below 200°C, the excess NO_2 reacts with NH_3 and forms ammonium nitrate (10.8), that is, it further gets decomposed into N_2O as described in reaction (10.9). (ii) At higher temperatures above 400°C, N_2O forms because of NH3 oxidation (10.10). Besides this, due to the variable engine operating conditions, the precise supply of urea is difficult. This leads to the occurrence of undesirable NH_3 emissions in the exhaust. To remediate this NH_3 and N_2O formation, the use of a high activity ammonia slip catalyst (ASC) and the optimal design of the injection system over a wide range of gas compositions, flow rates, and temperatures is required in NH_3-SCR. Table 10.1 shows the general chemical reactions of NO_x decomposition for the different reductant fluids. To

Table 10.1 The general chemical reactions of NO_x decomposition for different reductant fluids.

Reductant	Reduction reaction
Hydrocarbon (Propane)	$2NO + C_3H_8 + 4O_2 \rightarrow N_2 + 3CO_2 + 4H_2O$
Carbon monoxide	$2NO + 2CO \rightarrow N_2 + 2CO_2$
Hydrogen	$2NO + 4H_2 + O_2 \rightarrow N_2 + 4H_2O$
Alcohol (isobutonal)	$(CH_3)_2CHCH_2OH + O_2 \rightarrow (CH_3)_2CHCHO + H_2O$ $x(CH_3)_2CHCHO + yNO_2 + (x-y)\ O_2 \rightarrow N_2 + 4xCO_2 + 4xH_2O$

achieve the highest rate of reaction, it is essential to have the right pairing of the catalyst with respect to the reductant solution, along with a robust and reliable control unit for the precise supply of the reductant solution.

10.4 An assortment of reductants used in SCR

In the NO_x reduction process, the reductant is essential during the dissociation of NO_x from the catalyst because, during NO_x dissociation, it oxidizes the catalyst surface, which impedes the reduction process further. The various reductants utilized in the SCR system, including ammonia, hydrocarbons, carbon monoxide, hydrogen, and alcohol, are discussed.

10.4.1 Ammonia reductant

The most common reductant medium used for NO_x abatement is NH_3, available in both anhydrous and aqueous forms. The onboard storage of urea solution requires a special storage tank to resist freezing below $-11°C$ and urea decomposition in the tank above 50°C. It even crystallizes easily, so a highly precise engineered storage system is required. Instead, as an alternative approach, adding additives such as ammonia formate to the urea solution can make it an appropriate property to the desired operating conditions without freezing or the production of harmful emissions. Even, the ammonia can be generated from ammonia precursor compounds such as ammonia carbonate, methanamide, and guanidinium formate which eliminates the storage issues faced by urea.

10.4.2 HC reductant

Besides NH_3 as a reductant in the SCR system, hydrocarbons (HC) could also be used as an SCR reductant to remove NO_x from oxygen-rich diesel engine exhaust gas. The HC-SCR system is an attractive option for automobile applications because it uses a reductant that is a similar species to that in the exhaust gas mixture. Moreover, it is a convenient and less-expensive process where unburned HCs in the exhaust can also be used as a reductant. Various HC reductants such as methane, isobutane, octane, propene, and propane are utilized by researchers. From various studies, it is seen that oxygenated HC such as ethers, alcohols, aldehydes, esters, and ketones showed a higher NO_x conversion compared to nonoxygenated HC. The NO_x conversion rates for the various HCs were found to be in the following order: ethers > alcohols > aldehydes > esters > ketones > propane. The

HC reductant is a practically feasible reductant that has no slippage problem, is noncorrosive, independent to the NO/NO_2 ratio, and has a simple storage tank compared to an ammonia reductant. However, it requires a powerful catalyst to execute the effective reduction of NO_x for the commercial implementation of the HC reductant.

10.4.3 Other reductants

In addition to ammonia and hydrocarbon reductants, other reductants such as H_2 and CO can also be used in the SCR system. Compared to conventional ammonia, the H_2-SCR system offers higher NO_x conversion at a low temperature below 200°C. The main drawback with the H_2 reductant is that during the reduction reaction, undesired N_2O compounds are formed, and the presence of H_2 in the exhaust gas is small. To prevent this, a high N_2 selective catalyst needs to be used, and a separate onboard reliable H_2 source tank is required in the H_2-SCR system. As the CO reductant is considered the most practical reductant because it is generally contained in the exhaust, no separate reductant is needed. Moreover, the CO-SCR system is most suitable for the HCCI engine, where the CO concentration in the exhaust is very high and can be used as a reductant for NO_x. So, the simultaneous elimination of CO and NO_x occurs. Here, the only drawback is CO gets easily reacted with O_2 to form CO_2; a very active catalyst is needed to restrict this. On the whole, based on the choice of reductant, it is very essential to select a suitable SCR catalyst.

10.5 An assortment of catalysts for various SCR

To assist NO_x decomposition, a catalyst is required that can reduce the activation energy. The selection of the SCR catalyst is centered on the type of reductant adopted by the SCR system. In this section, an attempt to explore the various SCR catalysts utilized for different reductants as well as their per formances are explained.

10.5.1 Catalyst for NH_3 SCR system

Many reports have been published on the NH3-SCR catalyst, most of which used a metal catalyst, metal oxide catalysts, and hybrid catalysts. In 1988, Bosch et al. [1] suggested a catalyst supported with metal oxides, which are the active catalytic component for the reduction of NO_x in NH3-SCR. The main reason for doping with the supporting metal oxide is generally that the catalyst below its lights-off temperature (the temperature at which the catalytic reaction begins) shows a drop in catalytic activity. Also, with a higher temperature, instead of reacting with NO_x, it oxidizes the NH_3, thus yielding a lower NO_x conversion efficiency. Added to this, catalyst poisoning is another trouble area due to the sulfur and HC content in the exhaust gas. So, to improve a wide temperature range of activity and resist catalyst poisoning, it is typically supported on oxides of alumina, zirconium, silica, tungsten, cerium, etc. Based on these aspects, a discussion on vanadium, zeolite-based catalysts, and various composite metal oxide catalysts utilized by various researchers in the NH_3-SCR system will be presented in this section.

10.5.1.1 Vanadium-based catalysts

Many studies reported a vanadium catalyst as one of the active SCR catalysts; it contains V_2O_5 as the active component. But when it is exposed to the exhaust of sulfur-bearing fuels, it oxidizes the SO_2 to SO_3, which has a higher affinity to react with NH_3 to form ammonia sulfate, leading to the blocking of catalyst pores. To address this undesired oxidation, the vanadium is supported on a TiO_2 anatase form, which is an effective additive among other ceria, zirconia, alumina, and silica metal oxide supports. The main reason why the TiO_2 anatase form is the best support is because it has a low turnover frequency, which leads to providing a weak and reversible sulfated condition on the surface for SO_2. Further, there was an attempt to synthesize V_2O_5 with mixed oxides ($Al_2O_3/SiO_2/TiO_2$) as a supporter, but the results are unexpected, as the coexistence of hetero atoms lowers the titania crystallinity. For active vanadia species, crystalline titania is essential; hence, mixed oxides attain lower activity than the titania-supported catalyst. To alleviate the activity of V_2O_5/TiO_2, the promoting effect of V_2O_5 on sulfated TiO_2 by wet impregnation increases the acidity of Bronsted acid sites more than Lewis acid sites on the surface and provides greater conversion of NO—around 98% at low-temperature SCR reaction. But at a high temperature above 500°C, sulfate starts to decompose by the metal oxides present in titania. Similarly, V_2O_5 on sulfated Ti-pillared clay offers lower NO conversion at higher temperatures. Moreover, the metal oxide composition also influences the catalytic activity when there is an increase in V_2O_5 concentration from 0.78% to 1.4%, which triggers the catalytic reaction. As Fig. 10.6 illustrates, increases in the vanadia content of the V_2O_5/TiO_2 catalyst increase the active sites for ammonia adsorption. It visualizes the interaction of ammonia occurrences on the V—OH and V=O of the

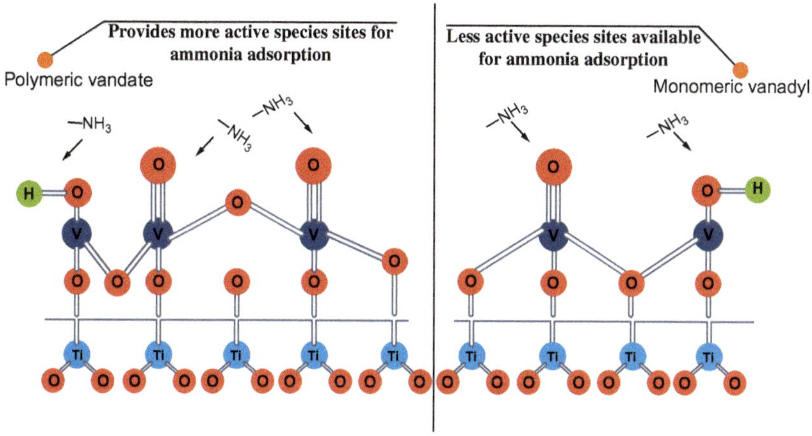

Fig. 10.6 Schematic structure model for the active sites of the V_2O_5/TiO_2 catalyst.

catalyst. Also Fig. 10.6 pronounces, the formation of high NO conversion on polymeric vanadate is more than the monomeric vanadyl. When the concentration is increased above 3%, a complete transformation of the TiO_2 anatase phase to the rutile phase at a high temperature above 400°C causes the undesired formation of N_2O as well as a reduction in the selectivity and thermal stability of the catalyst.

To improve thermal stability, quite active and selective oxides of tungsten (WO_3) and molybdenum (MoO_3) were tried with the support of TiO_2. The MoO_3/TiO_2 is highly reactive and has low selectivity, whereas WO_3/TiO_2 is vice versa but lower than V_2O_5/TiO_2. Thus, WO_3 or MoO_3 is added with V_2O_5 supported on TiO_2. As Fig. 10.7 shows, the ternary catalysts V_2O_5-WO_3/TiO2 and V_2O_5-MoO_3/ TiO_2 are both highly active and selective compared to V_2O_5/TiO_2. Further, compared to the ternary catalyst, V_2O_5-MoO_3/TiO_2 is more active than V_2O_5-WO_3/TiO_2. Moreover, the role of tungsten does not influence the vanadia catalytic activity; it only supports transporting the NH_3 to the active sites of vanadia. The addition of 9 wt.% WO_3 in V_2O_5/TiO_2 enhances the thermal durability up to 900°C by restricting the conversion of the TiO_2 phase change from the anatase to the rutile phase. It also inhibits TiO_2 sintering, lowers the surface area loss of anatase, and increases the Lewis sites of the catalyst. Another reason to add WO_3 and MoO_3 in V_2O_5/TiO_2 is because this covers the basic site of the TiO_2 surface for SO_2 oxidation and tends to reduce sulfation. Altogether, the WO_3 and MoO_3 in V_2O_5/ TiO_2 act as a promoter, stabilizer, and an SO_2 oxidation inhibitor in the NH_3 SCR reaction. The

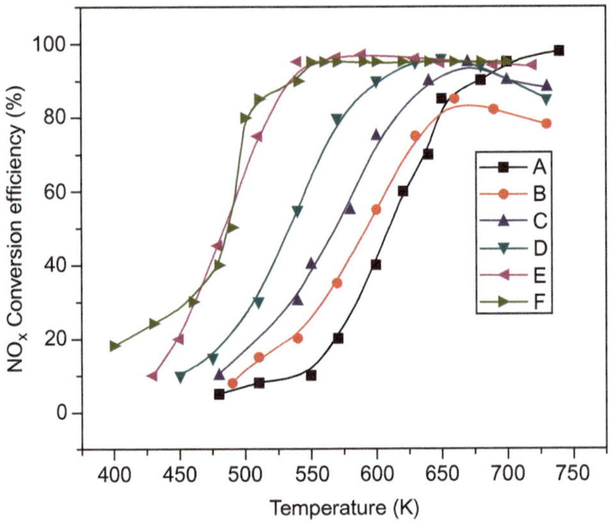

Fig. 10.7 Catalyst (A) 9%WO_3/TiO_2; (B) 0.78% V_2O_5/TiO_2; (C) 1.4% V_2O_5/TiO_2; (D) 0.78% V_2O_5—9% WO_3/TiO_2; (E) 1.4% V_2O_5—9% WO_3/TiO_2; (F) 1.5% V_2O_5—6% MoO_3/TiO_2 NO conversion efficiency for different temperature conditions.

performances of the various vanadium-based catalysts for the NH_3 reductant SCR system utilized by researchers are summarized in Table 10.2. Overall, the vanadium-based catalyst provides high sensitivity to alkali metal and HC poisoning at high-temperature exposure. But robust resistance to SO_2 and oil poisoning offers a way to recommend vanadium-based catalysts where there is a risk of exposure to sulfur fuels.

10.5.1.2 Zeolite-based catalysts

Zeolite -based catalysts are attractive toward high activity in wide operation temperature and high resistance to HC poisoning compared to vanadium-based catalysts. These catalysts are hydrothermally stable and have high activity at high temperature compared to the vanadium-based catalyst. Moreover, upon the installation of DPF with the SCR system, the zeolite catalyst is the best choice because, during DPF regeneration, the superior thermal durability of zeolite resists high-temperature exposure. Most commonly, the zeolites are characterized based on their morphological natures and compositions. Some used prominently in the SCR system are BEA, ZSM-5, MOR, FER, and CHA. Further, these catalysts are promoted with other metals to enhance their activity and NH_3 storage capacity. The performances of the various zeolite-based catalysts for the NH_3 reductant SCR system utilized by researchers is summarized in Table 10.3. Cu and Fe are the

Table 10.2 Performances of various vanadium-based catalysts for the NH_3-SCR system.

References	Catalyst	Catalyst preparation			Reaction conditions					De-NO_x performance		
		Method	Calcination temperature (°C) in air	Calcination time (h)	NH_3	NO_x	O_2	SV (h^{-1})	Reaction temperature (°C)	NO_x conversion	N_2 selectivity	N_2O selectivity
[2]	5.5 wt.% V_2O_5/Al_2O_3	Impregnation	500	4	600–1000 ppm	1000 ppm	4 vol%	102,000	180	78%	–	4 ppm
[3]	10 wt.% V_2O_5-La_2O_3/Al_2O_3	Impregnation	500	6	1000 ppm	1000 ppm	1 vol%	–	330	87%	–	–
[4]	6 wt.% MoO_3–12 wt.% V_2O_5/Al_2O_3	Impregnation	500	3	500 ppm	500 ppm	10 vol%	100,000	400	98%	–	–
[5]	6.5 wt.% V_2O_5-SiO_2	Sol-gel method	500	6	650 ppm	650 ppm	3 vol%	39,800	250	75%	–	–
[6]	2.35 wt.% V_2O_5/CNT (20–40 nm)	Incipient wetness technique	350	2	800 ppm	800 ppm	5 vol%	35,000	190	82%	–	–
[7]	3 wt.% V_2O_5/Carbon-coated monolith	Equilibrium adsorption	400 in Ar	–	800 ppm	700 ppm	3 vol%	17,000	180	78.60%	–	–
[8]	30 wt.% FeVOP/TiO_2	–	450°C in He	12	700 ppm	700 ppm	27,000 ppm	–	400	90%	–	15% at 450°C
[9]	2 wt.% V_2O_5–7 wt.% WO_3/TiO_2-SiO_2-SO_4^{2-}	Coprecipitation	550	5	300 ppm	300 ppm	4 vol%	12,000	380	92%	–	38 ppm
[10]	1 wt.% V_2O_5-SO_4^{2-}/TiO_2	Sol-gel method	500	12 h	5000 ppm	5000 ppm	5 vol%	13,500	450	99%	–	–
[11]	20 wt.% CeO_2-V_2O_5/TiO_2-ZrO_2	Wet impregnation	500	4	0.08 vol%	0.08 vol%	5 vol%	24,000	450	99%	–	–
[12]	1 wt.% V_2O_5–8 wt.% CeO_2–10 wt.% WO_3/TiO_2	Sol-gel	500	3	500 ppm	500 ppm	5 vol%	150,000	300	90%	97%	–

Table 10.3 Performances of various zeolite-based catalysts for the NH_3-SCR system.

		Catalyst preparation			Reaction conditions					DeNO$_x$ performance		
Reference	Catalyst	Method	Calcination temperature (°C)	Calcination time (h)	NH_3	NO_x	O_2	SV (h^{-1})	Reaction temperature (°C)	NO_x conversion	N_2 selectivity	N_2O selectivity
[13]	1 wt.% Fe-zeolite socony mobil–5	Ion-exchanging method	500	3	1000 ppm	1000 ppm	2 vol%	460,000	500	98%	–	–
[14]	3.1 wt.% Fe-mordenite	Ion-exchanging method	500	6	1000 ppm	1000 ppm	2 vol%	460,000	450	93%	–	–
[15]	1 wt.% Cu- 4 wt.%Fe-zeolite socony mobil–5	Subsequent ion-exchanged	500	4	500 ppm	500 ppm	10 vol%	–	450	90%	97%	–
[16]	Co-Cu-zeolite socony mobil–5	–	550	4	1000 ppm	1000 ppm	5 vol%	12,000	300	89%	–	–
[16]	Mn-Cu-zeolite socony mobil–5	–	550	4	1000 ppm	1000 ppm	5 vol%	12,000	300	83%	–	–
[17]	5 wt.% Cu-beta-1.5 wt.% Ce	Liquid-phase ion exchange	500	2	500 ppm	500 ppm	10 vol%	26,000	450	80%	–	–
[18]	Cu-zeolite socony mobil–5-2 wt.% ZrO_2	Liquid-phase ion exchange	500	2	500 ppm	500 ppm	10 vol%	28,000	400	80%	–	–

most promising metals for zeolite. The functionality of copper ion exchanged with the ZSM-5 catalyst relying on the SCR system under the reactant mixture (0.6% NO, 0.6% NH_3, 3.3%O_2) provides above 90% conversion to N_2 over the wide operating temperature of 200–500°C. To understand the conversion of NO to N_2, the reaction mechanism of NO_x with NH_3 over the Cu-zeolite catalyst is schematically expressed in Fig. 10.8. The active catalyst for the decomposition of nitrogen oxide depends on the ionic exchange level of the Cu^{2+} species on ZSM-5 and their Si/Al ratio. The Cu^{2+} ions are dispersed as counter-cations on the zeolite cation exchange sites. This provides a higher rate of specific activity. Also, another reason to achieve high activity is by providing a lower Si/Al ratio in the zeolite because a higher aluminum concentration gives more Bronsted acid sites as well as provides a way to introduce more Cu^{2+} ions at an exchangeable site. But these overloaded Cu^{2+} ions aggregate the

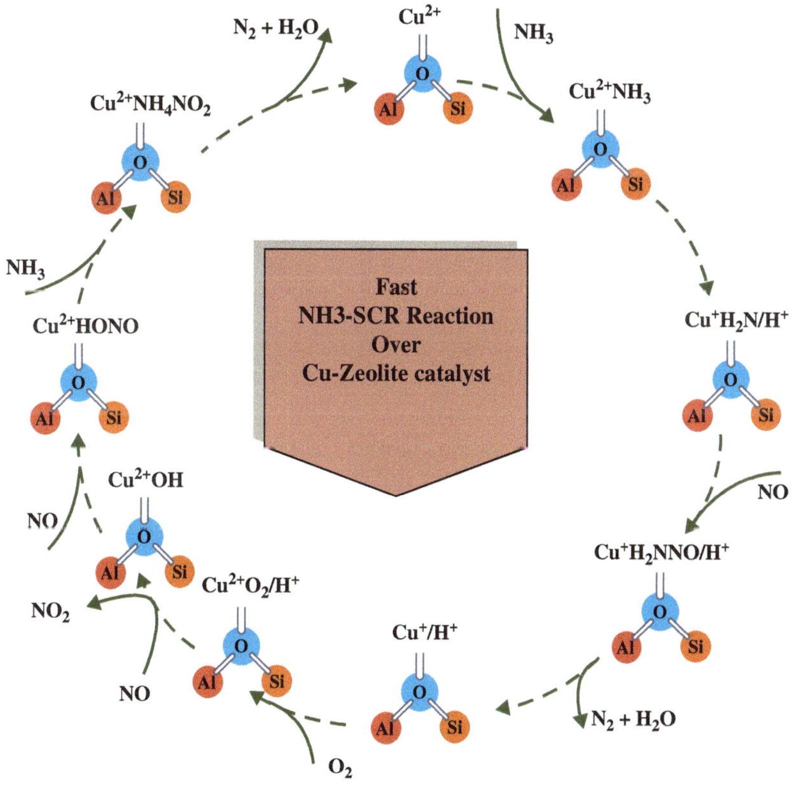

Fig. 10.8 SCR reaction mechanism of NO_x with NH_3 over the Cu-zeolite catalyst.

formation of CuO, which blocks zeolite pores, resulting in the deterioration of specific activity. Therefore, the Cu^{2+} ion concentration, being the crucial factor, drives the specific activity of Cu-ZSM-5. Moreover, the influence of the gas composition of O_2, H_2O, NH_3 on Cu-ZSM-5 also governs the catalytic activity. The increase in O_2 concentration offers higher NO conversion at lower temperatures for two reasons. First, O_2 reacts with NO to form active nitrites and nitrate intermediate species. Second, it promotes the reuse of Cu^{2+} ions by oxidizing the Cu^{+}. Whereas at the higher temperature, the presence of O_2 oxidizes more NH_3, resulting in a reduction of NO conversion. Unlike O_2, the presence of water in the feed supports NO conversion during both high and low temperatures. Further, the investigation of a higher NH_3 concentration with a water feed study reveals that the presence of water restricts higher NH_3 adsorption, which lowers the blocking effect of NH_3 on Cu-ZSM-5 pores and consequently results in higher NO_x reduction. Usually, in IC engines, some amount of unburned HC will exist in the exhaust, which would be adsorbed over the Cu-ZSM-5 catalyst and lead to the blockage of active sites by undesired coke formation, particularly at low temperature. This HC poisoning purely depends on the zeolite pore structure in ZSM-5 from ESR, and TGA analysis recognized that the presence of large pores and cages forms a huge deposition of HC over the active sites of zeolite. To address this issue, a comparison of large-pore ZSM-5 with various zeolite structures of BEA, FER, MOR, ERI, and CHA was studied. Among these, the CHA structure (Cu-SAPO-34 and Cu-SSZ-13) possesses small pores and cages, which act as an appropriate structure to resist HC poisoning and enhance the hydrothermal stability. Moreover, the small-pore Cu-doped CHA structure exhibits excellent activity and selectivity compared to the Cu-BEA and Cu-ZSM-5 catalysts. As a whole, the Cu-zeolite catalyst significantly provides the highest NO conversion up to 300°C and also is extremely tolerant to high temperature exposure, which is helpful for the DPF integrated SCR system.

The zeolite exchanged with Fe is specifically for the high-temperature SCR reaction. It provides higher SCR activity above 350°C and even possesses superior durability to high-temperature exposure. Also, the Fe-zeolite during the entire NO conversion no detection of undesired N_2O formation was observed in the temperature range from 300 to 600°C while Cu-zeolite does [13]. Another difference between Cu and Fe zeolite is the NH_3 adsorption behavior. The amount of NH_3 stored over the Fe-zeolite is less than the Cu-zeolite, which would be the main reason for the lower activity of Fe-zeolite at lower-temperature conditions. But an investigation of the presence of NO_2 shows that Fe-zeolite provides higher NO conversion compared to the Cu-zeolite below 250°C under an optimum NO_2/NO_x ratio of 0.5 [19]. Further, like Cu-zeolite, a variation in the Fe activity was observed for different zeolite structures. The result [14] reveals that an increase in activity was observed in the following order: Fe-CHA < Fe-BEA < Fe-FER < Fe-HEU < Fe-MOR. Moreover, the Fe-zeolite catalyst is resistant to sulfur poisoning and hydrothermally stable up to 650°C, but it shows moderate HC poisoning. Overall, from the above discussion and Fig. 10.9, it is understood that the difference between the performance these two zeolites mainly depends on their respective operating conditions and the zeolite structure [20]. The Cu-zeolite typically provides the highest NO_x conversion at a lower temperature (<300°C) while Fe-zeolite is at a higher temperature (>350°C). Therefore, a combined Cu-zeolite and Fe-zeolite as a separate bed in an SCR catalyst zone would provide a broad range of NO_x conversion temperatures. But this requires a proper bed design to achieve a reliable NO_x reduction performance along with superior characteristics such as high thermal stability, less NH_3 slip, and desirable byproduct formation. Further, to remove the complexity in design, another initiative tried to develop a single zeolite catalyst with both Cu and Fe cations. The bimetallic catalyst, Cu and Fe co-exchanged with zeolite NO_x reduction performance depend on Cu/Fe ratio and their

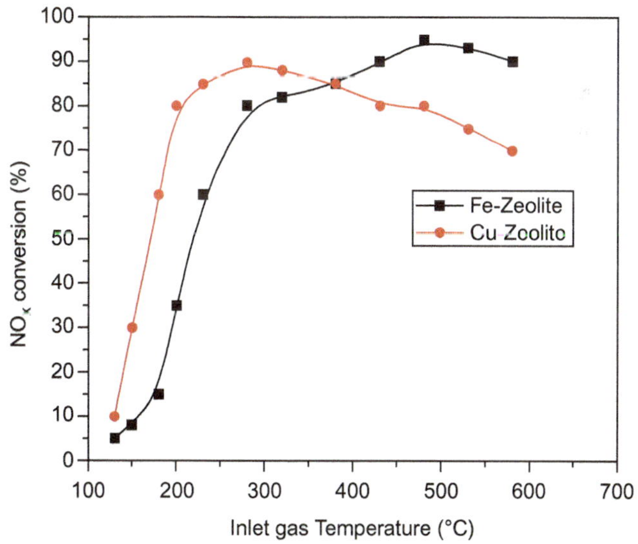

Fig. 10.9 NO_x conversion over Cu-zeolite and Fe-zeolite catalysts as a function of operating temperature.

preparation method. From the result [15], the catalyst prepared by the subsequent exchange of Fe-zeolite with copper or Cu-zeolite with iron yields better performance than the monometallic zeolite. Adding copper induces greater reducibility of Fe in the Cu-Fe-zeolite catalyst and also increases stronger acid sites. There are several combinations of bimetallic zeolite catalyst that were studied, including Co—Cu, Mn—Cu, Cr—Cu, Fe—Cu, Cu—Ce, and Cu-ZrO_2. Among these, Fe—Cu showed maximum activity [16–18]. Also, the high activity Cu—Fe pair was tested under different zeolite structures. The Cu-Fe-BEA showed the highest activity and N_2 selectivity among MOR, ZSM, and FER zeolite structures. This is due to the presence of a large pore structure over the BEA surface that facilitates the easy access of NO and NH_3 reagents to the active sites. But to resist HC poisoning, small pores are essential, as stated earlier. Hence, to better understand the characteristics of Cu-Fe-zeolite and to develop a commercial Cu-Fe-zeolite SCR catalyst, the effect of pore structure in the aspect of the activity, poisoning effect, and durability is critically important to find the optimum pore structure. This still would be the gap present over the Cu-Fe-zeolite catalyst. Overall, the combined Cu and Fe zeolite catalyst compared to the vanadium-based catalyst has the ability to operate in wide operating temperatures as well as higher selectivity toward N_2 along with lower N_2O formation and superior thermal durability.

10.5.1.3 Various composite metal oxide catalysts

Among the traditional vanadium- and zeolite-based catalysts, some of the earth metal oxide catalysts, particularly Mn, Ce, Mo, Co, Fe, and Cu, were doped or supported with other metal oxides to prepare a multimetal oxide catalyst. This would improve the catalytic activity and poisoning resistance while increasing the thermal stability for a wide range of operating temperature. These attribute enhancements on catalysts were achieved through the synergistic effects offered by the multimetal oxide catalyst. In the NH_3-SCR reaction, the injected ammonia in the form of NH_4^+ and coordinated NH_3 is typically adsorbed on the supporting species of SCR catalyst, which could react with the monodentate nitrate species attached to the active metal to form intermediate species. Further, it would react with gaseous or weakly adsorbed NO to form N_2. This implies that a proper supporting metal oxide is essential to provide active sites for NH_3 adsorption in the SCR catalyst. Moreover, the addition or doping of active metal oxides with base metal specifically enhances the activity and stability of the SCR catalyst. Table 10.4 summarizes the discussion of base metals with different metal

Table 10.4 Significance of various supporting or doping species on Ce- and Mn-based NH_3-SCR catalysts.

Base catalyst	Support/ doped metals	Significance
Mn	Ce-doped	It promotes more absorption of NO_3^- over the catalyst surface. t enhances the N_2 selectivity and promotes oxidation of NO to NO_2. For the easy decomposition of formed $(NH_4)_2SO_4$, Ce reduces the bonding between ammonia and sulfate ions. It provides more Lewis acid sites for the low-temperature SCR cycle. It enhances the surface characteristics.
	Ce-supported	Increase in the Mn^{4+}/Mn^{3+} ratio by promoting the large dispersion of Mn^{4+} species over the surface. It enhances the chemisorbed oxygen on the surface.
	Ti-supported	It increases surface acidity. It provides more Lewis acid sites. It offers a larger surface area and higher redox capability while resisting SO_2 poisoning.
	Al-supported	It provides a large surface area, pore-volume, and more hydroxyl group than Ti-support. It provides more Bronsted acid sites. The acidic nature support shows higher surface acidity. Promotes a high dispersion of manganese oxide in an amorphous state.
	Carbon-supported	Increases the BET surface area three times more than the Ti-supported. Enhances the adsorption and oxygen storage capabilities. Reduces the stability of formed $(NH_4)_2SO_4$.
	Fe-doped	On the surface, it provides better dispersion of Mn oxides. It enhances the oxygen adsorbed on the surface and provides higher surface acidity. Formation of an amorphous structure and facilitates the increase in catalyst oxidation-reduction capability on the surface. Improves the H_2O and SO_2 poisoning resistance. Strengthens both Lewis acid sites and Bronsted acid sites.

Continued

Table 10.4 Significance of various supporting or doping species on Ce- and Mn-based NH_3-SCR catalysts—cont'd

Base catalyst	Support/ doped metals	Significance
Ce	Cu-doped	It provides more oxygen storage capability. Enhances the surface redox capability at low temperatures. Greater enhancement of SO_2 resistance. Restricts the formation of cerium sulfate by interacting with SO_2.
	Ti-supported	It provides a better redox property. It enhances the good dispersion ofbase metal species. Increases surface acidity and thermal stability while resisting SO_2 poisoning.
	W-doped	It provides strong interaction among the other metal oxides. It decreases the thermal stability of undesired cerium sulfate formation. It generates more Bronsted acid sites. Inhibits the occurrence of surface oxygen reduction.
	Mo-doped	It reduces the growth of CeO_2 crystallite particle size. It provides improved oxidation-reduction ability and increases the surface acid sites. Promotes more Bronsted acid sites and strengthens the Lewis acid sites. Improves the chemisorbed oxygen and oxygen storage ability at the surface. Improves the resistance of H_2O and SO_2 poisoning. Inhibit more stable nitrate species adsorption and provides more ammonia adsorption.
	Zr-doped	It reduces the lights-off temperature. Improves the N_2 selectivity and enhances the oxidation NO to NO_2. It provides electronic interaction, thus inducing the transfer of Ce4+ to Ce3+. Promotes more acid Lewis acid sites and strengthens the Bronsted acid sites. Acts as an inhibitor for crystal transformation. It enhances the surface area, oxygen storage capability, active surface oxygen species, and sulfur resistance.

oxides acting as supporting or doping species that influence the performance of various SCR catalysts for the NH_3-SCR system. It is well known that noble metal catalysts have excellent activity at low temperatures, but they are too expensive. So, the development of transitional metal oxide catalysts can be used as an alternative for low-temperature SCR catalysts.

10.5.2 Catalyst for HC-SCR system

The HC-SCR system is an attractive option for automobile applications because it uses a reductant, which is one of the similar species present in the exhaust gas mixture. To accomplish the reduction mechanism of NO_x from the lean-burn engine exhaust gas effectively, the use of powerful catalysts is essential. The performances of various catalysts utilized by researchers in the HC-SCR system are summarized in Table 10.5. Iwamoto et al. [29] first developed a Cu-ZSM-5 catalyst for HC-SCR. Added to this, numerous active transition and noble metal catalysts are Co-ZSM-5 [30,31], Fe-ZSM-5 [32,33], Pt/Al [34,35], Pt/SiO_2 [36], Pt-ZSM-5 [37,38], and Rh-ZSM-5 [38]. Various studies show that although zeolite- supported HC-SCR catalysts are very active, the exposure of H_2O and SO_2 in the exhaust gas stream modifies the zeolite structure, which leads to severe deactivation [39,40]. Metal oxide-supported HC-SCR catalysts such as rare earth oxides or alumina-supported Pt, Ga, In, and Pd catalysts exhibit high hydrothermal stability. Especially, Pt-based catalysts have high activity and stability in low temperatures but are significantly more selective to undesired N_2O and limited by a narrow operating temperature. In Pd, Mn, and Co, the use of sulfated alumina or zirconia support alleviates both the stability and activity of the catalyst. Apart from these catalysts, the Ag/Al combination facilitates good activity and moderate resistance to H_2O and SO_2. Moreover, the Ag/Al-based catalyst compared to the HC reductant showed higher NO_x reduction and low sensitivity to H_2O and SO_2 in the oxygenated HC reductant, especially in ethanol and isobutanol. Ag/Al_2O_3 with an ethanol reductant provides a wide operating range of 310–610°C with above 90% NO_x conversion. Though Ag/Al_2O_3 with oxygenated HC has good activity and resistivity, undesired nitrogen-containing byproducts such as N_2O, NH_3, CH_3CN, and HCN are formed. Tatsuo Miyadera [41] employed a three-component composite catalyst Ag/Al_2O_3 + $CuSO_4$/TiO_2 + Pt/TiO_2 that was stacked in series. This composite catalyst effectively reduced NO_x to N_2 and the undesired N-containing byproduct compounds, even in the presence of water. However, this type of catalyst

Table 10.5 Performances of various catalysts for the HC-SCR system.

		Catalyst preparation			Reaction conditions					DeNO$_x$ performance	
Reference	Catalyst	Method	Calcination temperature (°C)	Calcination time (h)	Reductant type	NO$_x$	O$_2$	SV (h^{-1})	Reaction temperature (°C)	NO$_x$ conversion	N$_2$ selectivity
[21]	Cu-ZSM-5	–	–	–	C_2H_4 (250 ppm)	1000 ppm	2 vol%	–	350	70%	–
[22]	Co-ZSM-5	Ion-exchanging method	500	1	CH_4 (1000 ppm)	820 ppm	2.5 vol%	7500	400	65%	–
[23]	Fe-ZSM-5	Ion-exchanging method	–	–	C_4H_{10} (200 ppm)	2000 ppm	3 vol%	42,000	500	82%	–
[24]	Pt/Al_2O_3	–	–	–	C_3H_8(1000 ppm)	1000 ppm	10 vol%	–	400	98%	–
[25]	Pt/SiO_2	Impregnation method	500	16	CH_4 (1000 ppm)	1000 ppm	–	–	500	70%	–
[26]	Pt-ZSM-5	–	–	–	C_2H_4 (1000 ppm)	1000 ppm	2 vol%	72,000	212	54%	–
[27]	Rh-ZSM-5	Ion-exchanging method	500	1	C_2H_6 (1000 ppm)	1000 ppm	2.5 vol%	60,000	450	50%	–
[28]	Ce-Ag-ZSM-5	Ion-exchanging method	500	2	CH_4 (0.5% vol%)	0.5% vol %	2.5 vol%	7500	500	80%	–

stacking occupies more space, has a huge pressure drop, and also is not cost-effective. As a whole, it is understood that the development of a bimetallic catalyst would facilitate the required catalytic characteristics for the commercial implementation of the HC-SCR system in real-life applications. However, extensive research in bimetallic HC-SCR catalysts has not been done.

10.5.3 Catalyst for H_2-SCR system

In the H_2-SCR system, most researchers utilized Pt group metals as the active catalytic component. Among the Pt group metals (Pt, Pd, Rh, and Ir), the catalytic activity of Pt has the highest NO reduction. Though the Pt-based catalyst has the highest NO conversion, a large amount of undesirable N_2O was formed. The support compound on the Pt catalyst plays a major role in N_2/N_2O selectivity. So, the investigation of Pt catalyst effectiveness on various supports was found to be in the order of zeolite > SiO_2 > Al_2O_3. The metal oxide support on a Pt catalyst should have strong acid sites to adsorb NH^{4+} species, leading to more N_2 formation than N_2O. Moreover, during the NO conversion over the Pt catalyst, the byproduct NH_3 is formed to remove this zeolite-supported catalyst, especially Pt/H-ZSM-5, as it has the potential to convert NH_3 to N_2. Even the support compound for the Pt catalyst was tried with rare earth oxides. Only Pt/CeO_2 provided good NO_x reduction, as the remaining oxides possess high solid basicity, which reduces the NO_x reduction activity. The Pt loaded with multioxide support such as Pt/ La-Sr-Ce-FeO_3 [42], Pt/La-Ce-Mn-O [43], or Pt/MgO-CeO_2 [44] alleviates more activity than the single oxide support. Like the Pt group metal, the Pd catalyst performance was also influenced by the support species: TiO_2 support facilitates two reduction peaks, one at 100°C and the other at 300°C, with high NO conversion and N_2 selectivity compared to the Pd/Al_2O_3 catalyst. The reason for the second reduction peak is the reaction between in situ generated NO_2 and H_2 [45]. While for MFI zeolite support, provides 70% NO_x conversion to N_2 at 100°C with no other byproduct like N_2O formation and the bimetallic support La-$CaCO_3$ excite the Pd catalyst to achieve 100% NO conversion at 150°C with 78% N_2 selectivity. Moreover, an automobile exhaust temperature will be in the range of 150–400°C. Hence, to increase the wide operating temperature of the Pd/TiO_2-Al_2O_3 catalyst, the addition of V_2O_5 would increase the range of operating temperature up to 250°C as well as the NO conversion. Therefore, from all the research results, it is clear that choosing the right supportive species for the Pt group catalysts is very essential to synthesize an effective H2-SCR catalyst. The performances of various catalysts utilized by researchers in the H2-SCR system are summarized in Table 10.6.

10.5.4 Catalyst for CO-SCR system

For the CO-SCR system, the first catalyst was developed by Tauster et al. [49], who measured the performance of Ir/Al_2O_3; NO conversion of 90% was obtained at 400°C. After the first investigation on the Ir catalyst, much research has been done on Ir catalysts with various material supports such as Ir/silicate, Ir/SiO_2, Ir/TiO_2, and Ir/ZSM-5.IBetween Ir/SiO_2 and Ir/silicate, the Ir/SiO_2 showed higher NO conversion in the absence of SO_2. The presence of SO_2 in the feed decreases the Ir/SiO_2 activity but the activity of Ir/silicate was not influenced by the SO_2, although Ir/SiO_2 has the potential to restore its activity when the elimination of SO_2 in the feed occurs. Even Ir/SiO_2 with the presence of SO_2 provided the maximum NO_x conversion compared to Ir/TiO_2, Ir/Al_2O_3, and Ir/ZSM-5, which has a conversion efficiency below 15%. In addition, Ir supported with high oxidation number metal oxides such as WO_3, Ta_2O_5, and Nb_2O_5 also plays an important role in NO_x reduction using CO-SCR. Some researchers have also tried noble metal catalysts such as Pt/Al_2O_3 [50], Pt/SiO_2 [46], Pt/TiO_2 [47], Pt/WO_3/CeZrO [21], and Rh/Na-Beta zeolite [48]. Unfortunately, it is complicated to compare all the catalysts with one another under the same reaction conditions, but their results seem to indicate that Ir-based catalysts are more highly active than the other metal-based catalysts. The reason for this high activity is the presence of two NO adsorption sites on the Ir surface that facilitate this specific property. Hence, Ir-based catalysts, especially Ir/SiO_2, would be the most favorable catalysts for the CO-SCR system. The performances of various catalysts utilized by researchers in the CO-SCR system are summarized in Table 10.6.

10.6 SCR controller

The control of reductant fluid dosing dominates the overall performance of the SCR system. Typically, the operation of the control unit is done with the help of both hardware and software modules. To design and develop the dynamic SCR controller, the effects of the SCR system happening in different time scales have to be considered. The reductant dosing, sensor measurement, engine operation, exhaust gas transport, NH_3 adsorption, and their reactions are the fast time-scale action while the catalyst temperature

Table 10.6 Performances of various catalysts for the H_2-SCR system and CO-SCR system.

		Catalyst preparation			Reaction conditions					DeNO$_x$ performance	
Reference	Catalyst	Method	Calcination temperature (°C)	Calcination time (h)	Reductant type	NO_x	O_2	SV (h^{-1})	Reaction temperature (°C)	NO_x conversion	N_2 selectivity
[42]	0.1 wt.% Pt/$La_{0.7}Sr_{0.2}Ce_{0.1}FeO$	Wet impregnation method	400	2	H_2 (1 vol%)	0.25 vol%	5 vol%	80,000	180	80%	90%
[44]	0.1 wt.% Pt/Mg-Ce-O	Sol-gel method	400	2	H_2 (1 mol%)	0.25 vol%	5 mol%	80,000	180	85%	83%
[45]	Pd/TiO_2	Wet impregnation method	500	5	H_2 (3000 ppm)	1000 ppm	5 vol%	20,000	302	45%	–
[46]	Pt/SiO_2	Wet impregnation method	450	4	CO (500 ppm	500 ppm	10 vol%	12,000	20	78%	–
[47]	3 wt.% Pt/TiO_2	Sol-gel method	500	2	CO 1.5 vol%	0.5 vol%	1.5 vol%	–	350	75%	–
[21]	Pt/W-Ce-Zr	Precipitation method	–	–	CO (5000 ppm)	5000 ppm	2 vol%	40,000	380	14%	–
[48]	Rh/Na-Beta zeolite	Hydrothermal deposition	500	5	CO (1.5 vol%)	500 ppm	9 vol%	50,000	350	55.80%	–

and NH_3 surface coverage are the medium time-scale actions. The sensor drift, urea quality, catalyst aging, and poisoning are as long-term time-scale actions. Based on the different levels of time-scale response in the SCR system, from the control point of view, the relevant control strategy has to be constructed for precise and reliable control. In Fig. 10.10, the schematic representation for the design of the SCR controller is shown. The first thing is the NO_x emission characteristics study on various speed and load conditions of an engine. From an engine system study, the obtained input ($M_{air}+Mass_{fuel}$) and output (NO_x concentration) data help estimate the amount of reductant fluid to be injected with regard to the selected SCR catalyst. Further, the estimated reductant was mapped with the respective operating conditions of an engine system in the control model. This means that the design of the control model architecture purely depends on the input state parameters. Typically, the inlet gas temperature, NO_x, and reductant concentration before and after the catalyst system are the input state parameters. But the use of inlet gas velocity, NO_2/NO_x ratio, and the reductant surface coverage on the catalyst are used as input parameters to the controller. The use of the inlet gas velocity parameter provides the time at which the gas surges through the SCR; this would be beneficial

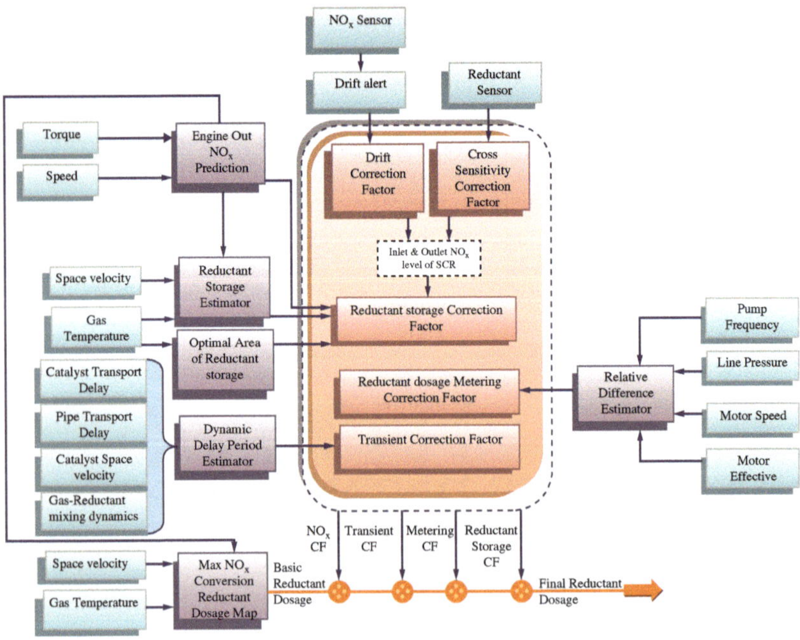

Fig. 10.10 Block scheme diagram for the design of the SCR controller.

for the precise supply of reductant dosage. The NO_2/NO_x ratio input signal helps to promote fast NO_x conversion SCR reaction (10.5) along with lower NH_3 slip. It also provides tracking of the NO_2 slip from the tailpipe, which is a major environmental issue. Moreover, the reductant coverage ratio on the catalyst is considered another important input control parameter in the SCR controller. To measure the reductant coverage ratio, the use of any measuring device inside the catalyst is impossible. Thus, the controller has to be designed in such a way as to accurately estimate the reductant coverage over the catalyst through measurable factors such as the amount of reductant supplied, NO_x, and reductant concentration before and after the SCR catalyst. Overall, these additionally added parameters could be helpful to achieve the required tailpipe NO_x concentration, which is less than 50 ppm (Euro 6), and an ammonia slip of less than 20 ppm (Euro 6) during the dynamic condition of the SCR system. Further, for the accurate measurement of input state parameters, the precise hardware system plays a significant role. But due to the rapid dynamic condition of the engine and also after a lifetime, uncertainty in hardware measurements arises, which has a direct effect on the misevaluation of the amount of reductant to be dosed in the SCR system. Some of the common imprecise function areas are: the NO_x sensors at the upstream and downstream of the SCR catalyst generate nonlinearity in measurements due to sensor drift (over a period of time, the output changes from its correct value for the same input) and crosses sensitivity toward NH_3 species. Second, a cause for uncertainty in the dosing unit is by the pump metering error, and also the rapid change of torque makes it very difficult for the pump to immediately respond, which creates a transient error. Thus, to remove this uncertainty in the measurement of input state parameters, additional sensors or appropriate error estimators and correction factors must be infused into the controller. Therefore, for the NO_x sensor drift problem, the NO_x emission level prediction block in the controller has to be encompassed with drift alert and drift correction factors. In the drift alert feature, when the sensor providing the signal apart from the user-defined drift limit, triggers the alert signal that intimates the maintenance needed. The drift correction feature diagnoses the deviation compared to the initial set point for adjusting back into the right specifications. Next, installing the NH_3 sensor at the downstream of the SCR catalyst provides data about the concentration of NH_3 to the NO_x emission level prediction block. This would help to estimate the NH_3-cross-sensitivity correction factor for the NO_x sensor, which has unwanted cross-sensitivity toward NH_3. Further, to reduce the metering error in the dosing unit, the estimation of the correction factor has to be proposed based on the relation between the pump frequency and the line pressure along with the relation between the effective voltage of the motor and the motor speed. Then, to estimate the transient correction factor, the best choice is to determine the gas flow period at each stage of the SCR with respect to the torque. This period can be categorized into four dynamic blocks: pipe transport delay, catalyst space velocity, exhaust gas-reductant mixing dynamics, and catalyst transport delay. Overall, with the infusion of all the stated errors and correction factors, the final reductant dosage is calculated by the controller using Eq. (10.11).

$$V_{\text{Final}} = V_{\text{base}} \times CF_{\text{N}} \times CF_{\text{ammonia storage}} \times CF_{\text{P}} \tag{10.11}$$

where V_{Final} is the final reductant dosage after the inclusion of the correction factor (mL/h), CF_{N} is the desired NO_x reduction correction factor, $CF_{\text{ammonia storage}}$ is the ammonia storage block correction factor for the reductant dosage emendation, and CF_{P} is the correction factor to reduce the perturbation $CF_{\text{p}}=f$(NO_x sensor drift correction factor, NH_3 cross-sensitivity correction factor, transient correction factor). Moreover, in the conventional SCR system, the sensing system is used at each stage to actively monitor the active control, self-diagnostics, and error correction. But due to cost limitations as well as the long term of the run, there are reliability issues that force researchers and car manufacturers to develop a virtual sensor-based SCR controller. To accomplish such a control system, a robust control strategy has to be designed by considering the above-discussed system input characteristics.

10.7 Conclusion

A broad overview of the state-of-the-art De-NO_x SCR technology role in NO_x reduction in IC engines has the greater potential to meet the increasingly stringent NO_x emission regulations in automobile applications. In the SCR system, ensuring a high NO_x conversion rate is done by the precise quantity of reductant agent supply based on the engine parameters such as exhaust gas temperature, level of NO_x, and engine speed. However, there are some complications to achieve high-level NO_x conversion efficiency in real-time driving conditions. First, sudden acceleration or deceleration occurs in milliseconds, which makes it difficult for the SCR system to react spontaneously. Second, the frequent load conditions require an immediate suitable dosing strategy. Third, the SCR catalysts have low activity at low temperature (during idling and starting stage) and also at high temperature (during high load and high speed); they work well only at the mid-temperature range. Even poisoning and aging of the catalyst by the formation of undesirable byproducts lead to a reduction in catalytic activity. So, efforts are being progressively made by numerous researchers to develop a positive and promising De-NO_x SCR system with a wide operating temperature, superdurability, and a precise reductant injection control algorithm. Still, many opportunities are available for further improvements in various areas of the SCR system to meet the new NO_x emission regulations as well as to promote good environmental standards for the future.

References

[1] Bosh H, Janssen FJ. Catalytic reduction of NO – a review of fundamentals and technology. Catal Today 1988;2:369.

[2] Blanco J, Avila P, Suárez S, Martin JA, Knapp C. Alumina-and titania-based monolithic catalysts for low temperature selective catalytic reduction of nitrogen oxides. Appl Catal Environ 2000;28(3–4):235–44.

[3] Centeno MA, Carrizosa I, Odriozola JA. In situ DRIFTS study of the SCR reaction of NO with NH_3 in the presence of O_2 over lanthanide doped V_2O_5/Al_2O_3 catalysts. Appl Catal Environ 1998;19(1):67–73.

[4] Koh HL, Park HK. Characterization of $MoO_2-V_2O_5/Al_2O_3$ catalysts for selective catalytic reduction of NO by NH_2. J Ind Eng Chem 2013;19(1):73–9.

[5] Caraba RM, Masters SG, Eriksen KM, Parvulescu VI, Fehrmann R. Selective catalytic reduction of NO by NH_3 over high surface area vanadia-silica catalysts. Appl Catal Environ 2001;34(3):191–200.

[6] García-Bordejé E, Lázaro MJ, Moliner R, Galindo JF, Sotres J, Baró AM. Structure of vanadium oxide supported on mesoporous carbon-coated monoliths and relationship with its catalytic performance in the SCR of NO at low temperatures. J Catal 2004;223(2):395–403.

[7] Huang B, Huang R, Jin D, Ye D. Low temperature SCR of NO with NH_3 over carbon nanotubes supported vanadium oxides. Catal Today 2007;126(3–4):279–83.

[8] Bagnasco G, Galli P, Larrubia MA, Massucci MA, Patrono P, Ramis G, Turco M. - Fe-vanadyl phosphates/TiO_2 as SCR catalysts. In: Studies in surface science and catalysis, Vol. 130. Elsevier; 2000. p. 653–8.

[9] Kobayashi M, Hagi M. V_2O_5-WO_3/TiO_2-SiO_2-SO_4^{2-} catalysts: influence of active components and supports on activities in the selective catalytic reduction of NO by NH_3 and in the oxidation of SO_2. Appl Catal Environ 2006;63(1–2):104–13.

[10] Baraket L, Ghorbel A, Grange P. Selective catalytic reduction of NO by ammonia on V_2O_5–SO_4^{2-}/TiO_2 catalysts prepared by the sol–gel method. Appl Catal Environ 2007;72(1–2):37–43.

[11] Zhang Y, Zhu X, Shen K, Xu H, Sun K, Zhou C. Influence of ceria modification on the properties of TiO_2–ZrO_2 supported V_2O_5 catalysts for selective catalytic reduction of NO by NH_3. J Colloid Interface Sci 2012;376(1):233–8.

[12] Ma Z, Wu X, Feng Y, Si Z, Weng D, Shi L. Low-temperature SCR activity and SO_2 deactivation mechanism of Ce-modified V_2O_5–WO_3/TiO_2 catalyst. Prog Nat Sci Mater Int 2015;25(4):342–52.

[13] Long RQ, Yang RT. Superior Fe-ZSM-5 catalyst for selective catalytic reduction of nitric oxide by ammonia. J Am Chem Soc 1999;121(23):5595–6.
[14] Long RQ, Yang RT. Selective catalytic reduction of NO with ammonia over Fe3+−exchanged mordenite (Fe–MOR): catalytic performance, characterization, and mechanistic study. J Catal 2002;207(2):274–85.
[15] Sultana A, Sasaki M, Suzuki K, Hamada H. Tuning the NO_x conversion of Cu-Fe/ZSM-5 catalyst in NH_3-SCR. Catal Commun 2013;41:21–5.
[16] Panahi PN, Salari D, Niaei A, Mousavi SM. NO reduction over nanostructure M-Cu/ZSM-5 (M: Cr, Mn, Co and Fe) bimetallic catalysts and optimization of catalyst preparation by RSM. J Ind Eng Chem 2013;19(6):1793–9.
[17] Zhao Y, Choi B, Kim D. Effects of Ce and Nb additives on the de-NO_x performance of SCR/CDPF system based on Cu-beta zeolite for diesel vehicles. Chem Eng Sci 2017;164:258–69.
[18] Seo CK, Choi B, Kim H, Lee CH, Lee CB. Effect of ZrO_2 addition on de-NO_x performance of Cu-ZSM-5 for SCR catalyst. Chem Eng J 2012;191:331–40.
[19] Bartley GJ, Chadwell CJ, Kostek TM, Zhan R. SCR deactivation kinetics for model-based control and accelerated aging applications. SAE Tech Pap 2012.
[20] Cavataio G, Girard J, Patterson JE, Montreuil C, Cheng Y, Lambert CK. Laboratory testing of urea-SCR formulations to meet tier 2 bin 5 emissions. SAE Tech Pap 2007.
[21] Zhu HO, Kim JR, Ihm SK. Selective catalytic reduction of NO with CO on Pt/W–Ce–Zr catalysts. React Kinet Catal Lett 2009;97(2):207–15.
[22] Li Y, Armor JN. Catalytic reduction of nitrogen oxides with methane in the presence of excess oxygen. Appl Catal Environ 1992;1(4):L31–40.
[23] Feng X, Hall WK. FeZSM-5: a durable SCR catalyst for NO_x removal from combustion streams. J Catal 1997;166(2):368–76.
[24] Hamada H, Kintaichi Y, Sasaki M, Ito T, Tabata M. Transition metal-promoted silica and alumina catalysts for the selective reduction of nitrogen monoxide with propane. Appl Catal 1991;75(1):L1–L88.
[25] Burch R, Ramli A. A comparative investigation of the reduction of NO by CH_4 on Pt, Pd, and Rh catalysts. Appl Catal Environ 1998;15(1–2):49–62.
[26] Hirabayashi H, Yahiro H, Mizuno N, Iwamoto M. High catalytic activity of platinum-ZSM-5 zeolite below 500 K in water vapor for reduction of nitrogen monoxide. Chem Lett 1992;21(11):2235–6.
[27] Burch R, Scire S. Selective catalytic reduction of nitric oxide with ethane and methane on some metal exchanged ZSM-5 zeolites. Appl Catal Environ 1994;3(4):295–318.
[28] Li Z, Flytzani-Stephanopoulos M. Effects of water vapor and sulfur dioxide on the performance of Ce–Ag-ZSM-5 for the SCR of NO with CH_4. Appl Catal Environ 1999;22(1):35–47.
[29] Iwamoto M, Yahiro H. Novel catalytic decomposition and reduction of NO. Catal Today 1994;22(1):5–18.
[30] Li YJ, Armor JN. Selective reduction of NO_x by methane on Co-ferrierites: I. Reaction and kinetic studies. J Catal 1994;150(2):376–87.
[31] Li Y, Armor JN. Catalytic reduction of nitrogen oxides with methane in the presence of excess oxygen. Appl Catal Environ 1992;1(4):L31–40.
[32] Feng X, Hall WK. FeZSM-5: a durable SCR catalyst for NO_x removal from combustion streams. J Catal 1997;166(2):368–76.
[33] Joyner R, Stockenhuber M. Preparation, characterization, and performance of Fe–ZSM-5 catalysts. J Phys Chem B 1999;103(29):5963–76.
[34] Hamada H, Kintaichi Y, Sasaki M, Ito T, Tabata M. Transition metal-promoted silica and alumina catalysts for the selective reduction of nitrogen monoxide with propane. Appl Catal 1991;75(1):L1–8.

[35] Burch R, Watling TC. The effect of promoters on Pt/Al_2O_3 catalysts for the reduction of NO by C_3H_6 under lean-burn conditions. Appl Catal Environ 1997;11(2):207–16.
[36] Burch R, Ramli A. A comparative investigation of the reduction of NO by CH_4 on Pt, Pd, and Rh catalysts. Appl Catal Environ 1998;15(1–2):49–62.
[37] Hirabayashi H, Yahiro H, Mizuno N, Iwamoto M. High catalytic activity of platinum-ZSM-5 zeolite below 500 K in water vapor for reduction of nitrogen monoxide. Chem Lett 1992;21(11):2235–6.
[38] Burch R, Scire S. Selective catalytic reduction of nitric oxide with ethane and methane on some metal exchanged ZSM-5 zeolites. Appl Catal Environ 1994;3(4):295–318.
[39] Captain DK, Roberts KL, Amiridis MD. The selective catalytic reduction of nitric oxide by propylene over Pt/SiO_2. Catal Today 1998;42(1–2):93–100.
[40] Li Z, Flytzani-Stephanopoulos M. Effects of water vapor and sulfur dioxide on the performance of Ce–Ag-ZSM-5 for the SCR of NO with CH_4. Appl Catal Environ 1999;22(1):35–47.
[41] Miyadera T. Selective reduction of NO_x by ethanol on catalysts composed of Ag/ Al_2O_3 and Cu/TiO_2 without formation of harmful by-products. Appl Catal Environ 1998;16(2):155–64.
[42] Costa CN, Savva PG, Andronikou C, Lambrou PS, Polychronopoulou K, Belessi VC, Stathopoulos VN, Pomonis PJ, Efstathiou AM. An investigation of the $NO/H_2/O_2$ (lean de-NOx) reaction on a highly active and selective Pt/La_0. $7Sr_0$. $2Ce_0$. $1FeO_3$ catalyst at low temperatures. J Catal 2002;209(2):456–71.
[43] Costa CN, Efstathiou AM. Transient isotopic kinetic study of the $NO/H2/O_2$ (lean de-NOx) reaction on Pt/SiO_2 and Pt/La– Ce– Mn– O catalysts. J Phys Chem B 2004;108(8):2620–30.
[44] Costa CN, Efstathiou AM. Pt/Mg-Ce-O catalyst for $NO/H_2/O_2$ lean de-NO_x reaction. Environ Chem Lett 2004;2(2):55–8.
[45] Ueda A, Nakao T, Azuma M, Kobayashi T. Two conversion maxima at 373 and 573 K in the reduction of nitrogen monoxide with hydrogen over Pd/TiO_2 catalyst. Catal Today 1998;45(1–4):135–8.
[46] Li MS, Seshan K, Lefferts L. Influence of NO on the reduction of NO_2 with CO over Pt/SiO_2 in the presence of O_2. Chin J Chem 2007;25(4):435–8.
[47] Wang JA, Cuan A, Salmones J, Nava N, Castillo S, Morán-Pineda M, Rojas F. Studies of sol–gel TiO_2 and Pt/TiO_2 catalysts for NO reduction by CO in an oxygen-rich condition. Appl Surf Sci 2004;230(1–4):94–105.
[48] Nakatsuji T, Yamaguchi T, Sato N, Ohno H. A selective NOx reduction on Rh-based catalysts in lean conditions using CO as a main reductant. Appl Catal Environ 2008;85 (1–2):61–70.
[49] Tauster SJ, Murrell LL. The NO CO reaction in the presence of excess O_2 as catalyzed by iridium. J Catal 1976;41(1):192–5.
[50] Abu-Jrai A, Tsolakis A. The effect of H_2 and CO on the selective catalytic reduction of NO_x under real diesel engine exhaust conditions over Pt/Al_2O_3. Int J Hydrogen Energy 2007;32(12):2073–80.

CHAPTER 11

Effects of fuel reformulation techniques in NO_x reduction

Ashwin Jacob and B. Ashok
Engine Testing Laboratory, School of Mechanical Engineering, Vellore Institute of Technology, Vellore, Tamil Nadu, India

11.1 Introduction

Energy security and climate change are two major issues faced by governments globally. It is estimated that more than half of global primary consumption is from fossil fuels such as gasoline and diesel for combustion. Consequently, the emissions generated from combusting these unaltered conventional fuels result in a large accumulation of toxic gases such as nitrogen oxides (NO_x), hydrocarbons (HC), carbon monoxide (CO), and carbon dioxide (CO_2) in the atmosphere. These toxic gases not only affect human health but also heat up the Earth's atmosphere, thereby causing global warming and climate change. Furthermore, the alarming rise in the cost of fossil fuels has paved the way to find new alternatives for conventional fuels. Hence, in recent years, advances in fuel production have led to the exploration of suitable feedstocks to extract surrogate fuels relative to conventional diesel and gasoline. There are three criteria followed in fuel production: long-term renewable feedstock alternatives, molecular and chemical compositions similar to fossil fuels, and the ability of the fuel to be used in internal combustion engines (ICE) without many modifications to the existing design. In recent years, feedstock generation such as first-, second-, and third-generation flora and fauna-based fuel sources have been established. Alcohols and biodiesel alternative fuels can be extracted from these feedstocks using specific extraction techniques based on fermentation, hydrothermal extraction, transesterification, etc. Processes that involve fermentation usually deliver alcohols such as methanol, ethanol, butanol, and fusel oil. Similarly, processes that involve transesterification deliver biodiesel with a strong fatty acid methyl ester profile [1]. The physicochemical properties of these surrogate alternative fuels should be able to emulate the roles and targets of conventional fossil fuels when they are used in ICE at various operating conditions. Hence, reformulating strategies are required that cater to

NOx Emission Control Technologies in Stationary and Automotive Internal Combustion Engines
https://doi.org/10.1016/B978-0-12-823955-1.00011-5

the conversion of these alternative fuels as optimal operating fuels similar to conventional fossil fuels in ICE. In order to reformulate conventional gasoline fuel, lower- or higher-chain alcohols are homogenized after the alcohols meet specific requirements such as research octane number, motor octane number, vapor pressure, density, lower heating value, aromatic content, olefin content, and distillation criteria. Almost all alcohols have similar properties to gasoline and can be reformulated by homogenizing different concentrations of alcohols. The effectivity and quality of the newly formulated blend can be assessed by extensive testing of the formulated fuel in a spark ignition (SI) engine at various operating conditions and analyzing the engine output characteristics. Similarly, to reformulate conventional diesel fuel, biodiesel fuels are homogenized after their cetane number, heating value, kinematic viscosity, flash point, and fatty acid methyl profile are analyzed in various concentrations for operation in a compression ignition (CI) engine to quantify its quality and effectivity based on the volume of NO_x emissions [2]. Apart from these alternatives, additive boosters such as nano metal-based additives and ethers are used as cetane improvers, oxygenates, octane boosters, etc., and are incorporated with fossil fuels to achieve multiple target functionalities apart from controlling NO_x emissions. Nanoadditives such as ferric oxide, cerium oxide, copper oxide, manganese oxide, magnesium oxide, graphene, alumina, carbon nanotubes, titanium oxide, etc., are made miscible with gasoline or diesel with techniques such as direct evaporation, laser ablation, and vacuum evaporation with an oil substrate one-step method or a two-step method such as ultrasonic agitation with a dispersant or by manual milling and grinding. This is done to avoid agglomeration and facilitate molecular and chemical stabilization to form a superior surrogate fuel compared to the base fossil fuel. In most cases, improvements in thermochemical fuel properties and mixture stabilization are evident in the homogenization of these nanoadditives at specific concentrations. Also, fuel reformulation that incorporate emulsion techniques and nanoadditives have significant benefits in acquiring superior engine output characteristics. This is because emulsion techniques and nanoadditives facilitate finer atomization and induce microexplosion which improves the combustion process. Similarly, ethers such as methyl tertiary butyl ether (MTBE) and diethyl ether (DEE) are oxygenated additives that facilitate low combustion noise and soot-free combustion, which improve the NO_x emissions positively in ICE [3]. Reformulations of fossil fuels employing ethers as additives are homogenized at multiple concentrations depending upon the ratio of the base fuel concentration. Apart from formulating base fuel-additive combinations, tertiary blends can be incorporated for advanced and more effective NO_x control in ICE. For instance, a fuel formulation involving different concentrations of gasoline-alcohol-nanoadditive-ether can be homogenized and stabilized to target and negate certain trade-offs to achieve optimal NO_x control in SI engines for various operating conditions. Similarly, a formulation involving miscible combinations of diesel-alcohol-nanoadditive-ether can be used in CI engines as an alternative fuel to control NO_x emissions without much compromise on other engine output characteristics. Therefore, in this chapter, basic concepts involved in fuel reformulations such as important factors, general composition, and fuel refining techniques are briefed and discussed in relation to ICE. Furthermore, the application and outcomes of various combinations of formulated surrogate fuels are highlighted using multiple studies and are tabulated for easy understanding. Apart from highlighting the role of nanoadditives, ethers, and alcohols in conventional fossil fuel reformulations, certain target-specific additives such as corrosion inhibitors, combustion enhancers, oxidizers, etc., are briefed based on their uses in their respective engine type with types and examples. Overall, this chapter will provide an in-depth understanding of controlling NO_x emissions using fuel reformulation techniques along with the basic nuances and concepts involved in the process.

11.2 Common factors that are crucial for fuel reformulations

In general, the nature of the fuel and its constituents such as composition and properties play an important role when they are used in ICE. These fuel constituents directly influence the nature of the emissions produced at different engine operating conditions. Ideal expectancies for any fuel used in an ICE is to provide good combustion quality, higher heat of combustion, minimized deposit formation, suitable latent heat of vaporization, low foaming tendency, material compatibility, stability, and ideal performance and emission characteristics at low and high temperatures. In order to achieve these abilities, fuel is carefully reformulated and tailored by manipulating certain internal and external methods. The internal method is altering the fuel composition and properties while producing the fuel. Similarly, the external method is mainly the addition of chemicals and additives to enhance performance or to prevent the degrading effects of the fuel. This section offers an in-depth analysis of the internal methods of fuel reformulation and its respective constituents.

11.2.1 General compositions of fuels

All forms of combustible fuels are predominantly composed of a complex mixture of different hydrocarbon arrangements. As the name suggests, the majority of these hydrocarbon chains are made of carbon and hydrogen bonds depending upon the ideal atomic and molecular arrangement. Typically, the carbon atom is a quadrivalent element that is capable of combining using single to triple bonds with a hydrogen atom. During fuel combustion in an engine, the bonds between the carbon and hydrogen atoms break, giving rise to chemical energy and new bonds with oxygen atoms. Apart from oxygen atoms, there will also be traces of sulfur and nitrogen atoms that mix with free radical oxygen to form NO_x emissions after combustion. Therefore, the idea is to manipulate these hydrocarbon chains in such a way that polluting compounds do not stick to the bonds while producing fuels. The stability of the hydrocarbon chain depends upon the chemical bond strength, which is in turn dependent on the structure and nature of various groups within the fuel composition chain [4]. The general compositions of gasoline or alcohol fuels can be classified based on their hydrocarbon combinations such as classes of olefins, aromatics, paraffin, and cycloparaffin, or naphthenes, as shown in Fig. 11.1. Paraffin belongs to the alkanes group, which can generally be represented by C_nH_{2n+2}. They have single bonds that fall under the class of saturated hydrocarbons, so they possess high heat content and low density. The single bond in paraffin allows it to have

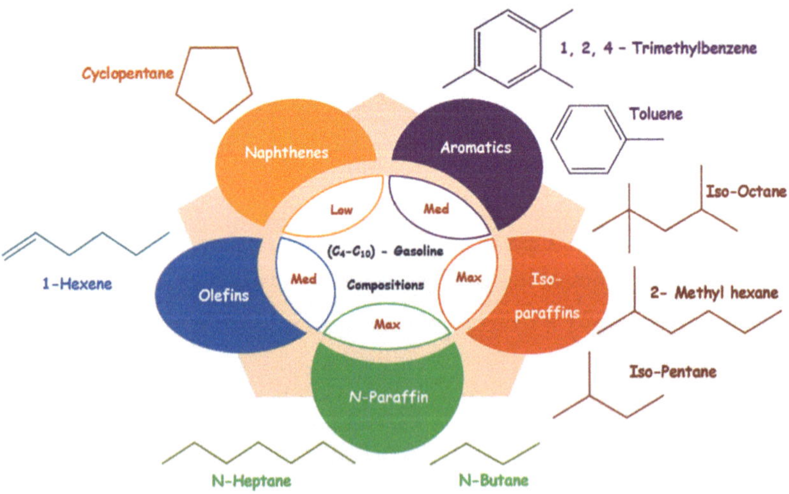

Fig. 11.1 General classification of hydrocarbon compositions with suitable examples.

straight-chained and multiple combinations of branched isomers. However, the straight-chained normal kinds of paraffin have a lower octane quality than isoparaffins, making the isomer combinations perfect knock-resistant fuels. Similarly, naphthenes or cycloparaffins are a class of stable, cyclic-structured saturated hydrocarbons that is generally denoted by C_nH_{2n}. The stability is often attributed to reduced autoignition but they possess a lower octane quality that can be rectified by the secondary processing of these fuels. Olefins are another common form of fuel composition comprised of unsaturated hydrocarbons containing one or more double bonds. Even though they have a similar chemical formula to the naphthenes, olefins have entirely different behaviors and characteristics. Because they predominantly have double bonds, the oxidation stability is lower for olefins compared to saturated hydrocarbons, as they can be easily oxidized. Aromatics are another class of hydrocarbons that has benzenoid-based rings with three double bonds. These fuel compositions are usually denoted by the chemical formulae C_nH_{2n-6} and exhibit aromatic odors. Generally, aromatics have the tendency to pollute due to their benzene toxicity. However, these aromatics have excellent antiknock abilities due to their high octane quality. Of these fuel compositions, depending upon the usage, the bond nature and chain arrangement can be manipulated with different compounds apart from hydrogen and oxygen atoms while producing alcohols or gasoline. Reformulation in this case cannot be done without experimentation. Therefore, the tailoring of these fuels based on fuel composition can be done using the emission results from engine test facilities as a reference. The physical and chemical properties of these fuels are important factors to achieve the desired fuel composition. The successive sections will discuss in depth the role of fuel properties in fuel reformulations and NO_x control.

11.2.2 Fuel properties

This section deals with the importance of the physicochemical properties of fuel on engine emissions and efficiency. The fuel properties vary drastically for different compositions of fuel; these were discussed in Section 11.2.1. Achieving the appropriate physicochemical properties can be done by tailoring the fuel composition while producing the fuel. It is important to acknowledge that the fuel properties influence all aspects of engine operation and the corresponding outputs such as the performance, combustion, and emission characteristics. Also, reformulating fuels based on their physicochemical properties will improve the tribological aspects of engine

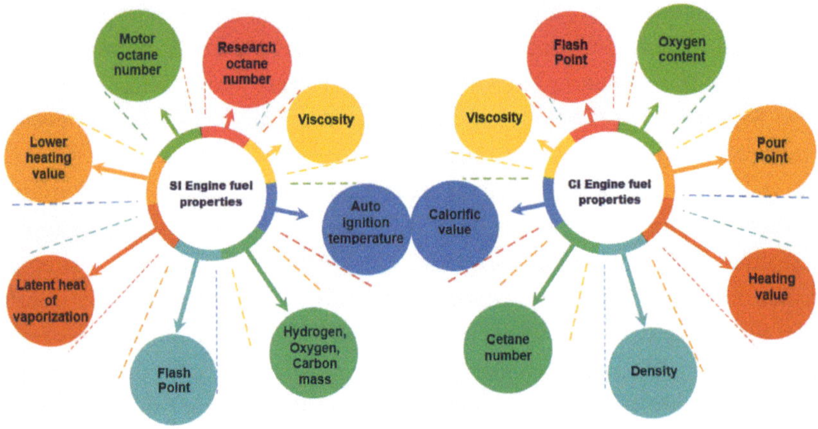

Fig. 11.2 Important physicochemical properties required for fuel reformulations in SI and CI engines.

operation. The most important common fuel properties for SI and CI engines include viscosity, surface tension, cetane number, flash point, volatility, octane number, calorific value, cloud point, and density, as shown in Fig. 11.2. Viscosity can be simply defined as a fluid's resistance to flow, where higher viscosity leads to greater resistance. Viscosity can be reformulated using the temperature exposure of the fuel and fatty acid carbon chain fuel composition modifications to achieve an ideal viscous range. Generally, lower viscosity leads to higher volatilization and excess fuel consumption due to vaporization along the cylinder wall, piston rings, and exhaust valves at higher temperatures. Likewise, higher viscosity results in improper fuel atomization and injection inaccuracies due to the higher diameter of the fuel droplets sprayed, consequently causing problems during a cold start. Also, higher viscous fuels cause excess smoke, NO_x, and hydrocarbon (HC) emissions [5]. Surface tension is another important parameter that facilitates the smaller-diameter fuel droplets during atomized spray inside the cylinder for combustion. Lower surface tension results in smaller droplet formation, which ultimately results in incomplete combustion and lowered emissions. For fuels such as diesel and biodiesel, the cetane number is an important criterion. It indicates the speed of fuel combustion and the required compression for ignition. Generally, higher cetane number fuels exhibit better ignition property. Also, fuels with poor cetane numbers tend to have unstable combustion and as a result, they produce huge amounts of NO_x, smoke, HC, and carbon monoxide (CO) emissions. The flashpoint is another important factor that is useful in determining the flammability of a hydrocarbon fuel chain under various engine operating conditions. Moreover, the quality and fatty acid composition of the fuel have no effect on the flashpoint. However, it is noteworthy that higher flashpoint fuels have easier storage, transportation, and handling amenities. Similarly, a correlated fuel factor is the volatility of a hydrocarbon chain. Fuels that have higher volatility result in a loss of fuel quantity at storage or while combusting even at lower temperatures. For gasoline and alcohols, the octane number defines the volume of a cycloparaffin composition in a specific volume of gasoline or alcohol, and this reduces the chances of knocking in engines. Generally, for all combustible fuels, the calorific value defines the total energy released in the form of heat when the fuel is burnt in the presence of oxygen. Higher calorific fuels tend to exhibit higher power output, whereas lower calorific fuels produce dense smoke and harmful pollutants such as NO_x. Some biodiesel variants with a longer ethyl ester fuel composition have a similar or higher calorific value to conventional diesel fuel, which is why they are broadly researched to replace or be blended with diesel. Similar to the flashpoint, another important parameter is the cloud point of the fuel. It is the minimum temperature at which the wax composition in the fuel turns into a cloudy form. Higher cloud point fuels generally result in injector clogging and improper fuel atomization, mainly during cold weather conditions. Typically, longer-chain aromatics or paraffin fuel compositions exhibit a higher cloud point and flashpoint. Finally, an important physical property is a mass per volume, also known as the density of the fuel. Changes in density have similar effects to the viscosity of the fuel, as they are interrelated. The idea here is to avoid higher-density fuels as they are sure to produce more NO_x and smoke emissions. As much as these physical properties affect the emission characteristics of fuel, fuel formulations also have to cater to the corresponding chemical properties of the fuel. Generally, common chemical properties include unsaturated fatty acids, oxygen percentage, oxidative stability, and water content in the fuel. Unsaturated fatty acids decide the per centage of NO_x emitted with engine operation mainly for biodiesel. Also, a higher composition of unsaturated fatty acids usually has a smaller cetane number, resulting in increased ignition delay with lesser emissions and thermal efficiency. The second most important factor for NO_x formation and the need for fuel reformulation is the oxygen percentage in fuels. Excess oxygen in a fuel results in reduced calorific value, higher fuel consumption, and higher NO_x emissions due to the combination of the excess oxygen atoms with the free nitrogen by oxidation [6]. However, combustion characteristics are fairly superior for high oxygen content fuels. In the shelf life and transportation of fuels, oxidation stability is an important parameter. Prolonged fuel storage results in changes in physicochemical properties that make the fuel unfit for use in engines. Factors such as high temperature or other environmental effects tend to separate the immiscible residues, which change the hydrocarbon chain stability, viscosity, and other important fuel properties. In case of higher moisture or water vapor levels in the atmosphere, fuels such as biodiesel get oxidized, therefore altering their fuel-like properties. These fuel properties can be altered or enhanced by tailoring the fuels during the production phases of these fuels. The next section sheds light on some production and processing techniques of conventional fuels and identifies areas where the fuels can be reformulated.

11.3 Methods of fuel refining and its role in tailoring fuel composition

Ideally, all conventional fuels in liquid and gaseous forms will have to go through a refinery phase during production of the fuel. Depending upon the physicochemical properties and the required hydrocarbon fuel composition, the refining can be altered. As shown in Fig. 11.3, the most common methods for gasoline, alcohol, and diesel are distillation, cracking, alkylation, polymerization, and isomerization. Cracking methods can be further classified into thermal and catalytic cracking. In the distillation process, crude oil can be processed into multiple forms of fuel. The required fuel form can be acquired by distilling the oil at different boiling temperatures. During distillation, gases can be acquired with lighter hydrocarbon chains and the heavier hydrocarbon chains are usually solid or liquid fuels at ambient temperatures. Therefore, the fuel formulation can be monitored based on the need. Cracking is another process that facilitates breaking hydrocarbon bonds by means of fluctuating temperature or using catalysts on the parent crude oil. Thermal cracking methods utilize high heat energy on the crude oil to break the carbon chain and create HC free radicals, therefore cracking it using a soaking chamber. The fuels formulated here are fractionated according to the required engine application. Catalytic cracking is a commonly used cracking process as it is flexible in providing lighter hydrocarbon fuels as well as heavier ones. A fluidized catalyst bed is made to run through the crude oil between the regenerator and reactor. The crude oil vapor is fractionated and simultaneously separated into heavier and lighter hydrocarbon fuels. A commonly used catalyst is zeolite due to its ability to control the formation

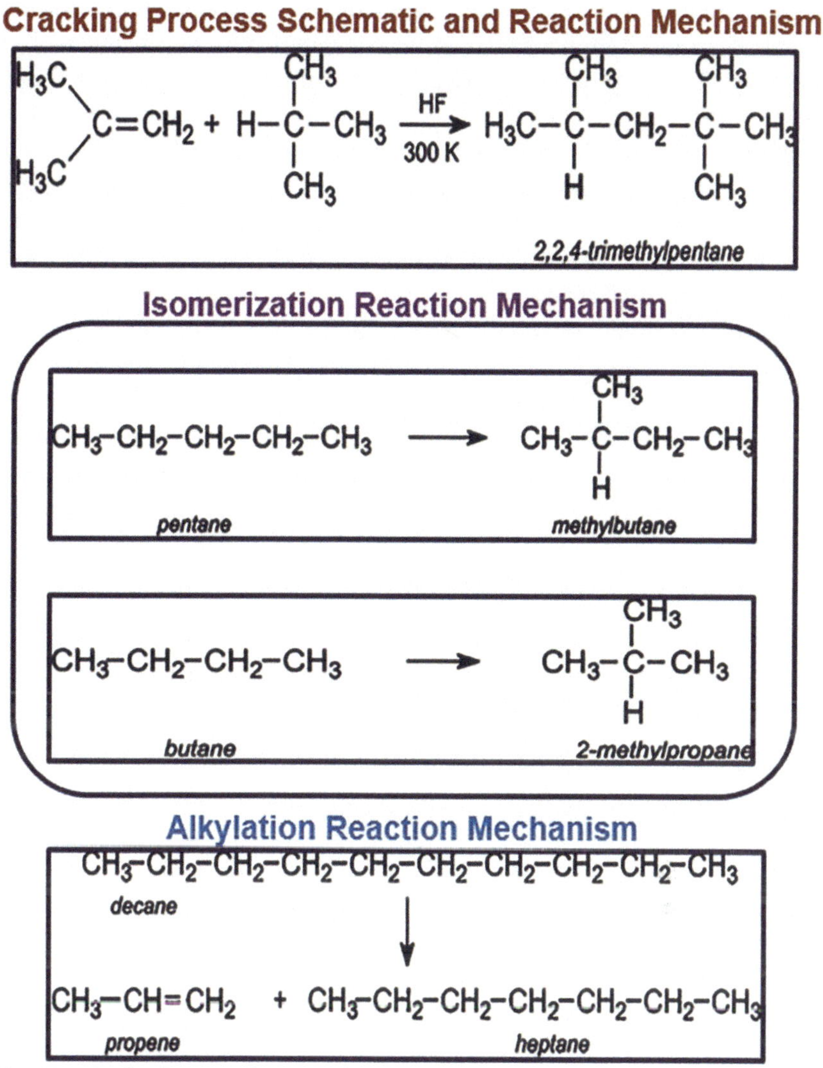

Fig. 11.3 Fuel refining techniques that contribute to the quality of the base fuel for reformulation.

of lighter olefins. Alkylation is another process for producing and reformulating high octane alcohols and gasoline by combining lighter olefins with cycloparaffin in the presence of a strong acidic catalyst [7]. Similarly, isomerization is another fuel-formulating production method that is mainly used to convert the low-octane quality of straight-chain paraffinic fuels to a superior branched HC molecule. For example, *n*-pentane with a research octane number (RON) of 62 can be converted into isopentane with a RON of 92. This process typically utilizes platinum- or zeolite-based catalysts to alter the straight fuel chain into a branched one. Polymerization is another form of fuel production that aims to combine lighter olefins such as alkenes together to give heavier olefins that usually display superior octane quality and low vapor pressure in alcohols and gasoline. This method usually uses phosphoric acid as its catalyst, as these pure olefin fuels have a low motor octane number (MON) compared to their RON. Biodiesel processing and formulation are done using transesterification, where the methyl ester component of the biodiesel is acquired using an alkaline catalyst, an alcohol, and an acid component. Upon heating the mixture, the biodiesel separates itself. The nature of the methyl ester can be reformulated by altering the heating duration and acid–base catalyst. These production techniques mostly offer a preliminary form of the fuel and their qualities need to be analyzed for making an ideal fuel. Therefore, the following section discusses the role of fuel quality in NO_x reduction and its importance in fuel formulation.

11.4 Formulation of fuels by blending to reduce NO_x emissions in IC engines

In order to curb harmful NO_x emissions from automotive engines, fuel reformulations that involve blending other biofuels, gases, and additives are widely being researched. Predominantly, NO_x is a combustion-based pollutant along with other sulfates, minerals, and engine oil. Some of the imperative factors that affect NO_x are cylinder pressure, air-fuel ratio, humidity, combustion temperature, and more importantly the oxygen content of the fuel. Also, the physicochemical composition of the fuel blend such as the cetane number, viscosity, bulk modulus, distillation temperature, and fuel aromatic content contribute to NO_x emissions. Therefore, it is important to formulate an ideal fuel blend with the target NO_x level in mind to curb the pollutant with the help of engine calibration and combustion strategies reinforced by after-treatment devices. Typically, the production of NO_x varies significantly with the blending nature. The common forms of fuel blend formulation usually incorporated by researchers for engine testing are biodiesel-diesel blends, gasoline-alcohol blends, additive-gasoline blends, additive-diesel blends, and biodiesel-additive blends, as shown in Fig. 11.4. When blend formulation is done with diesel-biodiesel, the resultant NO_x composition varies for different researchers due to the feedstock

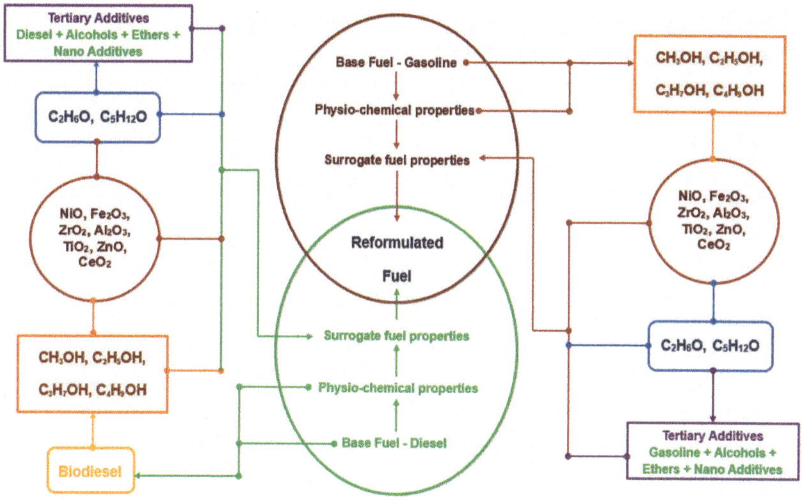

Fig. 11.4 Different possibilities of blend formulation pathways for gasoline and diesel with fuel additives.

and the physicochemical properties such as the degree of unsaturation, equivalence ratio, and oxygen bond within each methyl ester. Even though it is hard to predict the exact trend of the NO_x emitted based on the fuel composition, theoretically pure biodiesel should provide less NO_x compared to any blend ratio at specific engine operating conditions [8]. Similarly, alcohol-biodiesel blends are usually formulated as binary and ternary fuel blends because multiple factors such as the latent heat of evaporation, oxygen content, and combustion temperature play important roles in deciding the NO_x formed. For the most part, reformulating fuels with a small percentage of higher alcohols such as octanol might reduce the NO_x levels for some engine operating conditions due to their cooling effect, oxygenated characteristics, and in-cylinder temperature factor. Gasoline-biodiesel blends are an excellent combination in controlling NO_x emissions due to their combined fuel properties and their effects on engine output characteristics. Finally, the most effective form of fuel reformulation by fuel blending is achieved by the effective induction of different types of additives [9]. Advanced additives such as nanoadditives are becoming more prevalent in fuel reformulations, as they as more efficient in controlling NO_x emissions. Due to their nanoscale advantage over micron-scale additives, fuel miscibility is easily achievable without any residues or deposits. Nanoadditives can be classified into carbon nanotubes, nanoorganic additives, and metal, magnetic, and metal oxides. Metallic nanoadditives such as iron, ferric chloride, boron, and aluminum are mainly utilized as combustion catalysts to facilitate complete combustion while reducing fuel consumption and emissions. Metal oxide nanoadditives such as cupric oxide, manganese oxide, titanium oxide, zinc oxide, and cerium oxide are used mainly to reduce ignition delay and improve brake thermal efficiency. Generally, metal oxide additives such as zinc oxide, cerium oxide, cobalt oxide, and manganese oxide are added to the fuel blend with heightened mass fraction to control NO_x emissions for all engine operating conditions. Similarly, magnetic nanofluid additives are colloidal ferrofluids that improve performance characteristics such as fuel consumption and oxide-based emissions. Nanoorganic additives are unconventional in the field of engine testing; however, they have properties that might improve fuel volatility, which will improve the combustion and performance characteristics. Finally, carbon nanotubes are mainly used when fuel reformulation is done by water emulsion to improve the emission and combustion characteristics. All these additives have their functions interconnected as one factor influences another in many ways. However, these fuel reformulation additives can be categorized based on their primary functions. Fuel reformulation blends with ethers such as polyoxymethylene dimethyl ethers and dimethyl ether have significant properties that control NO_x without any trade-offs. Other additives such as ferrofluids, expanded polystyrene, acetone, and antioxidants are mixed with the above-mentioned additives to form effective combinations of hybrid additives, which might show high potential for NO_x control [10]. Finally, fuel reformulation by water emulsification is an emerging technique because the latent heat of vaporization and water heat might influence the in-cylinder temperature to reduce NO_x drastically. Emulsion techniques are also incorporated for biofuels for NO_x. However, prolonged exposure to excess water content might result in higher NO_x levels. Therefore, fuel formulations by blending are still being widely researched for diverse biofuels and additives. Through time, an ideal combination of biofuel-additive blend composition might be achieved to control NO_x more effectively. Certain additives that work well with gasoline might not suit for other fuels, such as diesel, to control NOx emissions.

11.5 Importance of additives on fuel reformulations for NO_x reduction in SI engines

In order to achieve high-quality fuel that caters to the current legislative norms, additives are used for reformulating conventional fuels and alternative biofuels. They play an important role in improving the fuel physicochemical properties, thereby providing competitive benefits on engine emission characteristics. Apart from the generic utilization of fuel additives, the criteria might be related to a specific role such as combustion improvement and oxidation stabilizing because the functions of the additives are interrelated. This section will delve deep into these types of fuel additive classifications using multiple research studies for formulating conventional gasoline using alcohols and ethers for lesser NO_x emissions.

11.5.1 Role of fuel additive combinations to reformulate gasoline for NO_x control

It is imperative that an appropriate fuel mixture with gasoline is formulated that can accurately emulate its target properties while operating in an SI engine. Alcohols have the potential to replicate conventional gasoline and can be blended with gasoline at any proportion for SI engine application without much modification to the existing design. However, numerous fuel properties are to be taken into account while reformulating gasoline to target NO_x reduction in SI engines. In addition to aromatic hydrocarbons in gasoline composition, additives such as alcohols and ethers are added in controlled proportions to reduce engine emissions [11]. These additives alter the physicochemical properties of the gasoline based on their miscible concentrations. Upon blending alcohols and ethers as additives for reformulating gasoline, there are significant changes in the blend viscosity, heating value, stability, and octane number. This change in the molecular structure of gasoline when it is blended with these additives plays a major role in achieving lower NO_x and significantly higher performance in SI engines. Alcohols and ethers have an innate molecular structure with adequate oxygen atoms to be blended with gasoline for controlling engine emissions such as NO_x, HC, and CO. Similarly, the latent heat of vaporization for alcohols is much more than gasoline, thereby facilitating a cooling effect within the cylinder that consequently hampers NO_x formation at high in-cylinder temperatures in combustion. More importantly, the lower heating values of alcohols and ethers are responsible for delivering lower exhaust temperatures when they are blended with gasoline. As a result, a significant reduction of NO_x is noted. Table 11.1 illustrates the significance of various concentrations of alcohol-ether-gasoline reformulated fuels on NO_x and other engine emissions. As described, NO_x and HC decrease upon blending with ethers and alcohols in most studies in SI engines. However, the NO_x levels vary with different operating conditions and blend concentrations, as the nature of HC bonds in gasoline varies with composition such as aromatics, olefins,

Table 11.1 Significance of fuel additive combinations on gasoline reformulations in SI engines with their effect on NO_x emissions.

Additive type and base fuel	Reformulated fuel	Effect on NO_x emissions	Effect on other engine emissions	Reference
Alcohol and Gasoline	Ethanol (5%, 10%, 15%)—Gasoline (95%, 90%, 85%)	Increased by 16.18% for higher ethanol percentages	CO, CO_2, and HC emissions decreased by 45%, 7.5%, and 40.15%, respectively	[11]
Alcohol and Gasoline	Gasoline (90%)—Ethanol (10%)—Methanol (10%)	Ethanol and methanol increased NO_x compared to gasoline	HC and CO are less for 10% ethanol addition	[12]
Alcohol and Gasoline	Ethanol (5%, 10%, 15%)—Gasoline (95%, 90%, 85%)	43% reduction for 15% ethanol addition	HC reduced by 6% along with CO and CO_2	[13]
Alcohol and Gasoline	N-Butanol (0%, 10%, 30%, 40%, 100%) Gasoline (100%, 90%, 70%, 60%, 0%)	n-butanol-gasoline blends and pure n-butanol reduced NO_x more than neat gasoline with exhaust gas recirculation	n-butanol-gasoline blends reduced HC and CO whereas neat n-butanol increased CO	[14]
Alcohol and Gasoline	Bioethanol (0%, 10%, 15%, 20%, 30%, 85%) Gasoline (100%, 90%, 85%, 80%, 70%, 15%)	15% NO_x reduction for 85% Ethanol 15% gasoline composition	Reduction of CO and HC by 15% and 20% for 85% ethanol 15% gasoline composition	[15]
Ether and Gasoline	MTBE—(0%, 10%, 15%, 20%), Gasoline (100%, 90%, 85%, 80%)	NO_x increased for rich gasoline mixture with MTBE. No change in NO_x during idling.	HC increased while CO decreased for higher ratios of MTBE	[16]

Table 11.1 Significance of fuel additive combinations on gasoline reformulations in SI engines with their effect on NO_x emissions—cont'd

Additive type and base fuel	Reformulated fuel	Effect on NO_x emissions	Effect on other engine emissions	Reference
Ether and Gasoline	DME (0%–30%), Gasoline (70%–100%)	NO_x reduced for increasing concentrations of DME	CO reduced by 10% for DME-gasoline blends along with HC.	[17]
Alcohol and Gasoline	Fusel oil (0%, 10%, 20%, 30%), Gasoline (100%, 90%, 80%, 70%)	NO_x reduced while using fusel oil-gasoline blend	HC and CO increased for fusel oil-gasoline blends	[18]
Alcohol Isomer and Gasoline	Isobutanol (0%, 3%, 7%, 10%), Gasoline (100%, 97%, 93%, 90%)	No change in NO_x emissions	UHC, CO, and CO_2 decreased at lower speeds	[19]
Nano-additive and Gasoline	Di-Methyl Carbonate (DMC)—Gasoline Blend	No change in NO_x emissions	UHC and PM reduced by 30% and 60%	[20]

paraffin, etc. It is also noteworthy to point out that a reduction of heating value and an increase in density and kinematic viscosity are observed as the concentrations of alcohol and ethers increase in the formulated gasoline. Combinations of methyl tertiary butyl ether with two and three carbon chain alcohols and gasoline at stoichiometric condition display the maximum reduction of NO_x emissions. Also, the increased water and oxygen content in a fusel oil-gasoline combination plays a major role in oxidizing most of the nitrogen oxides into free nitrogen upon combustion. Investigations on one and two carbon chain alcohol formulations with gasoline resulted in a significant reduction in CO_2 and NO_x emissions for multiple blend ratios. Certain formulations of alcohol-ether-gasoline blends show negative effects on NO_x emissions due to their molecular structure and their in-cylinder reactions upon combustion. For instance, formulations that involve increased concentrations of dimethyl ether in alcohol-gasoline result in higher NO_x and HC emissions at idle and stoichiometric conditions [16]. In recent years, nanoadditives have been incorporated in conventional gasoline fuel to control engine emissions and boost performance characteristics. Even though there is a dearth of studies incorporating gasoline reformulations with nanoadditives, metal oxide nanoadditives such as ferric oxide, titanium oxide, and manganese oxide are used to reduce NO_x emissions and enhance performance. Few of the studies reported a significant reduction of UHC and PM with minimal influence on NO_x emissions. Apart from these additives, there are multifunctional fuel additives that target different aspects of SI engine functionality while operating in the formulated fuel. A brief explanation is provided highlighting these additives in the next section.

11.5.2 Notable fuel additives with interrelated functionalities in SI engine outputs

Because the functions of the additives are interrelated, they can be broadly classified into two main categories: additives influencing fuel distribution and combustion characteristics and additives to improve oxidation stability and the fuel system. Most of the additives used for fuel distribution and combustion enhancement have multifunctional behaviors. Generally, antiknock, antimisfire, antipreignition, and antioctane requirement increase (ORI) as well as spark-aid additives, including some specific additives that enable smooth fuel distribution between cylinders, belong to the category of combustion enhancers. However, they also function as deposit control additives and indirectly affect NO_x and other engine emissions as they are a product of combustion. Due to the corrosive effects and chemical leaning effect of alcohols and ethers, only a fraction of their volume is used in fuels due to trade-off effects. Similarly, antioxidants operate by preventing chain reactions with free radicals tangled in hydrocarbon oxidation [20]. These additives are used based on the type of fuel composition and its application. Most of these additives fall under alkyl phenol and aromatic diamine compounds. As fuel processing and refining technologies improve each year, advanced fuel formulating additives for SI engines will always be researched for achieving low emissions and high engine performance. The next section deals with additives that are commonly used in CI engines to reformulate fuels for NO_x reduction and other implications.

11.6 Importance of additives on fuel reformulations for NO_x reduction in CI engines

Generally, CI engines produce much less HC and CO emissions; however, NO_x-related PM and soot emissions are higher than SI engines. Fuels such as diesel and biodiesel can be efficiently reformulated by adding suitable additives that cater to controlling emissions during CI engine operation. Also, these additives help reduce engine noise, and fuel production costs as well. Lately, nanoadditives and alcohols have gained traction in fuel surrogate and reformulation strategies of conventional diesel to meet legislative emissions norms, especially for NO_x emissions. The following sections will highlight the effects of using these additives in blend combinations while operating on CI engines.

11.6.1 Role of nanoadditives in conventional diesel fuel composition for NO_x reduction

The quality and physicochemical aspects of conventional diesel can be improved by homogenizing additives such as cetane improvers, oxygenated additives, and metal-based additives to target NO_x control while operating on CI engines [21]. In recent years, metal-based nanoadditives have been extensively exploited due to their multifunctional advantages over similar fuel additives in controlling engine emissions and enhancing performance. Table 11.2 highlights various studies that have reformulated diesel fuel with various metal-based nanoadditives at different proportions for operating in CI engines to assess their effect on engine output emission characteristics such as NO_x and other emissions. As described, aluminum oxide, copper oxide, carbon nanotubes, and cerium oxide are the most commonly used in these studies, which actively increased and altered the physicochemical properties of conventional diesel fuel such as the cetane number, density, heating value, and viscosity. Because diesel fuels possess much less inherent oxygen content, the metal based nanoadditives played a major role only in facilitating microexplosions for finer atomization, which improves the combustion characteristics and consequently induces complete combustion and reduced NO_x emissions. An emulsion of water with the reformulated diesel fuel with nanoadditives reduces all engine emission gases significantly at higher in-cylinder temperatures by the instantaneous vaporization of the water droplets. Most of the studies in the table used carbon nanotubes as a reformulating agent with diesel to control NO_x emissions. Carbon nanotubes act as a perfect deoxidizer during the combustion process at high

Table 11.2 Diesel fuel reformulations with various metal nanoadditives in CI engines with their influence on NO_x emissions.

Reformulated fuel-additive combination used	Effect on NO_x emissions	Effect on other engine emissions	Reference
Aluminum oxide-diesel fuel	10% increase	CO, UHC—17% decrease Soot/smoke/PM—26% decrease	[22]
Graphene nanoplatelets-diesel fuel	26.3% decrease	CO—30% decrease, UHC—23.2% decrease Soot/smoke/PM—29.2% decrease	[23]
Carbon nanotubes-diesel fuel	8.9% decrease	CO—9.6% decrease, UHC—11.4% decrease Soot/smoke/PM—15.2% decrease	[21]
Carbon nanotubes-diesel fuel	21% decrease	CO—20% decrease, UHC—22.6% decrease Soot/smoke/PM—5.5% decrease	[24]
Graphene-diesel fuel	29.9% increase	CO—14.2% increase	[25]
Aluminum oxide-diesel fuel	17% decrease	CO—6.92% decrease, UHC—32.7% decrease Soot/smoke/PM—65% decrease	[26]
Cupric oxide-diesel fuel	19% decrease	CO—13% decrease	[27]
Cerium oxide-diesel fuel	42.7% decrease	CO—increase, UHC—28.5% decrease	[28]

temperatures and consequently reduce NO_x into free nitrogen with the mechanism shown in Eq. (11.1):

$$C + 2NO \rightarrow CO_2 + N_2 \tag{11.1}$$

On the other hand, metal oxide nanoadditives such as aluminum oxide tend to increase NO_x emissions, as shown in some studies in the table, because aluminum oxide gets thermally dissociated at higher temperatures into free oxygen and a lower form of metal oxide, as shown in Eq. (11.2). This free oxygen reacts with the nitrogen oxides emitted during combustion and forms NO_x emissions. Similarly, cerium oxide nanoparticles work on the principle of nanocatalysis, where the metal oxide acts as a catalyst, absorbs the free oxygen, and breaks down the NO_x formed into free nitrogen by the mechanism depicted in Eq. (11.3) [28]. Therefore, water emulsion with carbon nanotube-diesel reformulations is ideal for NO_x control in conventional diesel fuel. The reaction of fuel additives differs with the molecular structure of the formulating diesel surrogate fuel. Therefore, the next section briefly discusses the role of additives, mainly nanoadditives, on biodiesel mixtures with conventional diesel fuel for controlling NO_x emissions.

11.6.2 Reformulations of biodiesel with nanoadditives for NO_x reduction

In order to improve and emulate conventional diesel fuel in CI engine operation, biodiesel alternatives were employed and assisted by metal-based nanoadditives and their oxides for controlling NO_x emissions. Table 11.3 describes the utilization of multiple nanometals as additives in diverse forms of biodiesel-diesel blends at different ratios for formulating novel fuels and their resultant emission outcomes while they are operated in CI engines. Most of the investigations revealed that NO_x emissions were slightly increased with increasing concentrations of nanoadditives. This is predominantly due to the thermal NO_x developed at near stoichiometric operation when the combustion flame increases. However, the results of NO_x emissions vary with different biodiesel-nanoadditive-diesel formulated fuels as their molecular composition and physicochemical properties vary after they are homogenized. Studies showed that metal additives such as carbon nanotubes drastically reduced NO_x emissions when formulated with biodiesel-diesel fuel rather than pure diesel fuel due to secondary atomization and a microexplosion before combustion [33]. It is noteworthy that a slight reduction of NO_x was observed with small amounts of ethers and alcohols in addition to the nanoadditives at lower and medium engine speeds. Studies that incorporated aluminum oxide as the biodiesel-diesel formulating additive resulted in a slight increase in NO_x emissions for increasing and decreasing concentrations of biodiesel and additive concentrations. The role of the additive is described by the reaction shown in Eq. (11.2) where the additive undergoes dissociation and forms nascent oxygen atoms. This free oxygen reacts with the nitrogen from the air-fuel mixture and forms NOx emissions at high in-cylinder temperatures. On the other hand, graphene additives in biodiesel-diesel blends showed a slight reduction in NO_x emissions for higher concentrations of graphene and a minimal concentration of biodiesel. In another study, graphene nanoparticles homogenized at lower concentrations

Table 11.3 Utilization of nanometal additives for reformulating biodiesel-diesel blends in CI engines.

Reformulated fuel-additive combination used	Effect on NO_x emissions	Effect on other engine emissions	Reference
Waste cooking oil-diesel blends with carbon nanotubes	27.5% ↑	CO—65.7% ↓ UHC—45% ↓ Soot/smoke/PM—29.4% ↓	[29]
Acacia Concinna biodiesel (B40) blend with titanium dioxide	Increased	CO—N/A UHC—38% ↓ Soot/smoke/PM—20% ↓	[30]
Tamarind seed methyl ester-diesel blends with multiwalled carbon nanotubes and alumina oxide	9% ↓	CO—56.6% ↓ UHC—68% ↓ Soot/smoke/PM—87.4% ↓	[31]
Emulsified soybean-diesel blends with zinc oxide	1.9% ↑	CO—25% ↓ UHC—11.5% ↓ Soot/smoke/PM—3.9% ↓	[32]
Calophyllum inophyllum biodiesel-diesel (B20) and titanium dioxide	Increased	CO—23% ↓ UHC—12% ↓ Soot/smoke/PM—Decreased	[33]
Jojoba methyl ester-diesel (B20) and aluminum oxide	70% ↓	CO—80% ↓ UHC—60% ↓ Soot/smoke/PM—35% ↓	[34]
Dairy waste biodiesel-diesel blends (B20) with multiwalled carbon nanotubes	2.3% ↓	CO—5% ↓ UHC—3.9% ↓ Soot/smoke/PM—N/A	[35]
Waste cooking oil biodiesel-diesel blend with emulsified SC1 novel nanoadditives	67.62% ↑	CO—62.05% ↓ UHC—60.94% ↓ Soot/Smoke/PM—N/A	[36]

reduced NO_x significantly at all engine loads and higher engine speeds due to a decrease in combustion duration. The use of cerium oxide as a reformulating agent at medium quantities with a diesel-biodiesel blend reduces NO_x significantly due to the principle of oxygen absorption, which converts the NO_x emissions into free nitrogen as shown in Eq. (11.3):

$$Al_2O_3 \rightarrow Al_2O + 2O \quad (11.2)$$

$$Ce_2O_3 + NO \rightarrow 2CeO_2 + N_2 \quad (11.3)$$

A few studies also incorporated nanometal additives such as manganese and nickel oxides in small quantities with diesel-biodiesel blends. They showed a reduction in NO_x emissions, even though a slightly better result was observed with manganese oxide due to its advanced catalytic effect as an oxygen absorbent. Some nanocatalyst additives such as magnesium oxide have shown a negative influence compared to manganese oxide on NO_x emissions, even though they have shown positive results with HC and CO emissions at lower engine loads compared to higher loads. However, unconventional additives such as cobalt oxide show significant improvement in NO_x emission control in similar engine operating conditions when they are homogenized with diesel-biodiesel blends [35]. Similarly, unconventional tertiary additives are used to formulate superior diesel fuel surrogates such as a combination of ethers-nanoadditives-alcohols. The following section briefly discusses the role of this unconventional combination of tertiary fuel additives in reformulating diesel fuel to control NO_x emissions in CI engines.

11.6.3 Tailoring of diesel fuel with tertiary additives and alcohols for NO_x reduction

Apart from individual additive reformulations with diesel, two or more fuel additives can be homogenized with diesel to enhance and target specific engine output characteristics based on their natural state and molecular composition. Table 11.4 describes the studies that have utilized tertiary fuel additives to reformulate diesel and biodiesel blends with alcohols, ethers, antioxidants, and nanoadditives in multiple combinations and concentrations to reduce NO_x emissions generated with engine operation. As highlighted, the alcohol-diesel-ether formulation is the most studied at different concentrations of each reformulating agent. Alcohols have the innate ability to boost thermal efficiency, improve combustion behavior, and ultimately reduce engine emissions. Due to its low viscosity, low fuel resistance

Table 11.4 Role of tertiary fuel additives in reformulating diesel fuel to reduce NO_x emissions.

Base fuel	Alcohol type and composition	Tertiary additive nature (Ethers/nanoadditives/antioxidants)	NO_x emissions produced	Effect on other engine emissions
Diesel	20% Ethanol	Methyl soyate	5.84 g/kWh	PM—0.21 g/kWh CO—0.7 g/kWh
Diesel	10% Ethanol	Nitro methane	674 ppm	PM—20.03
Diesel	10% Butanol	–	181 ppm	CO—380 ppm
Diesel	20% Ethanol	At. Peclete number 4	541 ppm	CO—±0.03 g/kWh
Diesel	10% Ethanol	CIZ emulsifier	842 ppm	PM—0.42 FSN CO—452 ppm
Diesel	20% Ethanol	Ethyl acetate and diethyl carbonate	14 g/kWh	CO—3.5 g/kWh
Diesel	50% Ethanol	Diethyl ether	101 ppm	CO—0.18% vol.
Honge Oil Biodiesel + Diesel	20% Methyl ester	Aluminum oxide	Increases by 11.27%	HC, CO, and smoke reduced by 27.72%, 48.4%, and 22.83%, respectively
Biodiesel + Diesel	45% Ethanol	Alumina-doped ceria-zirconia	NO_x emissions reduced for diesel-biodiesel-ethanol blends more than nanofuel blends	Nanofuel blends reduce CO and CO_2 emissions
Jatropha Biodiesel + Diesel	10% Ethanol	Alumina	Reduces by 22.53%	HC and CO reduced by 9.18% and 16.83%, respectively
Diesel	0–30% Methanol	Nanoemulsion	Drastic reduction at 15% water nanoemulsion blend	HC and CO increased
Fatty acid methyl ester + Diesel	3% Ethanol-methanol-butanol	Carbon nanotubes, aluminum oxide, titanium oxide	Reduced by 9.2%	CO, UHC, and smoke reduced by 26%, 7.5% and 36%, respectively

Diesel	20%–40% N-Heptanol	Multiwalled carbon nanotubes, graphene oxide, and nanoplatelets	Reduced by 12%	Soot reduced by 40%
Jojoba methyl ester + Diesel	20% Ethanol	Aluminum oxide and multiwalled carbon nanotubes	35% reduction	CO and UHC reduced 50% and 60%, respectively
Jatropha methyl ester + Diesel	40% N-Butanol	Multiwalled carbon nanotubes, graphene oxide, and nanoplatelets	Reduced by 45%	CO and UHC reduced by 55% and 50%, respectively
C. inophyllum methyl ester	100% Methyl Ester	Zinc oxide, ethanox	12.6% reduction by zinc oxide and 17.8% reduction by ethanox	–

is facilitated, which improves atomization upon injecting the formulated fuel [37]. Also, its oxygen content facilitates complete combustion and reduces the chances of NO_x formation. As described, higher concentrations of alcohols have resulted in lower NO_x emissions upon combustion due to the cooling effect of the alcohols and the quenching effect it produces, which hampers NO_x formation at high in-cylinder temperatures. Formulations that incorporated nanoadditives with alcohols and diesel showed a slight reduction in NO_x emissions, whereas the superior molecular structure achieved on formulating diesel-biodiesel-alcohol-nanoadditives showed significant reductions in NO_x emissions of up to 45%. This is due to the combined effect of oxygenated alcohols, ethers, and biodiesel fuels supported by the catalytic and oxidative effect of nanometal oxides, which ultimately results in the reduction of NO_x emissions at various operating conditions in CI engines. Formulations incorporating emulsion strategies using water inducted into the cylinder to be combusted with the base fuel and tertiary additive combinations have also proven effective in controlling thermal NO_x emissions. It is noteworthy to highlight that the reformulation combinations of tertiary additives have dominant control over reducing CO, HC, and CO_2 emissions over NO_x emissions as nanoadditives, alcohols, and ethers often thermally dissociate at higher in-cylinder temperatures. A few exceptions such as lower chain alcohols and carbon nanotube nanoadditive particles actively participate in NO_x reduction. Apart from these additives, there are multifunctional fuel additives that target different aspects of CI engine functionality while operating in the formulated fuel. A brief explanation is provided highlighting the distinctions in fuel reformulation techniques to mitigate NOx emissions.

11.7 Distinctions in fuel reformulation techniques to mitigate NO_x emissions

The effects of fuel reformulation approaches on engine output characteristics vary distinctly with respect to the combinations of additives used in conventional fuels such as gasoline and diesel and their respective engines. However, some combinations have significant improvements over other combinations in terms of their role in achieving positive engine output characteristics. For instance, reformulation techniques in CI engines show that carbon nanotube combinations with biodiesel or diesel have a significant reduction in fuel consumption and positive emission characteristics. Also, the required quantity of this nanoadditive to achieve the positive outputs from CI engine operation is much less compared to other metal oxide additives. Incorporating aluminum oxide as a biodiesel-diesel formulating additive shows a slight increase in NOx emissions for increasing and decreasing concentrations of biodiesel and additive concentrations in CI engines. Formulations that incorporated nanoadditives with alcohols and diesel showed a slight reduction in NO_x emissions, whereas the superior molecular structure achieved on formulating diesel-biodiesel-alcohol-nanoadditives showed significant reductions in NO_x emissions. Nanocatalysts such as magnesium oxide have shown a negative influence compared to manganese oxide on NO_x emissions while formulating with diesel or biodiesel, even though they have shown positive results with HC and CO emissions. However, cobalt oxide shows a significant improvement in NO_x emission control in similar CI engine operating conditions when homogenized with diesel-biodiesel blends. Moreover, small amounts of ethers and alcohols blended with nanoadditives slightly reduce NO_x at lower and medium engine speeds in CI engines. These effects are observed in CI engines due to the combined effect of oxygenated alcohols, ethers, and biodiesel fuels supported by the catalytic and oxidative effect of nano metal oxides, which ultimately results in the reduction of NO_x emissions at various engine operating conditions. Similarly in SI engines, alcohol and ether additives for reformulating gasoline significantly changed the blend viscosity, heating value, stability, and octane number. This change in the molecular structure of gasoline when blended with these additives plays a major role in achieving lower NO_x. For instance, combinations of methyl tertiary butyl ether with two and three carbon chain alcohols and gasoline at the stoichiometric condition displays the maximum reduction of NO_x emissions in SI engines. Also, formulations that involve increased concentrations of dimethyl ether in alcohols-gasoline results in higher NO_x and HC emissions at idle and stoichiometric conditions. Techniques that incorporated nanoadditives in SI engines reported a significant reduction of UHC and PM with minimal influence on NO_x emissions while reformulating gasoline with nanoadditives and metal oxide nanoadditives such as ferric oxide, titanium oxide, and manganese oxide. Hence, the different reformulating techniques discussed in this chapter influence the control of NO_x emissions in SI and CI engines distinctly.

11.8 Conclusion

Reformulations of conventional fossil fuels such as diesel and gasoline using alternative fuels can be an optimal solution for meeting the energy demand related to the automotive sector. Moreover, the utilization of additives such as ethers, alcohols, and nanoadditives has proven that they are potential boosters in the reformulated surrogate fuel blend in controlling NO_x emissions efficiently in ICE. From the studies analyzed in this chapter, certain assertions can be made that shed light on the effect of fuel reformulation for gasoline and diesel in controlling NO_x in ICE. Although alcohols are perfect additives in certain concentrations in SI engines, their oxygenated nature along with their inherent molecular composition, which has a substantial quantity of water molecules, results in excessive NO_x emissions upon combustion. However, this problem can be rectified if the alcohols are dehydrated or ethers can be added to address this problem.

It was noted that NO_x emissions slightly increased with increasing concentrations of nanoadditives. However, carbon nanotubes drastically reduce NO_x emissions when they are formulated with biodiesel-diesel fuel rather than pure diesel fuel due to secondary atomization and a microexplosion before combustion. It is noteworthy that a slight reduction of NO_x was noted with small amounts of ethers and alcohols in addition to the nanoadditives. Formulations that incorporated nanoadditives with alcohols and diesel showed a slight reduction in NO_x emissions, whereas the superior molecular structure achieved upon formulating a tertiary combination of diesel-biodiesel-alcohol-nanoadditives showed significant reductions in NO_x emissions of up to 45% in CI engines due to the combined effect of the catalytic and oxidative effect of nano metal oxides. To summarize, the formulation of gasoline and diesel fuels can be aptly emulated into surrogate fuels by utilizing suitable alternative fuels and additives to alter their molecular structure and chemical composition so that NO_x emissions can be controlled while they are operated in ICE.

References

[1] Jacob A, Ashok B, Alagumalai A, Chyuan OH, Le PT. Critical review on third generation micro algae biodiesel production and its feasibility as future bioenergy for IC engine applications. Energ Conver Manage 2020;228, 113655.
[2] Zaharin MS, Abdullah NR, Najafi G, Sharudin H, Yusaf T. Effects of physicochemical properties of biodiesel fuel blends with alcohol on diesel engine performance and exhaust emissions: a review. Renew Sustain Energy Rev 2017;79:475–93.
[3] Awad OI, Mamat R, Ali OM, Sidik NA, Yusaf T, Kadirgama K, Kettner M. Alcohol and ether as alternative fuels in spark ignition engine: a review. Renew Sustain Energy Rev 2018;82:2586–605.
[4] Sarathy SM, Farooq A, Kalghatgi GT. Recent progress in gasoline surrogate fuels. Prog Energy Combust Sci 2018;65:67–108.

[5] Cornejo A, Barrio I, Campoy M, Lázaro J, Navarrete B. Oxygenated fuel additives from glycerol valorization. Main production pathways and effects on fuel properties and engine performance: a critical review. Renew Sustain Energy Rev 2017; 79:1400–13.

[6] Fayyazbakhsh A, Pirouzfar V. Comprehensive overview on diesel additives to reduce emissions, enhance fuel properties and improve engine performance. Renew Sustain Energy Rev 2017;74:891–901.

[7] Rodríguez E, Gutiérrez A, Palos R, Vela FJ, Arandes JM, Bilbao J. Fuel production by cracking of polyolefins pyrolysis waxes under fluid catalytic cracking (FCC) operating conditions. Waste Manag 2019;93:162–72.

[8] Arbab MI, Masjuki HH, Varman M, Kalam MA, Imtenan S, Sajjad H. Fuel properties, engine performance and emission characteristic of common biodiesels as a renewable and sustainable source of fuel. Renew Sustain Energy Rev 2013;22:133–47.

[9] Nanthagopal K, Ashok B, Tamilarasu A, Johny A, Mohan A. Influence on the effect of zinc oxide and titanium dioxide nanoparticles as an additive with *Calophyllum inophyllum* methyl ester in a CI engine. Energ Conver Manage 2017;146:8–19.

[10] Danilov AM. Progress in research on fuel additives. Petrol Chem 2015;55(3):169–79.

[11] Shahad HA, Wabdan SK. Effect of operating conditions on pollutants concentration emitted from a spark ignition engine fueled with gasoline bioethanol blends. J Renew Energy 2015;2015. https://doi.org/10.1155/2015/170896, 170896.

[12] Kak A, Kumar N, Singh B, Singh S, Gupta D. Comparative study of emissions and performance of hydrogen boosted SI engine powered by gasoline methanol blend and gasoline ethanol blend. SAE Techn Pap 2015.

[13] Ghazikhani M, Hatami M, Safari B, Ganji DD. Experimental investigation of performance improving and emissions reducing in a two stroke SI engine by using ethanol additives. Propuls Power Res 2013;2(4):276–83.

[14] Gu X, Huang Z, Cai J, Gong J, Wu X, Lee CF. Emission characteristics of a spark-ignition engine fuelled with gasoline-n-butanol blends in combination with EGR. Fuel 2012;93:611–7.

[15] Costagliola MA, De Simio L, Iannaccone S, Prati MV. Combustion efficiency and engine out emissions of a SI engine fueled with alcohol/gasoline blends. Appl Energy 2013;111:1162–71.

[16] Al-Farayedhi AA, Al-Dawood AM, Gandhidasan P. Effects of blending MTBE with unleaded gasoline on exhaust emissions of SI engine. J Energy Resour Technol 2000;122(4):239–47.

[17] Ji C, Liang C, Wang S. Investigation on combustion and emissions of DME/gasoline mixtures in a spark-ignition engine. Fuel 2011;90(3):1133–8.

[18] Calam A, Solmaz H, Uyumaz A, Polat S, Yilmaz E, Içingür Y. Investigation of usability of the fusel oil in a single cylinder spark ignition engine. J Energy Inst 2015;88(3): 258–65.

[19] Elfasakhany A. Experimental investigation on SI engine using gasoline and a hybrid iso-butanol/gasoline fuel. Energ Conver Manage 2015;95:398–405.

[20] Chan JH, Tsolakis A, Herreros JM, Kallis KX, Hergueta C, Sittichompoo S, Bogarra M. Combustion, gaseous emissions and PM characteristics of di-methyl carbonate (DMC)-gasoline blend on gasoline direct injection (GDI) engine. Fuel 2020;263: 116742.

[21] Mei D, Zuo L, Adu-Mensah D, Li X, Yuan Y. Combustion characteristics and emissions of a common rail diesel engine using nanoparticlediesel blends with carbon nanotube and molybdenum trioxide. Appl Therm Eng 2019;162, 114238.

[22] Dhahad HA, Chaichan MT. The impact of adding nano-Al_2O_3 and nano-ZnO to Iraqi diesel fuel in terms of compression ignition engines' performance and emitted pollutants. Therm Sci Eng Prog 2020;18, 100535.

[23] Chacko N, Jeyaseelan T. Comparative evaluation of graphene oxide and graphene nanoplatelets as fuel additives on the combustion and emission characteristics of a diesel engine fuelled with diesel and biodiesel blend. Fuel Process Technol 2020;204, 106406.
[24] Zhang Z, Lu Y, Wang Y, Yu X, Smallbone A, Dong C, Roskilly AP. Comparative study of using multi-wall carbon nanotube and two different sizes of cerium oxide nanopowders as fuel additives under various diesel engine conditions. Fuel 2019;256, 115904.
[25] Ahmed A, Shah AN, Azam A, Uddin GM, Ali MS, Hassan S, Ahmed H, Aslam T. Environment-friendly novel fuel additives: investigation of the effects of graphite nanoparticles on performance and regulated gaseous emissions of CI engine. Energy Convers Manag 2020;211, 112748.
[26] Hasannuddin AK, Yahya WJ, Sarah S, Ithnin AM, Syahrullail S, Sidik NAC, Abu Kassim KA, Ahmad Y, Hirofumi N, Ahmad MA, Sugeng DA, Zuber MA, Ramlan NA. Nano-additives incorporated water in diesel emulsion fuel: fuel properties, performance and emission characteristics assessment. Energ Conver Manage 2018;169:291–314.
[27] Sahoo RR, Jain A. Experimental analysis of nanofuel additives with magnetic fuel conditioning for diesel engine performance and emissions. Fuel 2019;236:365–72.
[28] Saraee HS, Taghavifar H, Jafarmadar S. Experimental and numerical consideration of the effect of CeO_2 nanoparticles on diesel engine performance and exhaust emission with the aid of artificial neural network. Appl Therm Eng 2017;113:663–72.
[29] Hosseini SH, Taghizadeh-Alisaraei A, Ghobadian B, Abbaszadeh-Mayvan A. Performance and emission characteristics of a CI engine fuelled with carbon nanotubes and diesel-biodiesel blends. Renew Energy 2017;111:201–13.
[30] Saxena V, Kumar N, Saxena VK. Multi-objective optimization of modified nanofluid fuel blends at different TiO_2 nanoparticle concentration in diesel engine: experimental assessment and modeling. Appl Energy 2019;248:330–53.
[31] Raju VD, Kishore PS, Nanthagopal K, Ashok B. An experimental study on the effect of nanoparticles with novel tamarind seed methyl ester for diesel engine applications. Energ Conver Manage 2018;164:655–66.
[32] Vellaiyan S, Subbiah A, Chockalingam P. Multi-response optimization to improve the performance and emissions level of a diesel engine fueled with ZnO incorporated water emulsified soybean biodiesel/diesel fuel blends. Fuel 2019;237:1013–20.
[33] Praveen A, Rao GL, Balakrishna B. Performance and emission characteristics of a diesel engine using *Calophyllum inophyllum* biodiesel blends with TiO_2 nanoadditives and EGR. Egypt J Petrol 2018;27(4):731–8.
[34] El-Seesy AI, Attia AM, El-Batsh HM. The effect of aluminum oxide nanoparticles addition with Jojoba methyl ester-diesel fuel blend on a diesel engine performance, combustion and emission characteristics. Fuel 2018;224:147–66.
[35] Sunil S, Prasad BC, Kotresh M, Kakkeri S. Studies on suitability of multiwalled CNT as catalyst in combustion on a CI engine fueled with dairy waste biodiesel blends. Mater Today Proc 2020;26:613–9.
[36] Elkelawy M, Etaiw SE, Bastawissi HA, Marie H, Elbanna A, Panchal H, Sadasivuni K, Bhargav H. Study of diesel-biodiesel blends combustion and emission characteristics in a CI engine by adding nanoparticles of Mn (II) supramolecular complex. Atmos Pollut Res 2020;11(1):117–28.
[37] Gananamoorthi V, Devaradjane G. Effect of compression ratio on the performance, combustion and emission of DI diesel engine fueled with ethanol-diesel blend. J Energy Inst 2015;88:19–26.

CHAPTER 12

Influence of alcohol and gaseous fuels on NOx reduction in IC engines

C. Karthick[a], Kasianantham Nanthagopal[a], B. Ashok[a], and S.V. Saravanan[b]
[a]Engine Testing Laboratory, School of Mechanical Engineering, Vellore Institute of Technology, Vellore, Tamil Nadu, India
[b]Department of Mechanical Engineering, Asian College of Engineering and Technology, Coimbatore, India

12.1 Introduction

Various techniques such as fuel reformulation, the addition of oxygenated additives, octane boosters, and cetane improvers could improve the combustion and emission characteristics of conventional fuel in internal combustion (IC) engines. As a result of these techniques, some fuel characteristics will be improved to better the engine's performance. Moreover, these methods will influence particular fuel properties only. But the partial blending or complete replacement of alcohol and gaseous fuels with fossil fuels influences most fuel properties. Previous and ongoing studies on the use of alcohol and gaseous fuel in IC engines are effective due to their competence to reduce the reliance on conventional fossil fuels by increasing the percentage of renewable fuels. Nowadays, the emphasis on alcohols and gaseous fuels as potential fuels has been placed at the forefront of different policies enacted by emerging countries, with severe emission standards and decreasing fossil fuels [1]. Various types of emissions and their causes in automotive engines are shown in Fig. 12.1. Exhaust emissions from automotive engines have NOx (nitrogen oxides) emissions that are produced as a byproduct of the combustion of carbon-based fuels. Moreover, automotive vehicles around the world will contribute 50% of the overall NOx emissions. Various research works examined in this chapter indicate that NOx emissions varied considerably and increased at times owing to the combustion characteristics of the engine fuels [2]. A fuel's water content can substantially lower the engine exhaust emissions compared to alcohol content; nevertheless, this impacts the engine power and torque. In addition,

https://doi.org/10.1016/B978-0-12-823955-1.00012-7

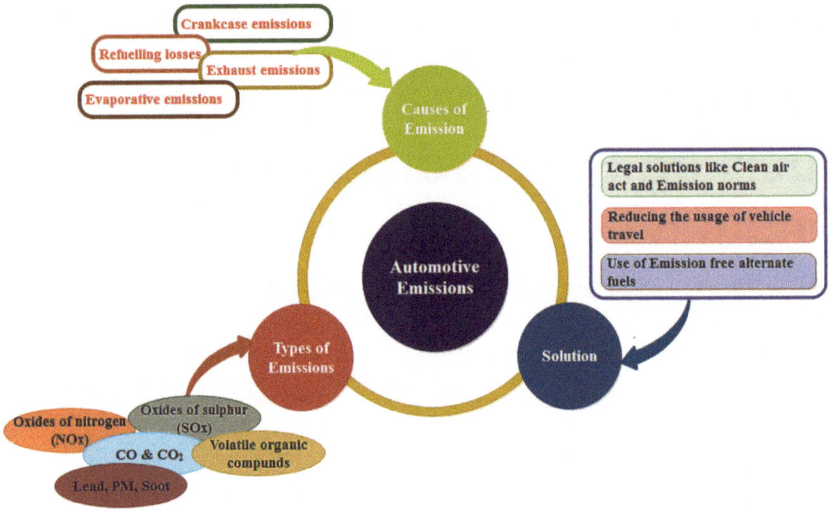

Fig. 12.1 Various kinds of automotive emissions, their causes, and possible solutions.

the maximum in-cylinder temperature in the cylinder is reduced. As previously stated, the interaction of nitrogen with engine fuel will cause higher NOx emission formation. It can also produce very hazardous greenhouse gases as they do not neglect oxidation in the combustion environment, which is rich in fuel, with lower temperatures. The exhaust gas temperature at different speeds of the test engine and the engine loads has offered important insight into how much NOx may be emitted in the fuel mixture. According to numerous research works, the blending of alcohol fuels such as methanol, ethanol, propanol, butanol, and pentanol in various proportions might lower NOx emissions. In contrast, many studies have found that blending alcohol with gasoline can raise NOx emissions. Although, increasing the alcohol concentration in the gasoline might result in NOx formation even at higher exhaust temperatures. This is a deciding parameter for the amount of NOx emitted by the engine [3]. NOx emissions are affected by the quantity of oxygen in the combustion zone as well as the temperature of combustion. They can be considerably decreased by utilizing alcohol fuel, which has a high latent heat of evaporation and therefore lowers the combustion temperature. However, because of the high oxygen concentration of the alcohol, NOx emissions may be enhanced. Furthermore, the cooling impact of the higher latent heat of evaporation does not considerably lower the in-cylinder temperature; the combustion process duration is reduced. Generally, there is a link between the amount of alcohol in a combination and the production of NOx [4]. Various fuel properties of alcohol and gaseous fuels influencing NOx emissions in IC engine are presented in Fig. 12.2. The fuel that contains

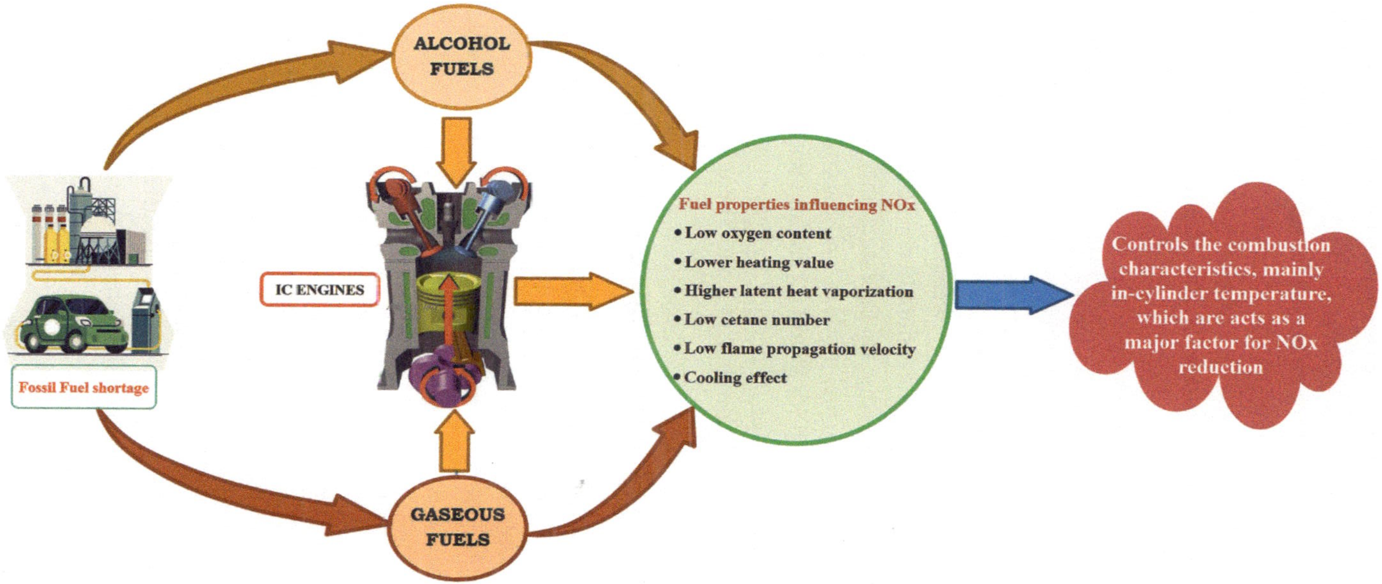

Fig. 12.2 Different fuel properties influencing NOx emissions.

nitrogen is also an essential source for NOx formation. NOx is considered a foremost source in an automotive engine. It is produced by the thermal formation of nitrogen oxides through the oxidation of two-atom nitrogen at the time of combustion with a temperature above 1800 K [5]. In recent years, various gaseous fuels have been researched as an energy resource to stabilize the effects of exhaust emissions from CI and SI engines. These gaseous fuels include hydrogen, compressed natural gas (CNG), biogas, etc., which are used as an alternate for diesel and gasoline in compressed ignition (CI) and spark ignition (SI) engines, respectively, due to the improved characteristics of the mixture proportions with air. Gaseous fuels have better octane value, thus they are an appropriate option for IC engines with higher compression ratios. Subsequently, gaseous fuel can restrict knocking better than fossil fuels. Moreover, gaseous fuels generate a lower amount of harmful engine exhaust gases when suitable conditions are identified for the mixture proportion and combustion [6]. This chapter discusses the suitability of alcohol fuels and gaseous fuels as blends and dual fuel mode as well as a complete replacement for conventional fossil fuels in both SI and CI engines without any engine modifications. Also, the influence of alcohol and gaseous fuel properties on the NOx emissions generated from both CI and SI engines is discussed.

12.2 Suitability of alcohol fuels for the engine application

Alcohol fuels are a suitable biofuel used as a partial and complete replacement for both diesel and gasoline fuels in IC engines. While blending the alcohol fuels with diesel increases the oxygen content, the heat of vaporization of the blended fuel subsequently reduces the emissions. Also, some properties of alcohol fuels such as cetane number, lower heating value, poor self-ignition, miscibility, and lubricating properties restrict its application as a complete diesel fuel replacement for CI engines. The properties of alcohol fuels depend on the number of carbons in the chemical structure. Further, alcohol fuels can be classified as lower alcohol and higher alcohol based on the number of carbon molecules. When the amount of carbon molecules is increased, the percentage of oxygen content (based on mass) is reduced and other properties such as density, cetane number, and calorific values are increased [7]. The physicochemical characteristics specify the quality of fuel to be used in an engine. The physicochemical properties of various alcohol fuels from carbon number C1 to C5 are compared in Fig. 12.3. Engine combustion characteristics, performance characteristics, and emission characteristics are highly reliant on the properties of the fuel. The

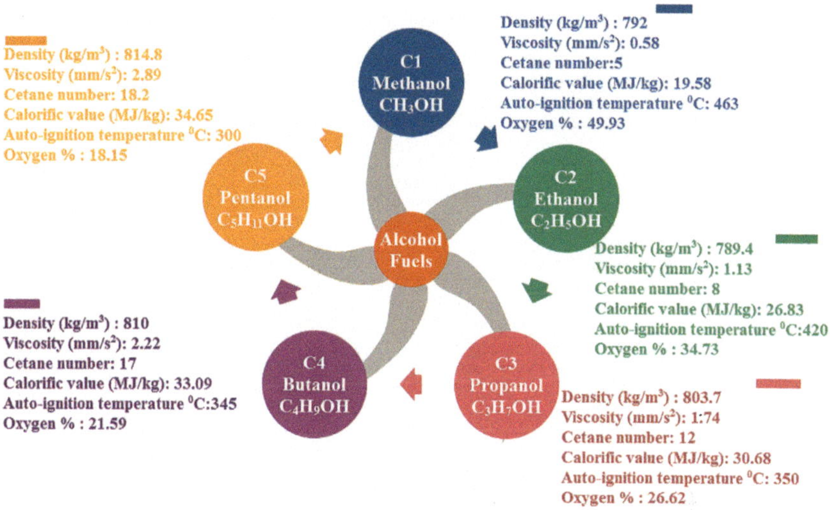

Fig. 12.3 Comparison of C1-C5 alcohol fuel properties.

superior characteristics that make alcohol fuels suitable for IC engine applications are provided in Table 12.1.

12.2.1 Methanol

Methanol is known as the first member of the alcohol family, having one carbon hydroxyl compound in its chemical structure. The existence of the hydroxyl compound in the chemical structure of methanol will contribute to the combustion process. Methanol has been recognized as a thriving additional oxygenated fuel for diesel engines. Also, methanol contains about 30% more oxygen than diesel, which allows CI engines to attain further complete combustion. Moreover, the use of methanol in SI engines is well known, but there are few studies on methanol use in CI engines [7]. Generally, a minor ignition delay at the combustion process is positive for the minimization of emissions. Moreover, the longer ignition delay and low cetane number will not be favorable to the 100% alternate for diesel in CI engines. So, instead of 100% replacement, the blending of methanol with diesel has been recommended for the CI engines. When methanol is blended with diesel, its low viscosity and greater volatility will reduce the viscosity of the blends and speed up vaporization. However, the in-cylinder temperature will be reduced with the use of methanol due to its higher latent heat of vaporization. Some drawbacks of methanol due to characteristics such as low cetane number, lubricity, heating value,

Table 12.1 Chemical and physical properties of various alcohol fuels [1, 8].

Properties	Gasoline	Diesel	Methanol	Ethanol	Propanol	Butanol	Pentanol
Chemical formula	C_8H_{15}	$C_{14}H_{30}$	CH_3OH	C_2H_5OH	C_3H_7OH	C_4H_9OH	$C_5H_{11}OH$
Carbon mass%	86	86.13	37.48	52.14	59.96	64.82	68.13
Hydrogen mass%	14	13.87	12.48	13.02	13.31	13.49	13.61
Oxygen mass%	–	–	49.93	34.73	26.62	21.59	18.15
Molecular weight (g/mole)	111.19	198.4	32.04	46.06	60.09	74.11	88.15
ΔH vap (kJ/kg)	351	232	1168	919.6	727.88	707.9	308.05
Vapor pressure (kPa)	–	0.0533	17	7.33	2.66	0.933	0.799
Density (kg/m^3)	737	849.2	792	789.4	803.7	810	814.8
Viscosity (mm/s^2)	0.5–0.6	2.72	0.58	1.13	1.74	2.22	2.89
Boiling temperature, ^0C	22–225	125–400	64.7	78	97.1	118	137.9
Flashpoint, °C	−45 to 38	>55	11	17	11.7	35–37	49
Self-ignition temperature, ^0C	420	254–300	463	420	350	345	300
Research octane number	88–98	–	109	109	–	98	–
Motor octane number	80–88	–	89	90	–	85	–
Cetane number	15-Oct	52	5	8	12	17	18.2–20
Calorific value, MJ/kg	43.5	42.49	19.58	26.83	30.68	33.09	34.65
Stoichiometric A/F ratio	14.58	14.95	6.46	8.98	–	11.17	–

difficulties in vaporization, and high self-ignition temperature will be reduced by blending with diesel. Because of the high octane number of methanol, blending it with diesel will minimize the different emissions produced during combustion [9].

12.2.2 Ethanol

Ethanol is known as the second member of the alcohol family, having two carbon hydroxyl compounds in its chemical structure. Ethanol requires the safest storage, handling, and transportation due to its lower flashpoint than fossil fuels. The high octane number of ethanol makes it possible for use in SI engines as a partial or complete replacement fuel for gasoline [2]. Moreover, ethanol has low toxicity because of its biological production process. Also, it has nearly 34% oxygen content by weight, lower density, lower viscosity, and better cold flow characteristics. This greater oxygen content of ethanol over gasoline can enhance the combustion efficiency and high combustion temperature. The lower C—H ratio and the high heat of evaporation can reduce the adiabatic flame temperature. The heating value of ethanol is lower than gasoline, which means an excess amount of fuel is required to attain the same output power in the engine. Moreover, ethanol necessitates more heat than gasoline owing to the higher heat of evaporation, which will increase the volumetric efficiency. Also, the high laminar flame speed of ethanol helps to end fuel combustion earlier, which improves engine performance [10].

12.2.3 Propanol

Propanol is another member of the alcohol family, and it has three carbon hydroxyl compounds in its chemical structure. There are two types of isomers available in propanol, n-propanol and isopropanol. Propanol is a better alternative fuel over the other alcohol members such as methanol and ethanol in some aspects with its lesser corrosive risks, high cetane number, and higher calorific value because of its higher carbon structure. Some other additional characteristics of propanol that can make it a perfect alternative fuel include greater energy density and a higher flashpoint than methanol and ethanol. Even though it has a considerably higher heat of vaporization than diesel, its low autoignition temperature will minimize the ignition delay time compared with the other alcohol members [3]. However, the higher production cost of propanol development restricts its possibility of using as an alternative fuel for automotive engines. Only a limited number of studies have been done on propanol, and research on the low-cost production of propanol is ongoing. Other than automotive engine applications, propanol has been used in the pharmaceutical industries, cosmetic production, and textile fabrication.

12.2.4 Butanol

Butanol is a well-researched biologically produced alcohol that has four carbon hydroxyl compounds in its chemical structure. Also, it has four different isomers–n-butanol, isobutanol, sec-butanol, and tert-butanol–with a similar chemical formula of C_4H_9OH. However, these isomers have similar heating values and slightly diverse combustion properties because of their different molecular structures. For instance, the combustion characteristics of n-butanol, isobutanol, and sec-butanol have the same burning rate and are differentiated by the structure of their flame [8]. In comparison with lower alcohols, butanol has various physicochemical properties that are similar to gasoline and diesel. Butanol has less toxicity and corrosiveness while being more hydrophilic than ethanol. Also, its higher flashpoint makes it highly companionable with the present fuel structure. Also, the viscosity of butanol is very similar to diesel, which is necessary for better lubrication. Moreover, butanol has some advantageous properties such as a high cetane number, higher heating value, lower latent heat of evaporation, lower viscosity, and better stability when blended with diesel. Due to the lower heat of evaporation, butanol needs low heat and low air temperature at the inlet compared with lower alcohol fuels. Moreover, the amount of butanol required is less than the lower alcohols because of its higher heating value. So, butanol is more effective than methanol and ethanol to attain better combustion and it can be a potential option for future biofuels [11].

12.2.5 Pentanol

Pentanol has five carbon hydroxyl compounds in its molecular structure, and it can be developed from renewable biomasses. Pentanol has a higher energy density than butanol as well as lower alcohols, which can further enhance the fuel economy rate. Pentanol has a high cetane number and can self-ignite better than methanol and ethanol. These characteristics will offer effective compatibility with CI engines. When pentanol is blended with diesel, it can enhance the fuel atomization value because of its high volatility and low kinematic viscosity. Also, pentanol has a higher oxygen content than lower alcohol fuels, which can minimize soot emissions [2]. In comparison with butanol and other lower alcohol fuels, pentanol attracts attention because of its high cetane number, greater energy content, and good stability when blending with diesel. Also, other characteristics such as viscosity, density, and latent heat of vaporization are similar to diesel properties. Like butanol, pentanol also has less corrosiveness, so that it acts as a corrosion inhibitor and protects the fuel supply line. Besides these benefits, pentanol shows no phase separation because of its highly hydrophobic nature over the lower alcohols [11]. Currently, there has been much research with the combustion and emission aspects of pentanol in both SI and CI engines.

12.3 Influence of alcohol fuels on NOx reduction in CI engines

The influence of various properties of alcohol fuels on NOx emission formation in CI engines is discussed in this section. Also, the effects on NOx influencing characteristics in lower alcohol fuels such as ethanol and methanol and higher alcohol fuels such as propanol, butanol, and pentanol due to different blend ratios is briefly discussed. The key sources of the variations in research comprise the kind of alcohol fuel and its characteristics, engine type, blending proportion, operating conditions, etc. The impacts of various alcohol fuels on NOx emissions in CI engines are shown in Table 12.2.

12.3.1 Lower alcohol fuels

NOx emissions vary critically because of the combustion characteristics of the lower alcohol fuels in the CI engine. In contrast, when engine combustion increases, NOx emission formation will be reduced. This condition will be applicable for all ranges of engine loads and blend ratios. Mostly, the NOx emissions formation changes depending on the temperature value at the combustion process. In that way, if the temperature attains a higher range at the end of combustion, then the NOx emissions will increase. The fuel properties influencing the in-cylinder temperature are the oxygen content of fuels, adiabatic flame temperature, latent heat of vaporization, laminar burning rate, air-fuel ratio, etc. In comparison with alcohol content, the water content can minimize the emissions. Among the emissions, the NOx emissions will not vary with compression ratio because the in-cylinder temperature differs in the engine. It's due to the water content, which will minimize the NOx emissions [3]. Generally, NOx emissions are formed when the nitrogen reacts with alcohol fuel. The temperature of the exhaust gas at various operating conditions such

Table 12.2 Impact of various alcohol fuels on NOx emissions in CI engines.

Fuel used	Injection	Operating conditions	NOx	Inference	Reference
Base fuel: Diesel, methanol, and ethanol each at 10, 15, 20, 25, 30, 35, 40%	Direct injection	Variable in-cylinder pressure	NOx ↑ when E % ↑, NOx ↑ when M% ↑, NOx of E > M	Excess amount of free oxygen	[12]
Base fuel: Diesel. Diesel 60% + tall oil 20% blended with M20, E20, B20, I20, F20. *E*-ethanol, M-methanol, B-butanol, I-isopropyl, F- fusel oil	CRDI	Variable load conditions	NOx ↑ when engine load ↑ for all blends	NOx ↑ due to the enrichment of oxygen in tall oil	[13]
Base fuel: Diesel. M10, M20. M-fethanol	Direct injection	BMEP	NOx ↑ for D100, M10 when load ↑. NOx ↓ for M20 when load ↑	NOx ↑ attributed to more oxygen content of methanol	[14]
Base fuel: Diesel. M10, M20, M10-nB10, M20-nB20. M-fethanol, nB-butanol	Direct injection	Variable load conditions	NOx ↑ when load ↑ for all blends	NOx ↑ due to more oxygen content	[15]
Base fuel: Diesel. B30, B20Bu10, B10Bu20, B15Bu15. B-biodiesel, Bu-butanol	Direct injection	Variable loads	NOx ↑ when torque ↑ for all blends	NOx ↑ due to slower combustion rate and higher heat released	[16]

Base fuel: Diesel. B10, B20, B30. B-butanol	Direct injection	Variable loads	NOx ↑ when load ↑ for all blends	NOx ↑ due to the higher oxygen concentration of butanol	[17]
Base fuel: Diesel. P20, B5P15, B10P10, B15P5. P-propanol, B-biodiesel	Direct injection	Variable engine loads	NOx ↑ for D80P20 when load ↑ for all blend	NOx ↑ attributed to latent heat of evaporation of propanol	[18]
Base fuel: Diesel. P10, P10EN2, P10EN4, P10EN6. P-propanol, EN-ethylhexyl nitrate	Direct injection	Variable loads	NOx ↑ when load ↑ for all blend	NOx ↑ due to the higher oxygen concentration of butanol	[19]
Base fuel: Diesel. Pentanol: 25%, 50%.	CRDI	BMEP	NOx ↓ when load ↑ for all blends	NOx ↓ due to the higher oxygen concentration and latent heat of evaporation of pentanol	[20]
Base fuel: Diesel. IP20, NP20. IP- isopropanol, NP- n-pentanol	CRDI	Variable loads	NOx ↓ when load ↑ for all blends	NOx ↓ attributed to lower peak in-cylinder temperature	[21]
Base fuel: Diesel. P10, P20, P30. P-pentanol	CRDI	Variable Loads	NOx ↑ when load ↑ for all blends	NOx ↑ lower heating value of pentanol	[22]

as different loads and speeds of the engine has led to important understandings of the range at which the blended fuel can release NOx. The use of lower alcohol fuels such as methanol and ethanol as alternate fuels in CI engines has been widely researched, and it has been discovered that they degrade engine performance, particularly when mixing levels surpass 10%. A few properties such as low cetane number, less calorific value, low flashpoint, bad miscibility with diesel, greater latent heat of evaporation, and high autoignition temperature restrict the use of lower alcohol fuels in CI engines [23]. Furthermore, by combining these alcohols with diesel, these blends are classified as Class I liquids, posing additional storage and handling challenges. The oxygenated characteristic of methanol and its poor cetane number outweigh the interactional effect of the robust quenching process instigated by its high heat of evaporation and low heating value, which will release higher NOx emissions. However, at low load conditions, the quenching process influence tends to decrease NOx at higher blend ratios of methanol with diesel. Moreover, if the methanol has been blended with any of the biodiesels, this leads to reduced NOx more than the diesel, owing to the cooling characteristics of methanol. Similarly, when ethanol is partially blended with diesel, the low calorific value and cooling effect balance the contrasting effect of the cetane value and oxygen content to produce a low amount of NOx emissions [24]. Besides, an increase in NOx value for methanol fuel has been observed due to the excess amount of free oxygen content. This leads to complete combustion, causing an increase in combustion temperature. Likewise, the same effect of an increase in NOx value has been identified while using the ethanol fuel in CI engines because of the greater oxygen concentration [12, 14].

12.3.2 Higher alcohol fuels

The application of various higher alcohol fuels such as propanol, butanol, and pentanol in CI engines and their properties on NOx emissions are discussed in this section. Higher alcohols with more than two carbon atoms have recently gained the attention of researchers due to the high cetane number and energy density, improved ignition rate, and blend stability. Furthermore, because of the increased carbon chain length, these alcohol fuels may be processed in transesterification, and the esters generated have a high cetane number, calorific value, and energy density. These higher alcohols mix well with diesel due to their low polarity, higher carbon percentage, and less hygroscopic nature. Also, they have less corrosiveness, higher flashpoint, and lower vapor pressure than diesel, which results in safe handling and reduced exhaust emissions [11]. Propanol blended with diesel has minimized NOx more than pure diesel. Also, the NOx value gets minimized while increasing the propanol percentage in the propanol-diesel blend. This is because of the cooling effect produced by the higher latent heat of evaporation of propanol, which decreases the combustion temperature and causes lesser NOx emissions [23]. Likewise, the butanol blended with diesel also results in slightly reduced NOx emissions more than the pure diesel, and they are further reduced with the increase in the percentage of butanol in the blend. The properties of butanol such as low calorific value and higher heat of vaporization lead to the cooling effect in the cylinder, which lowers the NOx emissions. This cooling effect turns against the high in-cylinder pressure produced at the premixed combustion stage due to low cetane value and longer ignition delay. Also, this is the reason behind the increase of NOx emissions in the ternary blends. However, lower NOx is also obtained for the ternary blends when the butanol percentage is exceeding the edge to counterbalance the shortcomings of prolonged ignition delay and lower cetane value [8]. Similar to propanol and butanol, pentanol also minimizes NOx emission formation when blended with diesel. Also, NOx generation has been reduced with the increase in the addition of pentanol percentage. As reported earlier, the higher latent heat of vaporization and lower cetane values of the pentanol reduce the in-cylinder temperature, lowering NOx emissions. For high load conditions, the more influential properties such as oxygen concentration and the lower cetane value of pentanol and the diesel blend prolong the ignition delay time, which causes bigger fuel drops in the premixed combustion process. This longer ignition delay will increase the combustion chamber temperature and NOx emission's thermal formation [25]. Additionally, a few other properties of pentanol such as the high atomic weight as well as the volatility and viscosity of the blend are also reasons behind the increase in NOx emission formation. Finally, the higher heat of vaporization, low energy density, and enhanced fuel atomization caused by the low viscosity of alcohol fuels influence the longer ignition delay to withstand the low temperature in the combustion chamber, thus reducing NOx formation.

12.4 Influence of alcohol fuels on NOx reduction in SI engines

The influence of various properties of alcohol fuels on NOx emission formation in SI engines is discussed in this section. The effects on NOx influencing characteristics in the lower alcohol fuels of ethanol and methanol and the higher alcohol fuels of propanol, butanol, and pentanol due to different blend ratios is briefly discussed. The impacts of various alcohol fuels on NOx emissions in SI engines is shown are Table 12.3.

12.4.1 Lower alcohol fuels

Several studies state that the addition of alcohol fuels with gasoline will increases the formation of NOx emissions. Meanwhile, greater alcohol concentration in the fuel composition can also generate NOx emissions at high exhaust gas temperatures. This alcohol content is the major influential factor to determine the amount of NOx emissions released from the engine. Many research works have shown that the addition of ethanol in the SI engine will increase NOx emissions. But some research works proposed the decrease of NOx emissions as well. Methanol at different ratios of 3%, 5%, 10%, and 15% has been tried with SI engines and the NOx emissions were compared with pure gasoline. From that, the methanol blend has an increasing trend on NOx when compared with gasoline due to more oxygen content in their molecular structure, and it achieves enhanced combustion efficiency. This more oxygen content in the fuel blend increases the cylinder pressure and temperature so that the NOx emissions have been increasing with methanol blends more than with gasoline. However, the decreasing trend of NOx emissions was also obtained with methanol blends of 5% and 10% because of the higher latent heat of vaporization than gasoline [37]. Similarly, 100% of methanol has been used instead of gasoline in SI engines and the NOx emissions produced are comparatively less than gasoline. This reduction is due to the higher latent heat of evaporation of methanol, which reduces the suction pressure and in-cylinder temperature during combustion. With the use of ethanol in SI engines, NOx emissions decrease with an increase in the ethanol content in the fuel blend. In comparison with pure gasoline and 85% ethanol in the ethanol-gasoline blend, the NOx emissions have been reduced due to a reduction in the flame temperature. Again, the ethanol percentage has been increased to 100%, so the NOx emission levels increase due to the advanced combustion. This advanced combustion leads to an increase of in-cylinder temperature and pressure when compared with 85% of ethanol. Also, the NOx emissions decrease when the ethanol percentage increases from 10% to 85% [38]. When using alcohol fuels in an SI engine, the combustion temperature is reduced, owing to the higher latent heat of vaporization and lower calorific value. This will lead to a decrease in NOx emissions. [39]. Sometimes, the NOx emissions of ethanol will increase because of the high pressure and high temperature at the

Table 12.3 Impacts of various alcohol fuels on NOx emissions in SI engines.

Fuel used	Injection	Operating conditions	NOx	Inference	Reference
Base fuel: Gasoline. Ethanol %: 25, 50, 75, 100	Electronic Injection System	Variable speed	NOx ↓ when E % ↑	NOx ↓ due to the higher latent heat of evaporation of ethanol	[26]
Base fuel: Gasoline. Ethanol %: 10, 20, 30, 40	Carburetor	Variable speed	NOx ↓ when E % ↑	NOx ↓ cooling energy flow effect	[27]
Base fuel: Gasoline. *E*-ethanol, B-butanol	Port Fuel Injection	Variable speed	NOx ↑ when EB ↑	NOx ↑ due to higher oxygen content	[28]
Base fuel: Gasoline. E-ethanol, B-butanol	Multiport Injection	Engine Speed	NOx ↑ for all blends	NOx ↑ due to the presence of oxygen in ethanol	[29]
Base fuel: Gasoline. Ethanol: 15, 30%	CRDI	Mean Effective Pressure	NOx ↓ when E15, NOx ↑ when E30	NOx ↑ due to higher oxygen content	[30]
Base fuel: Gasoline, Ethanol %: 10, 20, 30		Engine Speed	NOx ↓ when E ↑	NOx ↓ due to the higher latent heat of evaporation of ethanol	[31]
Base fuel: Gasoline. E20, E20L20, E20L40. E-ethanol, L-lemon peel oil	Port Fuel Injection	Mean Effective Pressure	NOx ↑ for all blends	NOx ↑ attributed to higher oxygen concentration	[32]
Base fuel: Gasoline. Ethanol + butanol%: 2,5,10,15,20	Port Fuel Injection	Engine Speed	NOx ↑ when blend ratio increases	NOx ↑ attributed to the presence of oxygen in ethanol	[28]

Continued

Table 12.3 Impacts of various alcohol fuels on NOx emissions in SI engines—cont'd

Fuel used	Injection	Operating conditions	NOx	Inference	Reference
Base fuel: Gasoline, ethanol	Port Fuel Injection	Engine Load	NOx↑ when load ↑	NOx↑ due to oxygen enrichment	[33]
Base fuel: Gasoline. methanol	MPFI	Methanol Fraction	NOx ↓ when load ↑	NOx↓ due to the low adiabatic flame temperature of the methanol	[34]
Base fuel: Gasoline. Pentanol %: 5, 10, 15, 20	Port Fuel Injection	Engine Load	NOx↑ when Pt↑	NOx↑ attributed to oxygen enrichment	[35]
Base fuel: Gasoline. Propanol %: 5, 10, 15, 20		Engine speed	NOx↑ when % P↑	NOx↑ due to higher oxygen content	[36]

combustion chamber. For example, while using 5% of ethanol (E5) with gasoline at high load conditions, the cylinder temperature tends to increase, which results in increased NOx emissions. Additionally, the more oxygen content of E5 results in a high NOx emission value. Also, NOx emissions of ethanol increase by nearly 25% when compared to gasoline. Up to 15% of ethanol blending in gasoline with the same combustion parameters will result in the maximum combustion temperature, which is a major factor for increased NOx formation. Additionally, ethanol blending will be the reason for the prolonged ignition delay due to more oxygen content. Consequently, the in-cylinder temperature increases because of the changes in combustion characteristics; the NOx emissions also increased. But, the NOx emissions have been reduced with an increase of ethanol concentration in the blend due to the low amount of heating value [7]. However, the lower energy density of ethanol will decrease the NOx emission level by increasing the ethanol concentration in the blend. Moreover, the NOx emission of the ethanol blend has been reduced due to the higher C—H molecule ratio of ethanol than gasoline. Furthermore, a low blend ratio of ethanol (up to 20%) shows that the NOx emission is greater than pure gasoline because of a greater heat release rate [40]. Likewise, another substantial factor for the increase of NOx emissions is the relative air-to-fuel ratio. Because the actual air/fuel ratio attains stoichiometric condition as the ethanol concentration in the blend is increased, subsequently complete combustion will be attained. This complete combustion will raise the combustion temperature; as a result, the NOx emissions also increase [37]. From this, we can state that the low blend ratio of ethanol will increase the NOx emissions due to enhanced combustion, resulting in a higher in-cylinder temperature. A high blend ratio of ethanol will lower the NOx emissions due to the decrease in in-cylinder temperature. This decrease of NOx emissions is because of the ethanol's high latent heat of evaporation, which reduces the in-cylinder temperature.

12.4.2 Higher alcohol fuels

In this section, the influence of various properties of higher alcohols in NOx formation is compared with gasoline and also with lower alcohols. The higher oxygen content of alcohol fuels leads to an increase in combustion temperature. At low-speed condition, the ethanol and isobutanol has lower NOx emission than gasoline. This is because of the higher latent heat of evaporation of ethanol and butanol than gasoline. But in high-speed conditions, NOx emissions will be increased because of the high fuel vaporization rate and in-cylinder temperature. This high fuel vaporization and complete combustion by butanol occurred because of the high oxygen content on their molecular structure. Some of the advantageous aspects for NOx generation are a higher combustion temperature and more oxygen content at peak temperature regions [29]. When comparing the methanol, ethanol, and butanol fuels blended with gasoline, the methanol-gasoline blend has less oxygen content than ethanol and butanol. So, methanol produces lesser NOx emissions than the other higher fuels. Likewise, NOx emissions of pure gasoline have been compared with the alcohol fuels of C1, C2, C3, and C4 each at 20% separately. The higher alcohol fuels such as propanol and butanol have lower NOx emissions than the lower alcohol fuels such as methanol and ethanol. Also, the lower alcohols resulted in high fuel consumption and higher cylinder temperature than the butanol and propanol at high engine speed. This increase in the cylinder temperature causes higher NOx emissions [41]. Sometimes, the isopropanol shows an increasing trend of NOx when compared with gasoline. Isopropanol blends of 10%, 20%, and 30% have been tried with SI engines, and they result in higher NOx emissions due to the high fuel consumption rate at high-speed conditions. Meanwhile, complete combustion has been achieved by the higher oxygen content of the isopropanol blend; due to this, NOx emissions also increase. Also, the NOx emission of isopropanol has been increased by the higher octane value of isopropanol, which results in shorter combustion [42]. Similarly, the higher NOx emissions of butanol blends than gasoline is due to the greater amount of oxygen concentration and lesser fuel-air ratio. Likewise, a similar increasing trend of NOx emission has been identified with 10%, 30%, and 100% of n-butanol blends. This increase is caused by the lower adiabatic temperature and thermal value of n-butanol than gasoline. On the other hand, NOx emission of butanol can be reduced by increasing the exhaust gas recycling process [43]. When the fuel is supplied with the carburetor-based method, the alcohols can reduce the in-cylinder temperature because of the higher latent heat of evaporation. This temperature reduction will reduce the NOx emission formation. Also, this is known as the major aspect for decreasing NOx emissions, particularly at high blend ratios as well as the pure condition of alcohol fuels. In lower blending ratios, the NOx emission may increase due to the improvement in the combustion process. Furthermore, the higher flame propagation speed of propanol and butanol will increase the combustion temperature. Then, more oxygen content of higher alcohols leads to the complete combustion of fuel, which causes more heat release. Finally, the higher antiknocking effect of propanol and butanol will permit higher compression ratio usage, resulting in increasing temperature and pressure after the compression process, which provides a favorable circumstance for Zeldovich NOx formation [44].

12.5 Suitability of gaseous fuels for engine applications

Among various gaseous fuels, hydrogen has promising characteristics for utilization in internal combustion engines. Some important characteristics of hydrogen include high calorific value, high diffusivity, wider flammability range, and high flame speed. The fuel properties of various gaseous fuels such as hydrogen, CNG, and biogas are compared in Fig. 12.4. Various important characteristics that make gaseous fuels suitable for IC engine application are provided in Table 12.4.

12.5.1 Hydrogen

Hydrogen (H_2) is a colorless, odorless, and flavorless gas. Comparatively, hydrogen is lighter than air by weight and fires with an invisible, unpolluted flame without soot and carbon deposits. Hydrogen is the only combustible fuel gas that does not have any carbon molecules. In comparison to widely used fuels, hydrogen also has a lower environmental impact. The exhaust emissions such as HC, CO, CO_2, and sulfur oxides are produced in less

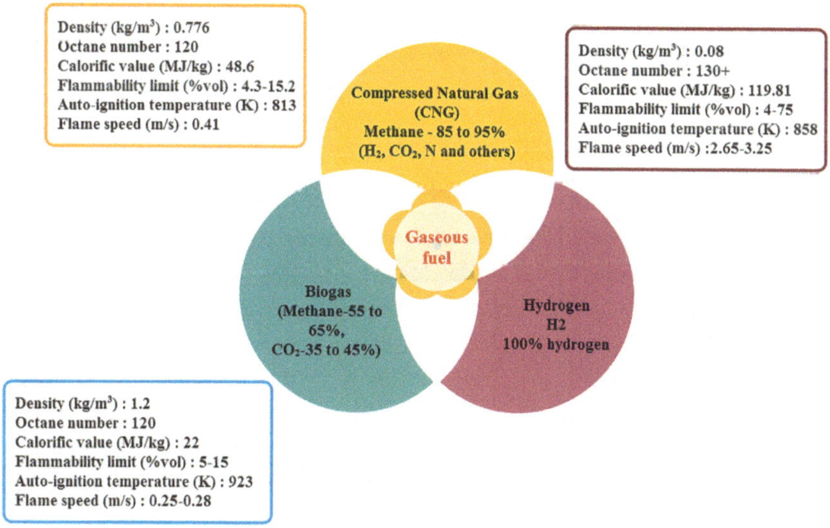

Fig. 12.4 Comparison of different gaseous fuel properties.

Table 12.4 Chemical and physical properties of various gaseous fuels [45–48].

Properties	Gasoline	Diesel	Hydrogen	Biogas	CNG
Chemical formula	C_8H_{15}	$C_{14}H_{30}$	H2	–	–
Carbon mass%	86	86.13	–	–	76.11
Hydrogen mass%	14	13.87	100	–	23.89
Molar mass (kg/mole)	109	204	2.02	–	17.3
Flammability limit (vol%)	1.2–6	0.7–5	4.0–75	5.0–15	4.3–15.2
Autoignition temperature, K	500–750	553	858	923	813
Adiabatic flame temperature, K	2470	2327	2379	2145–2199	2163
Density of gas at NTP (kg/m^3)	720–775	833–881	0.08	1.2	0.776
Lower heating value (MJ/kg)	44.79	42.5	119.81	22	48.6
Stoichiometric A/F ratio	14.58	14.95	34.3	6.95	17.2
Boiling temperature, ^{0}C	22–225	125–400	−252.9	–	–
Flame speed (m/s)	0.5	0.3	2.65–3.25	0.25–0.28	0.41
Gas constant	–	0.77	40.703	–	5.1147
Specific heat (Jg^{-1}K^{-1})	–	1.62	14.86	–	2.22

quantity when a hydrogen mixture is burned. However, even hydrogen has a high self-ignition temperature; it cannot be used in a CI engine without the use of a sparkplug. As a result, secondary fuel must be used as an ignition source to initiate hydrogen combustion in CI engines. Particularly, H_2 can be utilized in CI engines that run on dual fuel mode. The use of H_2 and biodiesel in CI engines is an effective option for reducing fossil fuel consumption. As a result, using biodiesel and H_2 as a dual fuel in CI engines improves the thermal performance, thereby lowering HC emissions. By eliminating the misfire phenomena, adding hydrogen will increase discharged emissions, efficiency, and fuel economy [49]. Furthermore, even though only a small volume of hydrogen is inducted into the cylinder, its high diffusivity results in stronger premixed petrol. Hydrogen fuel has calorific value of 119.81 MJ/kg, which is higher than the calorific value of diesel (42.5 MJ/kg). Also, hydrogen has a much lower density, so that hydrogen will be an unfavorable fuel to store at an adequate amount for driving range compared to other fuels. However, hydrogen burns in the existence of oxygen with a colorless (nearly light blue) flame at a high propagation speed. The capability to continuously ignite the mixture of air and hydrogen depends on the concentration of hydrogen in it. The peak interest in hydrogen as an automotive fuel is mostly owing to its high energy value and proenvironmental qualities. The higher energy density of hydrogen is the highest when compared with other familiar fuels used in IC engines. For example, 1 kg of hydrogen fuel can deliver nearly three times the energy of diesel fuel. [50].

12.5.2 Compressed natural gas

CNG, a nontoxic gaseous fuel, has numerous benefits for IC engine applications. Various characteristics such as the lower price of CNG than diesel and gasoline fuels due to greater availability make it suitable for IC engine applications. Natural gas has various components such as methane at nearly 90% and the remaining portion is occupied by the ethane, methane, and butane hydrocarbons. It also has some quantity of nitrogen and CO_2 gases. This nitrogen and CO_2 existence in CNG will lead to a diverse effect on the combustion characteristic such as a reduction of the heat release rate. Moreover, CNG fuel has a greater hydrogen percentage and a lesser carbon percentage in its composition when compared to gasoline. CNG has a lower heating value of 47.27 MJ/kg, which is more than gasoline. Also, the density of CNG fuel was significantly lesser than fossil fuels. This less density of CNG fuel will lead to a reduction in volumetric efficiency by the occupation of a huge volume in the combustion chamber. Due to the short carbon chain in its molecular structure, CNG has lesser gaseous emissions and the naturally purest exhaust emissions [49]. Moreover, CNG has high octane value presenting better resistance to knock, which makes the utilization of CNG in a higher compression ratio engine also possible. Probably, there are few difficulties in the application of CNG as a fuel in the CI engine. Because the self-ignition temperature of CNG is higher than diesel, it requires some ignition source such as a sparkplug or pilot fuel mode. Thus, CNG fuel is majorly applied in SI engines because of the similar properties when compared with gasoline. But CNG can be utilized in CI engines at dual fuel mode because of its other advantages [51]. For example, the cooling effect characteristics of CNG fuel cause a reduction in temperature at dual-fuel combustion. This low temperature in the combustion process minimizes the heat loss over the cylinder surfaces, which is a major factor to achieve high efficiency for diesel-CNG blends. CNG fuel has been extensively used for CI engines in dual fuel mode operation with pilot fuel injection. This dual fuel operation with pilot fuel will increase the thermal efficiency and simultaneously decreases the exhaust gas temperature so that the NOx emission formation is also lowered [52].

12.5.3 Biogas

Biogas is a widely accessible unconventional clean energy fuel source for IC engines due to its environmentally favorable character. Biogas, a sustainable fuel, is developed by the fermentation method in the absence of air from biological wastes in landfills and the digestion of sludge, agricultural wastes, municipal wastes, and living organic wastes. Over 15 days, the bacteria or microorganism converts the waste matter into biogas. Biogas has many constituent gases in it. The major constituent gases are methane (40%–70%), CO_2 (30%–60%), and other gases such as hydrogen and nitrogen, which are 1%–5% by volume. The major combustible part of biogas is methane (CH_4). Biogas has a comparatively lower carbon content than conventional fuels, which results in a decrease in exhaust emissions. Biogas has ignition temperatures in the range of 650–750°C. Also, biogas has less density that is 1/5 times lighter than air, and the calorific value lies between 18.6 and 26 MJ/m^3. But this calorific value and other characteristics are changeable when the composition of CO_2 has been varied. For the complete combustion of biogas, it requires an air -tomethane ratio of 10. Like CNG, biogas also has a better octane number to provide better resistance to knocking, so that it will be an appropriate option for engine applications even with a high compression ratio [49].

12.6 Influence of gaseous fuels on NOx reduction in CI engines

Different influential characteristics of gaseous fuels on NOx emission formation in CI engines are discussed in this section. Also, the effects on NOx influencing characteristics in gaseous fuels such as hydrogen, biogas, and CNG due to different mixture ratios are concisely discussed and summarized in Table 12.5. The key sources of variances in research comprise the varieties of gaseous fuel, fuel properties, mixing ratio, and test conditions (such as speed, load, etc.).

Table 12.5 Impact of various gaseous fuels on NOx emissions in CI engines.

Fuel used	Blend ratio	Injection mode	NOx	Inference	References
Hydrogen, Methane, Diesel	(H_2–30%, CH4–70%). Diesel–0%, 15%, 40%, 75%	PFI	NOx ↑	NOx increases 39.6% for the blend ratio, 75% due to higher in-cylinder temperature provided by hydrogen	[53]
Hydrogen, Diesel	H2–0.20, 0.40, 0.60, and 0.80 lpm	PFI	NOx ↑	NO_X increases due to hydrogen addition, which provides better combustion and leads to increased in-cylinder temperature at higher load	[54]
Hydrogen, Diesel, Biodiesel	H2–10, 20, 30, and 40 lpm	Hydrogen induction (Pilot injection)	NOx ↑	NO_X increases due to hydrogen addition at higher load	[55]
Hydrogen, Diesel	–	Hydrogen induction (Pilot injection)	NOx ↑	NO_X increases due to higher latent heat of vaporization of hydrogen	[56]
Diesel, Ethanol, CNG	–	CNG induction (Pilot injection)	NOx ↓	Absence of excess oxygen due to CNG replacing intake air also aids in lower NOx formation	[57]
Diesel, Biodiesel, CNG	15 lpm	Pilot injection	NOx ↑	NOx increases due to the higher availability of oxygen in biodiesel fuel	[58]

Continued

Table 12.5 Impact of various gaseous fuels on NOx emissions in CI engines—cont'd

Fuel used	Blend ratio	Injection mode	NOx	Inference	References
Diesel, Biodiesel, CNG	–	PFI	NOx ↑	NOx increases due to the higher flame temperature of CNG	[59]
Diesel, CNG, Biogas	–	PFI	NOx ↑, NOx ↓	NOx increases for CNG because of better and rapid combustion, NOx decreases for biogas due to lower peak combustion temperatures	[60]
Biodiesel, Diesel, Biogas		Biogas induction	NOx ↓	NOx decreases because the induction of biogas increases the specific heat capacity	[61]
Diesel, Methane	0–75%	PFI	NOx ↓	NOx decreases as methane concentration increases	[62]
Diesel, Methane	–	Direct injection	NOx ↑	NOx increases due to the higher intake charge temperature provided by methane	[63]

12.6.1 Hydrogen

High NOx emissions are one of the primary negative concerns of hydrogen combustion in an IC engine. These higher values of NOx emission occur at high engine load conditions. The major reason behind the generation of NOx emission in the combustion chamber is the reaction between the nitrogen and oxygen at higher temperatures produced during the combustion process. Also, some other compounds produced at the combustion chamber will cause the formation of smog and acid rain. Moreover, the major factors involved in NOx emission formation are in-cylinder temperature, air-fuel ratio, and rate of combustion. When hydrogen is utilized in dual-fuel mode, the high flame propagation velocity of hydrogen and little increase in ignition delay will lead to higher NOx emission formation than the diesel-operated engine. This trend is particularly applicable to the engine operated at high load conditions. Higher NOx formation because of an appropriate proportion of hydrogen is characterized by the higher flame velocity, which leads to an increase in combustion temperature and combustion rate. The increase of hydrogen energy share up to 30% will lead to an increase of NOx emissions by nearly 50% for dual fuel mode than diesel operated mode. Also, the cocombustion of hydrogen up to 40% with diesel fuel in a CI engine increases the NO emission at full engine load [46]. This drawback of hydrogen enrichment on NOx formation is comparatively low, and some research works reported a minor benefit in the reduction of NOx while the hydrogen percentage is not more than 10%. Furthermore, 50% of NOx formation has been reduced with the 30% of hydrogen share at a partial load operation. Meanwhile, the other emissions such as CO and smoke are also reduced at part load operations. Despite that, the reduction in NOx emission can be achieved because of the reduction of diesel combusted at a stoichiometric ratio near the spray fringe. But the hydrogen-to-air ratio is not noteworthy to proportionate the NOx formation. In addition, even a little quantity of hydrogen has a significant increase in NOx emission. This increase is mainly due to the unburned hydrogen that survives during the combustion phase and then oxidizes to increase the HO_2 level and encourages the NO to NO_2 conversion process. During the higher load operations, the hydrogen enrichment has a significant contrary influence on NOx emission formation. An intense rise of NOx emission with the increase in hydrogen proportion is achieved owing to the increase of pressure and the maximum rate of heat release. An exhaust gas recirculation (EGR) technique has been shown to be a quality tactic to reduce NOx formation and an applied technique for overwhelming the knocking of H_2 and diesel-based engines. The application of EGR in the combustion chamber countervails the increased heat release rate and cylinder pressure by hydrogen enrichment. Meanwhile, the EGR ratio increases, so the formation of NOx emission is reduced [64]. Like EGR, fuel injection techniques also play a vital role in NOx generation by hydrogen enrichment. The injection timing of fuel and the ratio of the hydrogen share have close reliance on the NOx emission formation and increase of cylinder pressure. However, the preinjection of fuel may cause a sudden increase in pressure in the cylinder, which causes higher NOx emissions. Contrarily, the advanced injection of fuel may offer an advantage toward the decrease of NOx emission. This reduction in NOx emission is due to the availability of enough timing for mixing the diesel molecules with hydrogen and air before the ignition. Also, the diesel becomes the ignition source for a CI engine cylinder. Consequently, perfect combustion with perfect heat release is attained. Delayed fuel injection after the top dead center may considerably decrease NOx formation. Due to this effect, a drawback of a decrease in thermal efficiency will happen. Though the increase in pressure and formation of NOx also depends on the quantity of hydrogen, these enriched hydrogen ratios lead to increased NOx [65].

12.6.2 Compressed natural gas

NOx emission formation is mainly dependent on the concentration of fuels and the in-cylinder temperature. For the utilization of CNG in the CI engine, the NOx emission is intensely linked to the air/fuel ratio and in-cylinder combustion temperature. Lean air/fuel mixtures and increased combustion temperature are the auspicious conditions for NOx formation. The lean-burn combustion process uses more excess air, generally requiring double the quantity for the complete combustion of fuels. On the other hand, a rich burn combustion engine operates at a stoichiometric air/fuel ratio, which is an accurately sufficient amount of air to burn the entire fuel quantity. This excess amount of air effectually cools down the maximum in-cylinder temperature during the combustion process, which decreases NOx emission formation. When the CNG burns at a low adiabatic flame temperature, it results in decreased NOx emissions. Even 50–80% of NOx emission reduction will be possible if the CNG fuel has been used in heavy-duty vehicles instead of diesel fuel [47]. Some research works report that the NOx formation rises with more diesel in the pilot mode and engine load. Meanwhile, the NOx emissions are reduced with higher speed at dual fuel mode operation. This decrease in NOx emission is because the specific heat capacity of NG is considerably higher than the air. In addition, the increase of natural gas percentage will increase the overall heat capacity of the combustion mixture. This will result in the decrease of temperature at the combustion chamber, which causes the decrease of NOx emission. This trend is more apparent at low load conditions of the engine because the combustion temperature is low at the low load conditions. Similarly, the long delay in the ignition of NG and diesel at dual fuel mode as well as the poor combustion quality of NG by the lean premixed condition will decrease the combustion temperature, thus resulting in a decrease of NOx emission. While increasing the ratio of natural gas in the air-fuel mixture leads to reduced oxygen content. The lower amount of oxygen content in the combustion chamber will cause decrease in NOx emission due to incomplete combustion. While increasing the engine speed, the residence time for the NOx emission formation will be less, leading to a decrease in NOx formation. An important reason for the increased NOx emissions while using natural gas is the better intensity of heat release at the premixed combustion process. This is due to the enhanced combustion of natural gas, which increases the peak cylinder temperature during combustion, resulting in increased NOx emission formation. Furthermore, a fuel with better combustion characteristics consumes a low amount of oxygen. This means that more oxygen will exist for NOx emission formation, thus increasing NOx emission formation [66].

12.6.3 Biogas

Various key parameters influencing the stimulation of NOx formation are high temperature, more oxygen concentration, and time of combustion. While increasing the load, the NOx emission also increases because of the increasing equivalence ratio for diesel and biogas-diesel mode. But in comparison with diesel, the NOx emission of the biogas-diesel mode is considerably lower. This reduction of NOx is due to the mixing of biogas and air, which leads to the reduction of oxygen content in the air/fuel mixture, thus decelerating the NOx formation rate. In addition, the CO_2 composition in biogas decreases the maximum combustion temperature because of the higher specific heat. Due to these effects, biogas utilization in CI engines has lesser NOx emission compared with diesel. At high load conditions, biogas undergoes rapid and complete combustion of comparably more biogas and diesel mixture than the low load conditions. This will lead to a higher temperature and pressure, which results in a higher NOx value [67]. For pure diesel, while increasing the load, the NOx formation starts to increase due to higher fuel consumption and higher temperature of exhaust gases. The enrichment of hydrogen and biogas with diesel reduces the NOx emission at high loads. The presence of biogas helps to reduce the oxygen content for the combustion process. Combustion of hydrogen increases the adiabatic flame temperature and hydrogen also contains water vapor content. Due to this water vapor, the increasing rate of exhaust gas temperature has been reduced. Also, the mixing of biogas substitutes the air and decreases the quantity of nitrogen in the engine cylinder. Thus, it leads to decreased NO and NO_2 emission values because of the reduced combustion temperature [68]. Sometimes, preheated biogas has been used in CI engines and resulted in a higher NOx concentration than the raw biogas. As a result of preheating the biogas, lower NOx emissions have been observed with raw biogas and minimum preheated temperatures. A higher NOx concentration was observed with the higher preheated temperature and also with diesel fuel. This preheating of biogas stimulates rapid chemical reactions and increases the temperature of the fuel in the combustion chamber. But this preheating of biogas will be beneficial in combustion efficiency because of the increase in combustion temperature [69]. In some cases, the oxygen concentration did not influence the formation of NOx emission. At a constant load and the same oxygen concentration, the NOx emission has been varied for the change of CO_2 proportion in the biogas mixture. When the mixture has higher CH4 and lower CO_2, then the NOx emission will be high. Otherwise, the increase of CO_2 ratio in the biogas mixture will reduce the NOx emission. The concentration of oxygen is not an influential factor on NOx emission when the ratio of CO_2 plays the role of controlling parameters [70].

12.7 Influence of gaseous fuels on NOx reduction in SI engines

Various influential properties of gaseous fuels on NOx emission in gasoline engines are discussed in this section. If the physicochemical properties of the gaseous fuel are not similar to gasoline, this will affect the combustion, performance, and emission characteristics of the SI engine. The influence of different properties of hydrogen, CNG, and biogas on NOx emissions due to different mixture ratios is briefly discussed. The impacts of gaseous fuels on NOx emissions in SI engines that are operated with various proportions of gaseous fuels are shown in Table 12.6.

Table 12.6 Impacts of various gaseous fuels on NOx emissions in SI engines.

Fuel used	Blend ratio	Injection mode	NOx	Inference	Reference
Gasoline, CNG, Hydrogen	10%, 18%	PFI	NOx ↑	NOx increases due to the addition of hydrogen, NOx of CNG is higher than gasoline	[71]
Gasoline, LPG	–	–	NOx ↑	NOx increases due to the higher burning speed and heating value of LPG at higher speed	[72]
Gasoline, LPG, Biogas	100%	PFI	NOx ↑, NOx ↓	NOx of LPG is higher than gasoline at higher load, NOx decreases due to the higher evaporation temperature of biogas	[73]
Hydrogen, CNG, Biogas	–	–	NOx ↑	NOx increases as hydrogen concentration increases	[74]
Methane, hydrogen	60%, 95%	–	NOx ↑	NOx increases due to the addition of methane	[75]
	–	DI	NOx ↓		[76]

Continued

Table 12.6 Impacts of various gaseous fuels on NOx emissions in SI engines—cont'd

Fuel used	Blend ratio	Injection mode	NOx	Inference	Reference
Methane, hydrogen				NOx emissions are lower for methane when compared to gasoline	
Biogas, syngas, hydrogen			NOx ↑	NOx increases due to the presence of hydrogen in syngas	[77]
Biogas, Hydrogen	–	–	NOx ↑	NOx increases due to the higher burning velocity of hydrogen	[78]

12.7.1 Hydrogen

The SI engine fueled with hydrogen releases NOx emission as a major emission because of its carbonless nature. Some CO and HC emissions were also seen in exhaust emissions due to the vaporization of the oil used for lubrication. Increased combustion temperature, oxygen content, and different types of chemical molecules in the combustion chamber are the major deciding factors for higher NOx formation in SI engines fueled with hydrogen rather than gasoline. As mentioned in the use of hydrogen in CI engines, the increased NOx emission is one of the important drawbacks in hydrogen-fueled SI engines. This increasing NOx formation will be reduced by the introduction of EGR and the injection of water at the intake manifold of the engine. Moreover, the injection of water at the intake manifold exponentially decreases the NOx emission. The primary reason for the extreme reduction of NOx emission is a decrease in in-cylinder temperature due to the mixtures in the cylinder being fully diluted with water. Also, the injection of water increases the number of chemical species present in the combustion process. Even nearly ultralow NOx emissions can be achievable with the injection of water in the intake manifold [74]. At various engine speeds, the NOx emission of hydrogen fuel has been compared with pure gasoline. The use of hydrogen in the SI engine resulted in an intense increase in NOx emission when compared with gasoline. The major factor influencing the rapid increase of NOx emission is the higher flame velocity of hydrogen than gasoline, which leads to maximum in-cylinder temperature. In this scenario, the remarkable developments in NOx emission reduction occur because of the excess air. The main reason for this control of NOx emission is the lean combustion of hydrogen and reduction in maximum in-cylinder temperature with an increase in excess air. Whenever the excess air ratio is kept constant, then the major factor influencing the NOx emission will be the in-cylinder temperature. Another way of reducing the NOx emission is by increasing the first injection proportion of hydrogen. The higher first injection proportion means the injection of hydrogen at an early stage, thus the distribution of hydrogen molecules in the fuel mixture is homogeneous. This means that the maximum in-cylinder temperature decreases, which leads to the reduction of NOx emission. Hence, the maximum in-cylinder temperature is less and the higher temperature region is small, which also causes lower NOx emissions than the direct injection of hydrogen. When hydrogen is added with methane at 10% and 30% load in the SI engine, the NOx emission is reduced and then again increases for the further addition of hydrogen. The addition of hydrogen up to 3% identified a decreasing trend of NOx emission because the introduction of a rich fuel mixture zone causes a lack of oxygen content. This leads to incomplete combustion of the hydrogen fuel. When the percentage of hydrogen increases more than 3%, the NOx emission also increases. This increase of NOx is due to the higher flame temperature of hydrogen [76].

12.7.2 Compressed natural gas

The constructive factors for higher NOx emission formation are more oxygen availability and maximum combustion temperature. As we know, the combustion temperature increases while adding the load, which leads to more NOx formation. When compared with the single fuel mode, the dual fuel mode has less NOx emission formation. This reduced NOx formation is due to the low premixed combustion process of CNG induction into the cylinder, which reduces the maximum combustion temperature. Additionally, this induction of CNG in dual fuel mode replaces the same quantity of air, thus reducing the other constructive factor of more oxygen content available for NOx formation by oxidation [79]. The influence of CNG fuel on NOx emission formation with different injection strategies of DI, PFI, and GDI systems shows lower NOx emission for the low speed and low load. Moreover, a linear increase in NOx formation is due to the poorer mixing of CNG with the air in the engine cylinder at low speed and low load conditions, thus forming a lean air/fuel mixture region. Under this lean air-fuel mixture environment, the in-cylinder temperature rises, thus generating higher NOx emission [71]. The varying trends of NOx emission with various CNG ratios and injection timings are studied. From that, the NOx emission starts to increase and then decreases due to the earlier injection. Under the condition of $\lambda = 1$, the NOx emission increases to the maximum when the CNG is injected at a crank angle of 120 degrees before TDC. At the same time, for $\lambda = 1.2$ and $\lambda = 1.4$, the NOx emission increases to the maximum when the CNG is injected at a crank angle of 105 degrees before TDC. In comparison with $\lambda = 1.2$ and 1.4, $\lambda = 1.4$ has low NOx emission due to the availability of more excess air. Some quantity of this excess air will not participate in the combustion process and carries more heat. The combustion temperature will be reduced and NOx emission will be decreased. Mostly, the in-cylinder temperature is the influential factor that affects NOx emission formation in a specific λ condition. The changes in in-cylinder temperature are due to the changes in the heat release rate during combustion because of various injection timing and mixing ratios. However, the delayed injection of CNG into the engine cylinder will lead to the accumulation of CNG around the sparkplug. This accumulation will inhibit NOx emission formation due to the nonconductive heat release during combustion. Also, the earlier injection inhibits NOx formation due to the homogeneous fuel mixtures, which restricts the heat release [79].

12.7.3 Biogas

The availability of oxygen in the gaseous fuel and the increase in temperature at the combustion process are openly related to NOx emission formation. While using biogas with 65% methane and 35% CO_2 in the SI engine, the NOx emission starts to increase from almost zero to maximum when the λ increases from 1 to 1.5. But the SI engine operated with other gaseous fuels such as natural gas has higher NOx at $\lambda = 1$ based on the Zeldovich mechanism, due to the increased temperature in the cylinder combustion chamber. For biogas, the existence of CO_2 in its mixture makes the fuel's heating value low. The in-cylinder temperature of the combustion chamber is reduced, which leads to reduced NOx emission. This CO_2 content in the biogas will generate a low amount of NOx emissions at both $\lambda = 1$ and 1.5 when compared with CNG [77]. The comparison of the biogas compositions such as 95% methane and 60% methane has resulted in more NOx emission generation at stoichiometric conditions. Whenever the engine is operated with methane, the NOx emission formation will be more due to the high temperature in the combustion chamber. In addition, if hydrogen has been added with the biogas, then the NOx formation will be further increased. This is due to the negative influence of hydrogen by its characteristics such as rapid burning speed and higher flame temperature. But this hydrogen addition with biogas will reduce the other emissions such as HC and CO because of the rise of the in-cylinder temperature. Besides, when the temperature exceeds 1400 K, then the rate of NOx emission formation will be more rapid [75]. However, the reduction of NOx emission is also achieved with biogas when compared to gasoline. At both full throttle and part throttle operations, the biogas can achieve a nearly 50% reduction of NOx emissions. Even though biogas consists of nitrogen in its composition, the latent heat of evaporation tends to decrease the maximum combustion temperature. This is due to the higher vaporization temperature of biogas compared with gasoline [73]. In contrast with gasoline, the biogas generates lesser NOx emission in exhaust gas while increasing the engine speed. This is because the mean gas temperature of a gasoline-operated SI engine was higher than the biogas. The NOx emission generated by the biogas was less than gasoline at all operating engine speeds. Generally, the NOx emission will be maximum with a lean mixture rather than the stoichiometric mixtures. The increase of oxygen concentration in the biogas will cause complete combustion and raise the combustion temperature. As a result, the increased oxygen concentration in biogas will provide increased NOx formation. Moreover, the enhancement of oxygen concentration raises the exhaust gas temperature, thus the combustion temperature will also increase. At any equivalence ratio, an increase of oxygen concentration is known as increasing the quantity of methane combusted [77].

12.8 Conclusion

A methodical assessment of the consequences of various fuel characteristics and their blending ratios that influence NOx emissions by the alcohol and gaseous fuels has been done, and the major conclusions are provided.

1. Most studies show lesser NOx emission compared with fossil fuels and NOx emissions are relatively increased when the engine load increases with both alcohol and gaseous fuels.
2. Numerous parameters are involved in the increase of NOx emissions in alcohol fuels such as high heat release rate, higher adiabatic temperature, complete combustion of fuel, more oxygen content, more cooling effect, etc.
3. Various properties related to the atomization process such as density, fuel viscosity, and heat capacity also have a substantial effect on NOx emission formation.
4. Numerous results show that more oxygen concentration in fuel composition will be the factor for higher NOx emission by increasing the temperature during the combustion reaction. But the higher latent heat of vaporization of alcohol fuels will reduce the in-cylinder temperature, thus reducing NOx formation.
5. Increasing the hydrogen proportion in the mixture has an impact of higher NOx formation due to its high burning velocity and higher latent heat of vaporization.
6. In most cases, CNG has lower NOx formation because it replaces the intake air in the cylinder, thus reducing the availability of excess oxygen content. Also, the higher flame temperature of CNG will cause increased NOx emission.

7. In general, biogas has lower NOx emission in dual fuel mode than single fuel mode. Moreover, the CO_2 concentration in the biogas will influence the NOx formation in that more CO_2 will reduce the in-cylinder temperature, thus reducing the NOx emission.

Alcohol and gaseous fuels are mixed with diesel and gasoline to attain various objectives of enhancing fuel properties, performance, combustion, and emission characteristics of the engine. But considering the NOx formation, the outcomes have been different. These differences may specify that the alcohol and gaseous fuels can lower the NOx emission when the appropriate operating parameters of the engine and suitable blending ratios are fixed. Moreover, the utilization of these alcohol fuels and gaseous fuels may cause some other issues such as increasing other emissions, affecting the performance, and corrosion or damage to engine parts. An endurance study should be carried out to ensure that the alcohol and gaseous fuels have no negative impact on engine life.

References

[1] Thiyagarajan S, Geo VE, Ashok B, Nanthagopal K, Vallinayagam R, Saravanan CG, Kumaran P. NO x emission reduction using permanent/electromagnet-based fuel reforming system in a compression ignition engine fueled with pine oil. Clean Techn Environ Policy 2019;21(4):815–25.

[2] Nanthagopal K, Kishna RS, Atabani AE, Ala'a H, Kumar G, Ashok B. A compressive review on the effects of alcohols and nanoparticles as an oxygenated enhancer in compression ignition engine. Energy Convers Manag 2020;203:112244.

[3] Mamat R, Sani MS, Sudhakar K, Kadarohman A, Sardjono RE. An overview of higher alcohol and biodiesel as alternative fuels in engines. Energy Rep 2019;5:467–79.

[4] Bragadeshwaran A, Kasianantham N, Balusamy S, Muniappan S, Reddy DM, Subhash RV, Pravin NA, Subbarao R. Mitigation of NOx and smoke emissions in a diesel engine using novel emulsified lemon peel oil biofuel. Environ Sci Pollut Res 2018;25(25):25098–114.

[5] Hill SC, Smoot LD. Modeling of nitrogen oxides formation and destruction in combustion systems. Prog Energy Combust Sci 2000;26(4–6):417–58.

[6] Sahoo BB, Sahoo N, Saha UK. Effect of engine parameters and type of gaseous fuel on the performance of dual-fuel gas diesel engines—a critical review. Renew Sust Energ Rev 2009;13(6–7):1151–84.

[7] Velvizhi G, Shanthakumar S, Das B, Pugazhendhi A, Priya TS, Ashok B, Nanthagopal K, Vignesh R, Karthick C. Biodegradable and non-biodegradable fraction of municipal solid waste for multifaceted applications through a closed loop integrated refinery platform: paving a path towards circular economy. Sci Total Environ 2020;731:138049.

[8] Karthick C, Nanthagopal K. A comprehensive review on ecological approaches of waste to wealth strategies for production of sustainable biobutanol and its suitability in automotive applications. Energy Convers Manag 2021;239:114219.

[9] Verhelst S, Turner JW, Sileghem L, Vancoillie J. Methanol as a fuel for internal combustion engines. Prog Energy Combust Sci 2019,70:43–88.

[10] Sakthivel P, Subramanian KA, Mathai R. Indian scenario of ethanol fuel and its utilization in automotive transportation sector. Resour Conserv Recycl 2018;132:102–20.
[11] Babu V, Murthy M. Butanol and pentanol: the promising biofuels for CI engines–a review. Renew Sust Energ Rev 2017;78:1068–88.
[12] Jamrozik A. The effect of the alcohol content in the fuel mixture on the performance and emissions of a direct injection diesel engine fueled with diesel-methanol and diesel-ethanol blends. Energy Convers Manag 2017;148:461–76.
[13] Özer S. The effect of diesel fuel-tall oil/ethanol/methanol/isopropyl/n-butanol/fusel oil mixtures on engine performance and exhaust emissions. Fuel 2020;281:118671.
[14] Huang J, Xiao H, Yang X, Guo F, Hu X. Effects of methanol blending on combustion characteristics and various emissions of a diesel engine fueled with soybean biodiesel. Fuel 2020;282:118734.
[15] Saiteja P, Ashok B. A critical insight review on homogeneous charge compression ignition engine characteristics powered by biofuels. Fuel 2021;285:119202.
[16] Thakkar K, Kachhwaha SS, Kodgire P, Srinivasan S. Combustion investigation of ternary blend mixture of biodiesel/n-butanol/diesel: CI engine performance and emission control. Renew Sust Energ Rev 2020;110468.
[17] Gawale GR, Srinivasulu GN. Experimental investigation of propanol dual fuel HCCI engine performance: optimization of propanol mass flow rate, impact of butanol blends (B10/B20/B30) as fuel substitute for diesel. Fuel 2020;279:118535.
[18] Bencheikh K, Atabani AE, Shobana S, Mohammed MN, Uğuz G, Arpa O, Kumar G, Ayanoğlu A, Bokhari A. Fuels properties, characterizations and engine and emission performance analyses of ternary waste cooking oil biodiesel–diesel–propanol blends. Sustain Energy Technol Assess 2019;35:321–34.
[19] Ravi S, Karthikeyan A. Assessment of performance and emission characteristics of diesel engine supplied with waste plastic oil propanol and ethylhexyl nitrate blends. Mater Today: Proc 2021;44:3642–6.
[20] Pan M, Huang R, Liao J, Jia C, Zhou X, Huang H, Huang X. Experimental study of the spray, combustion, and emission performance of a diesel engine with high n-pentanol blending ratios. Energy Convers Manag 2019;194:1-0.
[21] Chen H, Zhou Z, He J, Zhang P, Zhao X. Effect of isopropanol and n-pentanol addition in diesel on the combustion and emission of a common rail diesel engine under pilot plus main injection strategy. Energy Rep 2020;6:1734–47.
[22] Santhosh K, Kumar GN, Sanjay PV. Experimental analysis of performance and emission characteristics of CRDI diesel engine fueled with 1-pentanol/diesel blends with EGR technique. Fuel 2020;267:117187.
[23] Yusri IM, Mamat R, Najafi G, Razman A, Awad OI, Azmi WH, Ishak WF, Shaiful AI. Alcohol based automotive fuels from first four alcohol family in compression and spark ignition engine: a review on engine performance and exhaust emissions. Renew Sust Energ Rev 2017;77:169–81.
[24] Ghadikolaei MA, Wong PK, Cheung CS, Ning Z, Yung KF, Zhao J, Gali NK, Berenjestanaki AV. Impact of lower and higher alcohols on the physicochemical properties of particulate matter from diesel engines: a review. Renew Sust Energ Rev 2021;143:110970.
[25] Nanthagopal K, Ashok B, Saravanan B, Patel D, Sudarshan B, Ramasamy RA. An assessment on the effects of 1-pentanol and 1-butanol as additives with Calophyllum Inophyllum biodiesel. Energy Convers Manag 2018;158:70–80.
[26] Celik MB. Experimental determination of suitable ethanol–gasoline blend rate at high compression ratio for gasoline engine. Appl Therm Eng 2008;28(5–6):396–404.
[27] Mohammed MK, Balla HH, Al-Dulaimi ZM, Kareem ZS, Al-Zuhairy MS. Effect of ethanol-gasoline blends on SI engine performance and emissions. Case Stud Therm Eng 2021;25:100891.

[28] Mourad M, Mahmoud K. Investigation into SI engine performance characteristics and emissions fuelled with ethanol/butanol-gasoline blends. Renew Energy 2019;143:762–71.
[29] Zaharin MS, Abdullah NR, Masjuki HH, Ali OM, Najafi G, Yusaf T. Evaluation on physicochemical properties of iso-butanol additives in ethanol-gasoline blend on performance and emission characteristics of a spark-ignition engine. Appl Therm Eng 2018;144:960–71.
[30] Belgiorno G, Di Blasio G, Shamun S, Beatrice C, Tunestål P, Tunér M. Performance and emissions of diesel-gasoline-ethanol blends in a light duty compression ignition engine. Fuel 2018;217:78–90.
[31] Doğan B, Erol D, Yaman H, Kodanli E. The effect of ethanol-gasoline blends on performance and exhaust emissions of a spark ignition engine through exergy analysis. Appl Therm Eng 2017;120:433–43.
[32] Biswal A, Gedam S, Balusamy S, Kolhe P. Effects of using ternary gasoline-ethanol-LPO blend on PFI engine performance and emissions. Fuel 2020;281:118664.
[33] Wang X, Chen Z, Ni J, Liu S, Zhou H. The effects of hydrous ethanol gasoline on combustion and emission characteristics of a port injection gasoline engine. Case Stud Therm Eng 2015;6:147–54.
[34] Yilmaz İ, Taştan M. Investigation of hydrogen addition to methanol-gasoline blends in an SI engine. Int J Hydrog Energy 2018;43(44):20252–61.
[35] Yaman H, Yesilyurt MK. The influence of n-pentanol blending with gasoline on performance, combustion, and emission behaviors of an SI engine. Eng Sci Technol 2021.
[36] Mourad M, Mahmoud KR. Performance investigation of passenger vehicle fueled by propanol/gasoline blend according to a city driving cycle. Energy 2018;149:741–9.
[37] Masum BM, Masjuki HH, Kalam MA, Fattah IR, Palash SM, Abedin MJ. Effect of ethanol–gasoline blend on NOx emission in SI engine. Renew Sust Energ Rev 2013;24:209–22.
[38] De Simio L, Gambino M, Iannaccone S. Effect of ethanol content on thermal efficiency of a spark-ignition light-duty engine. Int Sch Res Not 2012;2012:01–8.
[39] Kumar AN, Ashok B, Nanthagopal K, Ong HC, Geca MJ, Victor J, Vignesh R, Jeevanantham AK, Kannan C, Kishore PS. Experimental analysis of higher alcohol–based ternary biodiesel blends in CI engine parameters through multivariate and desirability approaches. Biomass Convers Biorefin 2020;1–6.
[40] Chen R, Nishida K. Spray evaporation of ethanol–gasoline-like blend and combustion of ethanol–gasoline blend injected by hole-type nozzle for direct-injection spark ignition engines. Fuel 2014;134:263–73.
[41] Li Y, Gong J, Deng Y, Yuan W, Fu J, Zhang B. Experimental comparative study on combustion, performance and emissions characteristics of methanol, ethanol and butanol in a spark ignition engine. Appl Therm Eng 2017;115:53–63.
[42] Sivasubramanian H, Pochareddy YK, Dhamodaran G, Esakkimuthu GS. Performance, emission and combustion characteristics of a branched higher mass, C3 alcohol (isopropanol) blends fueled medium duty MPFI SI engine. Eng Sci Technol 2017;20 (2):528–35.
[43] Zhen X, Wang Y, Liu D. Bio-butanol as a new generation of clean alternative fuel for SI (spark ignition) and CI (compression ignition) engines. Renew Energy 2020;147:2494–521.
[44] Shirazi SA, Abdollahipoor B, Windom B, Reardon KF, Foust TD. Effects of blending C3-C4 alcohols on motor gasoline properties and performance of spark ignition engines: a review. Fuel Process Technol 2020;197:106194.
[45] Yan F, Xu L, Wang Y. Application of hydrogen enriched natural gas in spark ignition IC engines: from fundamental fuel properties to engine performances and emissions. Renew Sust Energ Rev 2018;82:1457–88.

[46] Jamrozik A, Grab-Rogaliński K, Tutak W. Hydrogen effects on combustion stability, performance and emission of diesel engine. Int J Hydrog Energy 2020;45 (38):19936–47.
[47] Khan MI, Yasmin T, Shakoor A. Technical overview of compressed natural gas (CNG) as a transportation fuel. Renew Sust Energ Rev 2015;51:785–97.
[48] Ashok B, Ashok SD, Kumar CR. LPG diesel dual fuel engine–a critical review. Alex Eng J 2015;54(2):105–26.
[49] Mirhashemi FS, Sadrnia H. NOX emissions of compression ignition engines fueled with various biodiesel blends: a review. J Energy Inst 2020;93(1):129–51.
[50] Gürbüz H. Analysis of the effects of multiple injection strategies with hydrogen on engine performance and emissions in diesel engine. Int J Hydrog Energy 2020;45 (51):27969–78.
[51] Thiyagarajan S, Sonthalia A, Geo VE, Chokkalingam B. Effect of waste exhaust heat on hydrogen production and its utilization in CI engine. Int J Hydrog Energy 2020;45 (10):5987–96.
[52] Khayum N, Anbarasu S, Murugan S. A role of the combined effect of fuel injection parameters on a dual fuel diesel engine. Mater Today: Proc 2021.
[53] Karagöz Y, Güler İ, Sandalcı T, Yüksek L, Dalkılıç AS, Wongwises S. Effects of hydrogen and methane addition on combustion characteristics, emissions, and performance of a CI engine. Int J Hydrog Energy 2016;41(2):1313–25.
[54] Koten H. Hydrogen effects on the diesel engine performance and emissions. Int J Hydrog Energy 2018;43(22):10511–9.
[55] Akcay M, Yilmaz IT, Feyzioglu A. Effect of hydrogen addition on performance and emission characteristics of a common-rail CI engine fueled with diesel/waste cooking oil biodiesel blends. Energy 2020;212:118538.
[56] Deb M, Paul A, Debroy D, Sastry GR, Panua RS, Bose PK. An experimental investigation of performance-emission trade off characteristics of a CI engine using hydrogen as dual fuel. Energy 2015;85:569–85.
[57] Paul A, Bose PK, Panua RS, Banerjee R. An experimental investigation of performance-emission trade off of a CI engine fueled by diesel–compressed natural gas (CNG) combination and diesel–ethanol blends with CNG enrichment. Energy 2013;55:787–802.
[58] Paul A, Panua RS, Debroy D, Bose PK. An experimental study of the performance, combustion and emission characteristics of a CI engine under dual fuel mode using CNG and oxygenated pilot fuel blends. Energy 2015;86:560–73.
[59] Ghaffarzadeh S, Toosi AN, Hosseini V. An experimental study on low temperature combustion in a light duty engine fueled with diesel/CNG and biodiesel/CNG. Fuel 2020;262:116495.
[60] Verma S, Das LM, Bhatti SS, Kaushik SC. A comparative exergetic performance and emission analysis of pilot diesel dual-fuel engine with biogas, CNG and hydrogen as main fuels. Energy Convers Manag 2017;151:764–77.
[61] Yoon SH, Lee CS. Experimental investigation on the combustion and exhaust emission characteristics of biogas–biodiesel dual-fuel combustion in a CI engine. Fuel Process Technol 2011;92(5):992–1000.
[62] Tripathi G, Sharma P, Dhar A. Effect of methane augmentations on engine performance and emissions. Alex Eng J 2020;59(1):429–39.
[63] Lee TK, Park H, Hyun J, Lee C, Song HH. Applicability of high-pressure direct-injected methane jet for a pure compression-ignition engine operation. Fuel 2019;251:428–37.
[64] Talibi M, Hellier P, Ladommatos N. The effect of varying EGR and intake air boost on hydrogen-diesel co-combustion in CI engines. Int J Hydrog Energy 2017;42 (9):6369–83.

[65] Dimitriou P, Tsujimura T. A review of hydrogen as a compression ignition engine fuel. Int J Hydrog Energy 2017;42(38):24470–86.

[66] Wei L, Geng P. A review on natural gas/diesel dual fuel combustion, emissions and performance. Fuel Process Technol 2016;142:264–78.

[67] Verma S, Das LM, Kaushik SC. Effects of varying composition of biogas on performance and emission characteristics of compression ignition engine using exergy analysis. Energy Convers Manag 2017;138:346–59.

[68] Khatri N, Khatri KK. Hydrogen enrichment on diesel engine with biogas in dual fuel mode. Int J Hydrog Energy 2020;45(11):7128–40.

[69] Prabhu AV, Avinash A, Brindhadevi K, Pugazhendhi A. Performance and emission evaluation of dual fuel CI engine using preheated biogas-air mixture. Sci Total Environ 2020;142389.

[70] Rahman KA, Ramesh A. Studies on the effects of methane fraction and injection strategies in a biogas diesel common rail dual fuel engine. Fuel 2019;236:147–65.

[71] Lather RS, Das LM. Performance and emission assessment of a multi-cylinder SI engine using CNG & HCNG as fuels. Int J Hydrog Energy 2019;44(38):21181–92.

[72] Çinar C, Şahin F, Can Ö, Uyumaz A. A comparison of performance and exhaust emissions with different valve lift profiles between gasoline and LPG fuels in a SI engine. Appl Therm Eng 2016;107:1261–8.

[73] Simsek S, Uslu S. Investigation of the impacts of gasoline, biogas and LPG fuels on engine performance and exhaust emissions in different throttle positions on SI engine. Fuel 2020;279:118528.

[74] Nadaleti WC, Przybyla G, Belli Filho P, de Souza SN, Quadro M, Andreazza R. - Methane–hydrogen fuel blends for SI engines in Brazilian public transport: potential supply and environmental issues. Int J Hydrog Energy 2017;42(17):12615–28.

[75] Ashok B, Jeevanantham AK, Prabhu K, Shirude PM, Shinde DD, Nadgauda NS, Karthick C. Multi-objective optimization on vibration and noise characteristics of light duty biofuel powered engine at idling condition using response surface methodology. J Energy Resour Technol 2021 Apr 1;143(4), 042301.

[76] Di Iorio S, Sementa P, Vaglieco BM. Analysis of combustion of methane and hydrogen–methane blends in small DI SI (direct injection spark ignition) engine using advanced diagnostics. Energy 2016;108:99–107.

[77] Nadaleti WC, Przybyla G. SI engine assessment using biogas, natural gas and syngas with different content of hydrogen for application in Brazilian rice industries: efficiency and pollutant emissions. Int J Hydrog Energy 2018;43(21):10141–54.

[78] Park C, Park S, Lee Y, Kim C, Lee S, Moriyoshi Y. Performance and emission characteristics of a SI engine fueled by low calorific biogas blended with hydrogen. Int J Hydrog Energy 2011;36(16):10080–8.

[79] Rai AA, Bailkeri NK, Shrinivasa Rao BR. Effect of injection timings on performance and emission characteristics of CNG diesel dual fuel engine. Mater Today: Proc 2021;46:2758–63.

CHAPTER 13

Impact of NOx control measures on engine life

Madhu Sudan Reddy Dandu[a], Kasianantham Nanthagopal[b], B. Ashok[b], Dhinesh Balasubramanian[c,d,e], and R. Sakthivel[f]
[a]Department of Mechanical Engineering, Sree Vidyanikethan Engineering College, Tirupati, Andhra Pradesh, India
[b]Engine Testing Laboratory, School of Mechanical Engineering, Vellore Institute of Technology, Vellore, Tamil Nadu, India
[c]Department of Mechanical Engineering, Mepco Schlenk Engineering College, Sivakasi, India
[d]Mechanical Engineering, Faculty of Engineering, Khon Kaen University, Khon Kaen, Thailand
[e]Center for Alternative Energy Research and Development, Khon Kaen University, Khon Kaen, Thailand
[f]Department of Mechanical Engineering, Amrita School of Engineering, Amrita Vishwa Vidyapeetham, Coimbatore, Tamil Nadu, India

13.1 Introduction

The previous chapter dealt with the use of alcohol and gaseous fuels such as ethanol, methanol, natural gas, liquefied petroleum gas, and hydrogen or syngas as fuel sources for internal combustion engines. Also, the composition of NOx emissions such as carbon monoxide, hydrocarbons, nitrogen oxides, and particulate matter in the exhaust emission as well as NOx reduction rates were reviewed. The previous chapter discusses alcohol fuel as a substitute for gasoline fuel in an SI engine because of its similar physical and chemical properties as well as the variations in fuel consumption and modifications in the engine for reducing NOx.

Better fuel economy and higher power with a lower maintenance cost have increased the popularity of diesel engine vehicles. The automobile vehicles fueled with diesel are used for the bulk movement of goods and powering stationary are more economically than any other engine in this size range. The anticipation of additional improvements in petroleum fossil fuels has forced engine manufacturers to upgrade technology in terms of power, fuel economy, emissions, and durability. Diesel emissions are categorized as carcinogenic [1]. The stringent emission legislations are compelling engine manufacturers to develop technologies to combat exhaust emissions. To meet these emission regulations with competitive fuel economy, exhaust gas treatments and optimized combustion are necessary [2]. Higher emissions of nitrogen oxides and particulate matter have been noticed as major emissions from diesel engines. Several factors such as fuel

NOx Emission Control Technologies in Stationary and Automotive Internal Combustion Engines
https://doi.org/10.1016/B978-0-12-823955-1.00013-9

properties, type of engine, and operating conditions of the engine affect the formation of NOx emissions in diesel engines. The self-ignition engines produced higher NOx compared to diesel engines due to their higher combustion temperature. The diesel engines produced lower NOx because of their lower combustion temperature. Moreover, the direct injection diesel engine combustion depends more on fuel atomization and vaporization condition compared with the indirect injection diesel engine. Using a lean fuel mixture with a higher combustion temperature may produce NOx emissions. This may happen for petroleum diesel fuel because of the lower heat release rate at the mixing controlled combustion phase and the higher heat release rate at the premix or rapid combustion phase [3].

A diesel engine fuel system has several types of components that can be divided into two classes: static and dynamic components. Static components include the fuel pump injector housing, fuel tank, fuel filter, fuel line, cylinder liner, exhaust system, etc. Dynamic components include the piston, piston ring, fuel pump, filter plunger, connecting rod, inlet and exhaust valves, etc. Dynamic components are usually metallic and are always in motion with each other as well as with static components during engine operation. Sliding contacts between these components are always accompanied by wear. In a diesel engine, the lubricating system wear particle is washed and suspended in the lubricating oil. The analysis of the concentration of metallic particles in lubricating oil provides sufficient information and prediction about the wear rate, element source, and engine condition. The tribological properties of the lubricating oil samples collected at regular intervals in the durability tests of the diesel engine play a key role in assessing the life of the engine. The properties influencing the wear assessment include viscosity, density, flashpoint, ash and moisture content, total base number, and pentane and benzene insolubles. The viscosity of the lubricating oil is the most important property that affects the wear rate of various engine components. Higher viscosity lubricating oil increases the friction loss and lower viscosity oil prevents the establishment of a protective film. When the engine is under operating condition at normal temperature, a small quantity of petroleum fuel is diluted in the lubricating oil. At the engine starting condition in low temperature, a large amount of fuel is diluted in the lube oil. These higher dilutions affect the lube oil properties such as reducing lubricating oil viscosity, flashpoint, and pour point, and decreasing the load-carrying capacity of the lubricating oil.

To meet stringent vehicular exhaust emission norms worldwide, several exhaust pretreatment and posttreatment techniques have been employed in modern engines. Exhaust gas recirculation (EGR) is a pretreatment technique being used widely to reduce and control the nitrogen oxide emissions from diesel engines. The engines using EGR emit a lower quantity of exhaust gases compared to non-EGR engines because part of the exhaust gas is recirculated [4]. Thus, even if the concentration of toxic substances in the exhaust gas remains unchanged, the total quantity of emissions of toxic substances is reduced for the same volumetric concentration. Diesel engines operating at low loads generally tolerate a higher EGR ratio because recirculating exhaust gases contain a high concentration of oxygen and a low concentration of carbon dioxide and water vapors. However, at higher loads, the oxygen in exhaust gas becomes scarce and the inert constituents start dominating along with increased exhaust temperature. Thus, as the load increases, diesel engines tend to generate more smoke because of the reduced availability of oxygen [5]. EGR controls the NOx because it lowers the oxygen concentration and the flame temperature of the working fluid in the combustion chamber. The reasons for reduced NOx emissions using EGR in diesel engines are reduced oxygen concentration and decreased flame temperature in the combustible mixture. At a part load, oxygen is available in sufficient quantity. However, at high loads, the oxygen is reduced drastically. Therefore, NOx is reduced more at higher loads compared to part loads. However, the use of EGR leads to a trade-off in terms of soot emissions. The higher soot generated by EGR leads to long-term usage problems inside the engines such as higher carbon deposits, lubricating oil degradation, and increased engine wear. The research revealed that the extent of wear of the top piston ring in the engine using EGR is lower than a normal operating engine. The possible reason for this may be the lower temperature of the engine combustion chamber using EGR. However, the wear rates of the second and third compression rings and the oil ring are comparatively higher for engines using EGR. The possible reason for this may be the presence of a higher amount of soot and wear debris in the lubricating oil of the engine using EGR.

This chapter focuses on the extensive literature from many researchers in evaluating the impact of NOx control measures on the life of the engine. This work presents the effects of the NOx reduction technique on engine performance and tribological characteristics such as compatibility, endurance, wear, and frictional properties. Also, we present various research and review articles on the effect of pretreatment techniques on the wear of piston rings, valves, the surface of the combustion chamber, and the cylinder head. We also review experiments conducted for exhaust gas recirculation on carbon deposits, brake power, exhaust gas temperature, and gas opacity.

13.2 Various methods for the determination of engine life

Primarily, it has remained with the focus of extensive amount of research because it is renewable and reduces the emission of pollutants. Some significant issues have not been adequately inspected, including the compatibility of different fuels including biodiesel, alcohol fuels with the crank case lubricating oil, the thermal stability of the lubricating oil with biodiesel, the change in the physical and chemical properties of lubricating oil with biodiesel, etc. These requirements are to be addressed to confirm the long-term suitability of different alternate fuels in the current family of compression ignition engines.

13.2.1 Long-term endurance study

An endurance test typically refers to a test carried out to determine whether the internal combustion engine can hold the operating load and withstand it for a long period [6]. The test cycle starts with examining the engine, while it can withstand a huge amount of load for a long period of time followed by measuring the reaction of the engine parameters under such operating conditions [7]. The endurance tests for internal combustion engines can be conducted according to standards IS: 10000 (Part IX)–1980. After the completion of initial performance, the endurance test on the engine shall be conducted. The long-term endurance test can be conducted on different engines such as constant speed and variable speed engines. This is briefly discussed in the following sections.

13.2.1.1 Long-term endurance test for constant speed internal combustion engines

The methodology and test cycle for conducting a long-term endurance test on constant speed internal combustion engines according to the standards is clearly discussed here. The long-term endurance test for the constant speed internal combustion engines shall be conducted after the initial performance tests specified according to the standards IS: 10000 (Part VIII)–1980. After completing the initial performance tests, then followed by operating the engine for 32h at rated speed in which each cycle is with 16h continuous running. At the end of each cycle of 16h, the engine should be stopped for necessary servicing and any required adjustments. Before starting the next cycle, the temperature of the lubricating oil in the engine must come down to room temperature [8]. The test cycle for the long-term endurance test for the constant speed internal combustion engines is presented in Fig. 13.1. Each running cycle consists of 16h made of the following sequence of engine running conditions. First, the engine is started and 100% load is applied on the engine; the engine is operated for 4h, which includes the warm-up period of 30min. After that, the load is reduced to 50% on the engine, which will run for 4h. Then, the load is increased to 110% and run for 1h. This is followed by bringing down the engine to no load condition and keeping it idle for 30min. Again, the engine load is increased to 100% operating for 3h. Finally, the engine load is reduced to 50%, which will be running for 3.5h. This completes the cycle for a duration of 16h and shall be followed to the next cycle; the same procedure is followed for the remaining cycles. The formation of NOx emissions from the constant speed engines was observed to be minimum because of the uniform temperatures in the engine cylinder.

Fig. 13.1 Long-term endurance test cycle for constant speed engine.

13.2.1.2 Long-term endurance test for variable speed internal combustion engines

The methodology and the test cycle for conducting the long-term endurance test on variable speed internal combustion engines according to the standards is clearly discussed here. The long-term endurance test for the variable speed internal combustion engines shall be conducted after the initial performance and the speed limiter or governor check as specified according to the standards IS: 10000 (Part VIII)–1980. After completing the initial performance tests, then followed by operating the engine for 100 h and shall consist of nonstop running periods of 10 h duration with not less than 2 h stoppage between consecutive running periods [9]. During the tests, the lubricating oil shall be used that is recommended by the manufacturer. Before starting the next cycle, the temperature of the lubricating oil in the engine shall come down to the room temperature.

The total cycle duration of the long-term endurance test on the engine is 100 h, and each running period shall consist of five cycles, each running for 2 h and made up of the following sequence of running conditions on the engine:

75% of full load at the declared maximum speed for 50 min.

Full load at speed corresponding to maximum torque for 45 min.

Idling condition for 5 min.

Full load at declared maximum speed for 20 min.

This constitutes the cycle duration of 2 h. This completes the cycle, and the next cycle will follow. The same procedure is repeated for the remaining cycles represented in Fig. 13.2. The formation of the NOx emissions in the variable speed engines was found to be higher because of the varying temperatures.

13.2.2 Material compatibility study

Material compatibility is a primary concern if there is a change in the composition of fuel in the fuel system. The designers take utmost care in incorporating the materials in the fuel system based on different testing methods available in laboratories [10]. Thus, changes in the composition of the fuel and the introduction of various alternate fuels frequently create some problems in the components of the engine. These considerations posed a great challenge with the compatibility of biodiesel to the tribologists. They have to provide an elucidation to decrease the tribological degradation of the various metals and the fuel being used. Some of the findings have already established the corrosion of materials and wear engine components by using

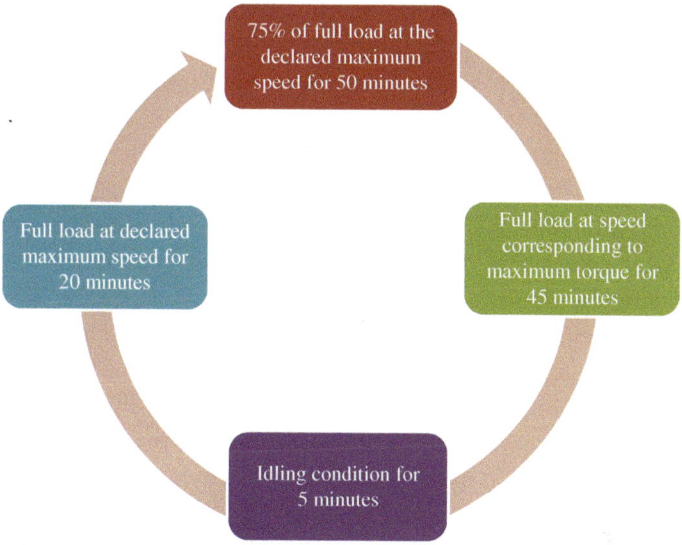

Fig. 13.2 Long-term endurance test cycle for variable speed engines.

fuels such as diesel and biodiesel. The results showed that a more corrosive environment happened with biodiesel when compared to diesel, but it provides better lubricity regarding wear and friction.

The importance in analyzing the various tribological parameters of biodiesel is decisive for its automobile applications [11]. After all the usage of biodiesel causes numerous environmental issues and the depletion of fossil fuel, however it levies certain thoughtful teething troubles to the researchers. Among the greatest issues is the material compatibility of automotive components with biodiesel. If biodiesel is used, then much importance must be given to certain engine components such as the cylinder, piston, piston rings, bearings, etc. Tribologists set a goal to address and overcome the difficulties related to various engine components as well as fuel interaction. In general, fuel passes across the engine components such as the fuel filter, fuel tank, cylinder liner, piston, piston rings, connecting rod, etc. [12]. The use of different alternate fuels in the engine definitely has an impact on the material, which in turn increases the formation of NOx emissions due to different properties possessed by the alternate fuels.

13.2.3 Impact of endurance study on lube oil degradation

The lubricity of any alternate fuel can be assessed by proper tribological investigations before adapting it as a fuel for regular use. The lubricity of internal combustion engine fuel is examined through the inherent lubricity characteristics such as wear and frictional characteristics, its corrosive nature, and its effect on lubricating oil contamination and degradation. The inherent lubricity of any alternative fuel is very important for its flow in the fuel pump, fuel injectors, and other places [13]. Different tribometer techniques are used to assess the lubricity of any fuel such as the four-ball tribotester, the pin-on disc tribotester, the high-frequency reciprocating tribotester, and the reciprocating wear tester under various operational conditions in terms of parameters such as load, sliding or rotating speed, oscillating frequency, operating temperature, etc. [14]. The wear and frictional characteristics of any fuel are very useful for understanding their lubricity behavior. Further, the wear and frictional characteristics of fuel are greatly influenced by their corrosive and hygroscopic nature, autooxidation, lower volatility and higher viscosity during sliding conditions. These properties may degrade the fuel quality during storage or use. In addition, fuel would be contaminated due to exposure with different metals [15]. Hence, the corrosion behavior of an alternative fuel also needs to be critically studied and investigated in various metals such as copper, aluminum, stainless steel, zinc, brass, and bronze under different operating conditions using an immersion test [16]. This could show the corrosion characteristics of fuel in terms of corrosion rate measurement, weight loss during the immersion test, and the changes in surface morphology [17].

The physical and chemical characteristics of fuel also greatly affect the lubrication behavior and wear of internal combustion engine components. Hence, it is evident that studies of the tribological parameters on lube oil also need to be investigated to assess the influence of any chemical changes in fuel on lube oil performance and durability [18]. The impact of engine fuel on lubricating oil contamination is of great importance, as it has an effect on fuel consumption, emissions, and engine life [19]. The increased trend of degradation of lube oil leads to the higher formation of NOx emissions. This can be attributed to the burning of lubricating oil in the engine cylinder, which results in higher temperatures generating much more NOx emissions. Some of the research studies shown that the lubricating oil performance and quality can be assessed on the basis of various analytical techniques in condition monitoring [20]. This is well established and effective, and is used to inspect the engine characteristics during operation at on- and offroad conditions along with the prediction of engine components during failures [21].

The condition monitoring of the analysis of lubricating oil can be evaluated in two different ways [22]:

(1) The measurement of lubricating oil properties such as viscosity, density, ash content, moisture content, flashpoint, pentane and benzene insoluble, etc., can determine the remaining lifetime of the lubricant and its effectiveness based on contamination and degradation. (2) Debris monitoring involves the measurement of wear particles deposited in the lubricating oil resulting from surface wear and monitoring the lubricating oil by different tests, including physical and chemical tests [23]. Fig. 3 shows that to monitor wear debris in the lube oil, ferrography [24] is one of the techniques used. It can be employed for diesel engines for their condition monitoring as well as measurements of wear in the engine parts [25]. To determine the wear metal content in the used lubricating oils, various methods are available, including atomic absorption spectrometry [26], x-ray fluorescence, and inductively coupled plasma-optical emission spectroscopy [27]. Similarly, for wear debris quantifying and analysis, different techniques are available, including the visual and microscopic examination as well as dimension and weight analysis.

13.3 Correlation of smoke and NOx emissions on engine life

Smoke and nitrogen oxide are two significant emissions from diesel engines. In accordance with the extended Zeldovich thermal NO mechanism, the higher temperature of the resident burning gas inside the combustion chamber of a diesel engine facilitates the reaction of the nitrogen within the inlet air with oxygen to generate high NOx emissions. In contrast to NOx formation, black smoke is generated by the incomplete burning of the diesel fuel. This incomplete burning may be caused by insufficient air or an excessive fuel supply. Understanding the correlation between these two emissions is an important step toward developing the technology for an appropriate strategy to control or eliminate them. However, there was no obvious correlation between the age of the tested diesel engines and the black smoke reflectivity. In addition, if the make and engine displacement volume of the tested diesel engines are not taken into consideration, then the correlation between the black smoke reflectivity and nitrogen oxide emissions weakens. It was observed that the emission indices of black smoke reflectivity decreased linearly with the increase of the emission indices of NOx for the tested diesel engines belonging to the same group of engine and engine displacement volume.

The mechanisms for controlling the formations of NOx and black smoke depend on various engine operating conditions that are not consistent with each other. For example, high-temperature burning gas would result in the larger formation of NOx [28] while black smoke is produced in a lesser amount due to the occurrence of a more complete burning process under the high-temperature burning environment. The optimization of engine operation conditions is a promising approach for the simultaneous reduction of both black smoke and NOX emissions. The correlation of NOx and black smoke can thus be applied to obtain optimal engine operating conditions, leading to pollution reduction from diesel engines.

13.3.1 Impact of smoke emission on engine durability

Most of the research studies revealed that the tested diesel engines emitted black smoke with low reflectivity. When the tested diesel engines were classified into various groups based on their makes and engine displacement volumes, the make of a tested engine became a dominant factor for the trend of black smoke reflectivity. A higher emission index of black smoke reflectivity was observed if the diesel engines were operated at low engine speed and full engine load conditions [29]. Moreover, the larger the displacement volume of the engine, the lower the emission index of black smoke reflectivity. The emission index of black smoke reflectivity of the tested diesel engines was also influenced by the make of the engine. The addition of nanoparticles to fuel blends tends to further decline the occurrence of metal oxides in the form of nano emulsion bounces thermal stability through the combustion period and augments the combustion rate owing to rapid evaporation, thus tumbling the emissions of smoke. On the other hand, the inclusion of an oxygenated additive showed increased emissions with respect to smoke. This can be attributed to the intensification in C—C bonds, which tends to diminish the accessibility of oxygen content and the augmentation in aromatic substances [30].

13.3.2 Impact of NOx emissions on engine life

Most of the research studies revealed that tested diesel engines produced low nitrogen oxide concentration. The age of the tested engines has a significant influence on NOx emissions. The older the tested diesel engines, the higher the NOx concentrations emitted. When the tested diesel engines were classified into various groups based on their makes and engine displacement volumes, the make of a tested engine became a dominant factor for the trend of

NOx emission concentration [31]. A higher emission index of nitrogen oxides was observed if the diesel engines were operated at low engine speed and full engine load conditions. Moreover, the larger the displacement volume of the engine, the lower the emission index of nitrogen oxides emitted. The emission index of nitrogen oxide emission of the tested diesel engines was also influenced by the make of the engine.

13.3.3 Effect of oil degradation on NOx emissions

Lubricating oil of good quality is crucial for better engine performance. Lubricating oil decreases the friction of moving parts, keeps the engine clean, reduces oxidation, and removes various particles. Water can be the potential component to degrade lubricating oil through moisture from the atmosphere inclusive of the combustion process and condensation. However, the condensation of moisture present in the exhaust increases corrosion in the combustion chamber. The major influence of NOx emissions is due to a change in temperature rather than oxygen availability [32]. Also, at the start of lubricating oil degradation, the oxidation forms lightweight materials such as acids, alcohols, and ketones. Some of the metallic elements that were deposited in the lubricating oil sample showed a wear rate in the diesel engines. A calcium compound is used as a detergent in the typical commercial lube oil. These compounds provide an alkaline reserve to neutralize acidic byproducts to reduce the formation of insoluble products and provide some measure of corrosion protection.

13.4 Effect of NOx reduction devices on SI engine life

The emissions from SI engines are comprised mainly of NOx. In order to control NOx emissions, some of the adopted methods will have a small impact on the life of the SI engine in terms of wear of engine components, tribological characteristics, etc. These are discussed briefly in this section.

13.4.1 Engine performance behavior

It is well known that biofuel with ethanol used in petrol engines has drawbacks such as material compatibility due to the corrosive influence on components of automobiles. The causes of corrosiveness with fuels containing alcohols are the improved water content because of the hygroscopic nature of ethanol as well as the partial existence of organic acids. In addition, alcohols are polar solvents leading to the corrosion of the metal components of the engine [33]. Many researchers presented better results by adding ethanol to gasoline in analyzing various parameters of engine performance. This includes the torque of the engine, engine brake power, thermal efficiency, etc. [34]. The blends of ethanol and gasoline have been widely analyzed for their adverse impact with emissions of gas - phase exhausting from the SI (Spark Ignition) engines and these emissions of numerous significant gas - phase kinds identified to be condensed by adding ethanol to petrol [35]. This gas-phase species includes carbon monoxide and hydrocarbon emissions, including toluene, benzene, and butadiene. The other emissions of the gas-phase species, including ethylene, unburned ethanol, aldehydes, and methane, may essentially be increased. By adding ethanol, there is a decrease in the emissions of particle mass as well as particle number [36]. When the engine is operating at optimized conditions with respect to the performance parameters, the formation of NOx emissions in the chamber is reduced due to proper combustion taking place in the engine cylinder.

13.4.2 Tribological behavior

The use of ethanol as a fuel in automotive engines results in certain tribological issues, including the possibility of polluting the lubricant oil. In addition, the latent heat of evaporation is greater for ethanol when compared with gasoline ethanol, with accumulation in the lubricant oil considered a severe issue [37]. The amassing of ethanol in the crank case causes austere wear by which many researchers reporting repeatedly with different models as well as sizes of flex - fuel engines, as it influences the lubrication and friction [38]. At the temperature of 100°C, if ethanol is being added to the formulated oil does not show any impact on the antiwear tribofilm formation; however revealed that at this temperature ethanol completely vaporized from the lubricant oil. Hence pointing the significance of analyzing the presence of content of ethanol in lubricant oils with monitoring as well as maintaining suitable amount of concentration of ethanol all over the testing time [39]. And revealed that acetaldehyde and ethanol have minimal effect with aging performance of the lubricant oils expressed in different parameters relating to common oil conditioning, but acetic acid progresses oxidation and persuades significant sludge formation [40]. The trend of increased NOx emissions in the engine cylinder result in burning lubricating oil, which in turn degrades the lube oil at a faster rate as well as leads to the formation of CO and HC emissions.

13.4.3 Wear on engine components

It was observed that the types of wear metals from the engine components in the lubricating oil include iron, molybdenum, aluminum, copper, lead, tin, nickel, etc. The sources of these wear metals can be grouped together into two categories: wear from engine components and environmental sources in the engine cylinder. The wear sources include the cylinder head, piston, bearing, vale seats, etc., and the environmental sources include the lube additive, fuel oil contaminant, coolant additive, etc. [41]. The concentration of traces of metal in the lube oil increases with its use because of accumulating wear debris. The concentration of wear metals in the lube oil could be used as an indicator of engine wear containing metals mentioned in the above table. This concentration of wear metals from engine components possesses some warning limits, and these wear metals exceeding the prescribed warning limits indicates severe wear in the engine cylinder. The main elements from engine cylinder wear include iron, aluminum, copper, lead, tin, chromium, silicon, and boron; each of these metals has its own warning limit [42]. Greater temperatures leads to increased NOx emissions, which also affects combustion in the SI engine due to hot spots. This leads to the greater wear rate of engine components.

13.5 Impact of NOx reduction devices on CI engine life

The emissions from CI engines are mainly comprised of NOx. Some of the methods being adopted to control these NOx emissions will have a small impact on the life of the CI engine in terms of the wear of engine components, tribological characteristics, etc. These are discussed briefly in this section.

13.5.1 Engine performance behavior

Many research studies revealed that there will be a decreasing trend in the brake thermal efficiency of the engine due to the impact of NOx reduction devices. This can be partially attributed to the lower calorific value of the fuel blends. It also leads to more consumption of fuel, increasing the BSFC. The decreasing trend of BSFC could be attributed to varying the properties of the blend by adding biodiesel; the major contributing factors are the density and viscosity of the blend [10]. On the other hand, the higher viscosity and boiling point of the diesel and biodiesel affect the values of BSEC. Further, the temperature of the exhaust gases also rises with intensification in the concentration of biodiesel in the blend. This is because of the viscosity of the blend, which in turn affects the atomization of fuel blends in the engine cylinder during the combustion process [43]. In general, the formation of NOx emissions in the diesel engine was found to be higher when compared to the petrol engine. The duration of the power stroke in the cylinder is reduced due to increased NOx emissions.

13.5.2 Tribological behavior

Generally, the lube oil of the engine fueled with biodiesel demonstrates greater change in its density when compared to the lube oil of the diesel engine in the entire duration of the endurance test. This may be due to the accumulation of wear debris, fuel, and moisture affecting the density of the lube oil. The density of the lube oil increases due to the higher wear of engine components. The dilution of the fuel reduces the density of the lube oil because the density of fuels is lower compared to lube oil. Also, because of the relentless heating and exposure to moisture, the density increases. The rate of increase in the viscosity of lubricating oil in biodiesel fueled engines becomes higher than that of diesel. This may be because the oxidation nature and polymerization of lube oil leads to increasing the viscosity while the dilution of the fuel tends to reduce the viscosity. There is a slightly higher ash content in the lube oil of an engine fueled with biodiesel when compared to diesel. Variations in the ash content of lube oil with use show metallic wear debris addition to the lube oil. The total base number (TBN) is reduced with the use of lube oil for diesel as well as biodiesel. The oxidation of the ester molecules of biodiesel in the lube oil improves the formation of organic acids while also leading to the reduction of the alkalinity reserves of lube oil. The level of fuel dilution for biodiesel might be a result of the variation of the flashpoint of the biodiesel [44]. After engine operation for several hours, the fuel dilution level of lube oil with biodiesel was becoming greater, leading to a change in the flashpoint.

Pentane insolubles replicate the amount of sludge formed by metallic wear debris, oil oxidation, and the carbon content of fuel. Benzene, aromatic in nature, may dissolve the resinous material of lube oil. Henceforth, a difference in the pentane and benzene insolubles designates the content of resinous material in the lube oil. Greater differences among the pentane and benzene insolubles show the higher oxidation nature of lube oil, leading to the higher polymerization of the lube oil base-stock. An increase in the fuel dilution of biodiesel leads to a greater amount of resinous material in the lube oil. It was also observed that with the increase in the use of lubricating oil, the density also increases due to several potential impurities such as wear debris, soot, etc. However, a minor increase showed less wear. Similar results were observed with viscosity and flashpoint. But the TBN decreases while using lubricating oil with respect to duration representing less wear and lower degradation [45]. The trend of increased NOx emissions in the engine cylinder results in the burning of lubricating oil, which in turn degrades the lube oil at a faster rate and also leads to the formation of CO and HC emissions.

13.5.3 Wear on engine components

It is well known that by increasing the concentration of biodiesel in the blend, both the friction and wear decrease. This is due to the occurrence of free fatty acid components, oxygenated moieties, degree of unsaturated molecules, etc. in biodiesel. In addition, the distortion of worn surfaces also declines with an increase in the concentration of biodiesel in the blend. With the increase of rotating speed, the lubricity also decreases with respect to wear and friction [46]. Biodiesel affords improved lubricity properties. Improved engine deterioration occurs due to the limited lubricity of diesel. Therefore, diesel is added to biodiesel to enhance the fuel lubricity properties by eliminating the sulfur compounds. The lubricity of biodiesel may be affected by many parameters. Among them, the oxidation process is the one to be considered because this process reconverts the esters into various fatty acids such as acetic, formic, propionic, caproic, etc. The oxidation of biodiesel was also influenced by factors such as the concentration of impurities present, the presence of antioxidants in biofuels, and their storage conditions [47]. Some studies indicated that increased lubricity occurred due to the presence of certain components in biodiesel such as the degree of unsaturated molecules, oxygenated moieties, free fatty acid components, long chain molecules, glycerides, etc. Similarly, some of the parameters in biodiesel such as degradation, absorption of moisture, autooxidation, corrosiveness, etc., resulted in decreased lubricity. In addition, improved lubricity was observed owing to the high viscosity of biodiesel, but this promotes poor atomization and also improper functioning of fuel injectors. The sound effects of biodiesel and its blends on engine tribological characteristics are also examined and presented through parameters including pump plunger surface, carbon deposits on injectors and in combustion chambers, and the nozzle discharge coefficient. Greater temperatures lead to increased

NOx emissions, which also affects the combustion in CI engines due to a reduction in the duration of the expansion stroke. This leads to the greater wear rate of engine components.

13.6 Effect of advanced technologies on engine durability

It is evident that the sensitivity of the NO formation rate relates to the temperature and oxygen concentration. Hence, in order to reduce the NOx formation inside the combustion chamber, the temperature and oxygen concentration in the combustion chamber need to be reduced. Even though certain cetane improving additives are capable of reducing NOx, the amount of reduction is reported to be inadequate. Moreover, most of these additives are expensive. Retarded injection is an effective method employed in diesel engines for NOx control. However, this method leads to increased fuel consumption, reduced power, and increased hydrocarbon emissions and smoke. Water injection is another method for NOx control, but this method enhances the corrosion of vital engine components. In addition, it adds to the weight of the engine system because of the requirement of a water storage tank. It is also difficult to retain water at a desired temperature during cold climatic conditions [48].

In general, pollutant control methods can be divided into two types: after combustion control and during combustion control. The former includes selective catalytic reduction (SCR) and DeNOx function to control NOx emissions [49]. The latter consists of a low-excess-air (LEA) burning mechanism, fuel modification, engine design modification, and exhaust gas recirculation (EGR). A famous in-cylinder method to reduce the NOx production rate is water injection into the air intake, directly into the cylinder or in an emulsion with the fuel [50]. A major benefit of water injection when compared with EGR is the potential reduction of both NOx and particulate matter emissions either at low or high loads. However, water injection puts liquid water in the cylinder to affect the lubricant, engine wear, and fuel consumption due to poor atomization and degraded combustion [51].

The removal of NOx from the lean exhaust gases of diesel engines is an important challenge. Selective catalytic reduction requires a supplementary substance or reducing agent that in the presence of catalysts produces useful reactions, transforming NOx into N_2 and H_2O. The combination of particle filtration using a diesel particulate filter (DPF) and the most efficient deNOx technology is called SCR. It is widely used as the best solution to minimize emissions of diesel engines. In light-duty diesel engines applied with a common rail injection system, the late injection events used as a regeneration strategy for diesel particulate filters (DPFs) increased the opportunity for unburned fuel to reach the cylinder walls, enter the lubricating oil, and promote its deterioration.

The engine using EGR emits a lower quantity of exhaust gases compared to non-EGR engines because part of the exhaust gas is recirculated. Thus, even if the concentration of toxic substances in the exhaust gas remains unchanged, the total quantity of toxic substances is reduced for the same volumetric concentration. Diesel engines operate at low loads and generally tolerate a higher EGR ratio because recirculating exhaust gases contain a high concentration of carbon dioxide and water vapors. However, at higher loads, the oxygen in the exhaust gas becomes scarce and the inert constituents start dominating along with increased exhaust temperature. Thus, as the load increases, diesel engines tend to generate more smoke because of the reduced availability of oxygen [5]. The rise in the smoke level of engine exhaust due to EGR affects the engine performance in various ways. Increased smoke or soot level causes a considerable increase in carbon deposits and the wear of the various vital engine parts such as the cylinder liner, piston rings, valve train, and bearings. The wear of the materials also increases due to chemical reactions taking place on the surface with respect to adsorption, corrosion, the abrasion of material, or the rupture of antiwear film by soot. The application of EGR also adversely affects the lubricating oil quality and engine durability. This is because EGR results in more soot formation, which in turn affects the lubricating oil by thickening the oil and also increases the wear debris in the lubricating oil. Increased soot level and wear debris in lubricating oil may adversely affect the piston rings because piston rings are used to scrape off the excess lubricating oil from the cylinder liner and return it to the oil sump [52].

13.7 Effect of fuels on engine durability

The different sources of fuels such as neat vegetable oils or their blends with diesel pose various long-term problems in compression ignition engines. These problems include poor atomization characteristics, ring sticking, injector coking, injector deposits, injection pump failure, dilution of lubricating oil, etc. Such problems are not usually experienced with short-term engine operations. Sometimes the engine fails catastrophically if run on neat vegetable oils continuously for a longer period. The properties of vegetable oils responsible for these problems are high viscosity, low volatility, and polyunsaturated character.

13.7.1 Desirable fuel properties for longer engine life

In the majority of short-term tests employing vegetable oil as a fuel, the peak power outputs developed by the engines could be comparable with those of the engines operated with mineral diesel fuel. The thermal efficiencies of the engine were generally reported to improve with vegetable oils. However, due to its energy content, vegetable oil consumption was higher than that of mineral diesel fuel. Remarkable attention has been focused on various biofuels, in particular biodiesel and bioethanol, due to their renewability and biodegradability [53]. Biodiesel derived from vegetable oil has been considered a partial substitute for diesel fuel and it has been commercialized in many countries. In a similar trend, ethanol fuel is being used as an alternative to gasoline in many parts of the world. Hence, the increasing demand of biofuels necessitates deep research in terms of tribological aspects. The desirable fuel properties favorable for longer engine life in terms of friction, wear, and lubricity are:

Enhanced scuffing protection performance compared to hydrocarbons of petroleum diesel fuel.

The observed ester molecules resulting from biodiesel will act as surfactants for the metal surfaces.

More effective film formation.

Oxygen content of biodiesel could reduce the friction between the metal surfaces.

The internal compounds such as C—C, COH, and COOH groups might reduce friction.

The presence of aliphatic fatty acids (stearic acids) can enhance the lubrication property due to the development of lubrication films.

The protecting film can decrease the thermal energy in terms of sliding contact, thus increasing the lubricity.

The presence of unsaturated fatty acid compounds could increase the lubricity behavior.

13.7.2 Influence of conventional fuels on engine life

The composition and structure of petroleum or conventional fuels play vital roles in the performance and life of engine lubricating oil. Variations in physical and chemical properties of diesel fuels are the source of dilution in engine lubricating oil. Moreover, the properties of engine lubricating oil vary with its usage, due to fuel thinning and the addition of impurities or contaminants through wear of the various moving components of the engine. The conventional fuels have good oxidation stability, and this nature particularly helps in long-term applications to avoid the characteristics of engine life such as fuel degradation, poor lubricity, enhanced corrosion, and degradation of materials. The moisture absorption by fossil fuels has a great influence on wear rate.

13.7.3 Effect of alternate fuels on engine life

Although many fuel properties of diesel and alternate fuels, particularly biodiesel fuels, are close to each other, the level of compositional differences is an alarming issue to understand tribology in biodiesel. The following points are against the tribological behavior of alternate fuels:

The oxidation of biodiesel in long-term applications causes fuel degradation, poor lubricity, enhanced corrosion, and degradation of materials.

Instability in the properties of biodiesel fuel due to several environmental factors and wide exposure of metals.

Moisture absorption and autooxidation of biodiesel have great influence on the wear rate.

Instability in oxidation as an alarm of varying temperature along with acquaintance to air.

Low volatility nature.

Reacting with unsaturated hydrocarbons of biodiesel during usage.

Filter plugging.

Injector coking.

Sticking of moving parts.

Biodiesel oxidation depends on unsaturated compounds, impurity concentration, etc.

Biodiesel properties could be affected by storage conditions such as light, temperature, atmospheric humidity, etc.

Biodiesel impurities such as free fatty acid, unused catalysts, moisture, and different glycerides are not compatible with automotive materials.

The catalysts used for biodiesel esterification are very aggressive on corrosion.

The acidic nature of biodiesel enhances the degradation of metal surfaces.

Higher biodiesel viscosity provides more deposition in the fuel injection system.

Alternate fuels have a great impact on engine life in terms of various characteristics, including lubricating oil degradation, tribological characteristics, wear of vital components, etc. Many researchers have carried out work on alternate fuels, particularly focusing on the impact on engine life. These are summarized and presented in Tables 13.1–13.3. The characteristics possessed by various alternate fuels used as fuels in diesel engines result in the promotion of growth in NOx emissions. Therefore, in order to explore the advantages of alternate fuels, they should be blended with diesel in certain limited proportions. The formation of NOx by using blends depends on the various properties such as self-ignition temperature and kinematic viscosity.

13.7.4 Effect of various additives on engine durability

The oxygenated additive inclusion to biodiesel and their blends diminish the emissions of nitrogen oxides in diesel engines with respect to various operating environments by impeding the formation of free radicals that backs NO development. The oxygenated additives added to various fuel blends compel the formation of hydroxyl radicals throughout the combustion process. The oxidation of carbon monoxide and unburnt HC is inhibited by the deterioration of OH radical development, which tends to decrease the temperature in the engine cylinder and henceforth lowers the NOx emissions. A similar trend is found in increasing the CO and HC emissions due to the oxidation effect by adding the oxygenated additive, resulting in incomplete combustion [76]. Furthermore, the reducing trend of NOx emissions was observed with the nanoparticle addition to the fuel blends because the nanoparticles behave as catalytic convertors by flouting down NOx into nitrogen and oxygen. Excessive lube oil dilution leads to several problems, such as reduced oil performance and durability. Usually, oil breakdown is promoted by the oxidation and polymerization of unsaturated fuel constituents. So, a serious danger is the possibility that unburnt fuel entering the oil sump may be oxidized, thus promoting lubricant degradation and thickening, sometimes by severe sludge precipitation and significant loss of its dispersion. On the other hand, oil thinning may be due to fuel dilution or the share of polymers used as viscosity modifiers. Apart from that, certain metals such as copper and lead become leached from bearings due to fuel in the lubricating engine oil. Because the state of oil deterioration depends also on operation conditions, the oil performance grade, the engine type, and its service, the life span of engine oil has not been well understood. There are limited

Table 13.1 Effects of biodiesel on lubricating oil degradation in CI engines on endurance tests (Duration: 100–500 h).

Fuels	Engine	Type of tribological analysis	Engine operation duration	Significant results	Reference
Diesel and cotton seed oil (B50 & B100)	Turbocharged diesel engine	Wear of engine components	200 h	Higher wear in combustion chamber Higher ash quantity More carbon deposits	Fort et al. [43]
Soy bean oil, biodiesel, and diesel	Four-cylinder direct injection, turbocharged diesel engine	Wear of engine components	200 h	Normal levels of wear metal in lube oil for biodiesel except aluminum and lead with slightly levels	Clark et al. [54]
15% palm biodiesel and 85% diesel	Four-stroke, water-cooled, indirect injection, and naturally aspirated engine/ SAE 40	Lubricating oil analysis Wear of engine components	300 h	Minimized wear with B15 than diesel Brake power improved Exhaust emissions reduced Changes in viscosity are similar for both fuels Reduction in total base number by increasing biodiesel blends	Kalam and Masjuki [55]
Rubber seed oil and diesel blends	Four-stroke direct-injection single-cylinder diesel engine	· Lubricating oil analysis	100 h	Reduction in viscosity of lube oil with rubber seed oil Higher carbon deposits with rubber seed oil due to incomplete combustion	Ramdhas et al. [56]

Continued

Table 13.1 Effects of biodiesel on lubricating oil degradation in CI engines on endurance tests (Duration: 100–500 h)—cont'd

Fuels	Engine	Type of tribological analysis	Engine operation duration	Significant results	Reference
ROME B20 (rice bran oil methyl ester), mineral diesel	Medium-duty direct-injection transportation diesel engine (CIDI)	Condition monitoring of lube oil Wear of engine components	100 h	Ash content, density, moisture content, viscosity increase with usage of lube oil Lower wear of engine components for biodiesel fueled engine	Shailendra Sinha and Avinash Kumar Agarwal [44]
Karanja oil methyl ester and diesel	Compression ignition direct-injection engine	Wear metal analysis	100 h	30% lower wear for biodiesel due to higher viscosity and lubricity	Pandey et al. [57]
Rice bran oil, methyl ester B20, and diesel	Medium-duty transportation compression ignition engine	Wear of engine components	100 h	Lower wear of all engine components except big end bearing for B20 Lower concentrations of all wear metals except lead for B20	Sinha and Agarwal [45]
Diesel and B100 biodiesel	Single-cylinder, direct-injection diesel engine	Wear of engine components	200 h	Structural changes observed on the surfaces of fuel injector nozzle and pump piston for B100 biodiesel	Celik and Aydin [58]
B20 Karanja biodiesel and diesel	Four-stroke, in-line, water-cooled, direct injection, naturally aspirated CI engine/SAE 40	Wear of engine components	250 h	Lesser wear of engine components for biodiesel than diesel fuel Higher carbon deposition on cylinder head, piston top, and injector tip with B20	Dhar and Agarwal [59]

B20 Jatropha biodiesel and diesel	Four-stroke DI diesel engine, naturally aspirated	Lubricating oil analysis	250 h	Viscosity of lube oil is decreased for B20. B20 showed more increase in density of lube oil Higher carbon deposits with B20 Concentration of wear metals for B20 were greater	Liaquat et al. [60]
B20 Rapeseed oi, biodiesel, and biolubricant	Four-stroke, single cylinder, water-cooled, direct-injection, variable compression ratio diesel engine/SAE20W40	Lubricating oil analysis Wear of engine components	150 h	Improved brake power, brake thermal efficiency, and mechanical efficiency with biolubricant Ferrogram showed decreased Fe, Al, and Cu wear for biodiesel and biolubricant	Arumugam et al. [9]
Diesel Pongamia biodiesel 20	Constant speed diesel engine/ SAE20 W40	Condition monitoring of lube oil Wear metal analysis in lube oil samples Physical measurement of cylinder surfaces · Loss of weight in piston rings	256 h	Wear rate was higher in engine with PME20 owing to autooxidation, hygroscopic nature, and more deposit formation	Nanthagopal and Thundil Karuppa Raj [61]

Table 13.2 Effects of biodiesel on lubricating oil degradation in CI engines on endurance tests (Duration: 500–1000 h).

Fuels	Engine	Type of tribological analysis	Engine operation duration	Significant results	Reference
LOME B20, mineral diesel oil	Four-stroke, water-cooled, single-cylinder vertical engine	Wear assessment of components of engine	512 h	AAS showed low concentration of wear metals Ferrography showed lesser concentration of wear debris with smaller size for the engine with biodiesel	Agarwal et al. [62]
20% linseed oil methyl ester and diesel	Single-cylinder, water-cooled, portable	Wear of engine components	512 h	Lower wear rate No filter plugging	Agarwal et al. [63]
Diesel and 100% Refined palm oil	KUBOTA Model ET80 single cylinder indirect injection diesel engine	Lubricating oil analysis Wear of engine components	1000 h	Wear in the compression rings is higher for the engine fueled by refined palm oil when compared with the engine fueled by diesel	Prateepchaikul et al. [7]
Diesel LOME B20 (linseed oil, methyl ester)	Compact, water-cooled single-cylinder diesel engine of portable type (4 kW capacity)	Condition monitoring of lube oil	512 h	The extent of durability problems such as injector coking, filter plugging, carbon deposits, and lubricating oil contamination were lower for a biodiesel engine when compared with a diesel engine	Agarwal et al. [6]

Diesel and 100% Karanja oil (straight vegetable oil)	Constant speed, four-stroke, single-cylinder, water-cooled, direct-injection diesel engine	Lubricating oil analysis Wear of engine components	512h	Wear of engine liner with Karanja oil is higher when compared with engine with diesel fuel Engine with Karanja oil has reasonable long-term performance when compared with engine with diesel fuel	Agarwal et al. [8]
RME (rapeseed oil, methyl ester), diesel	EURO IV engine	· Regulated emissions and particulate matter	500h	No difference in SFC through the test, but average temperature of exhaust improved by 15% Emissions of HC and CO are under regulatory level Emissions of NOx improved by 10% with biodiesel instead of diesel fuel	Juergen Krahl et al. [64]
Soy methyl ester (SME 100), castor oil, methyl ester (CME 100),and diesel	Single-cylinder Agrale M93ID direct-injection engine	· Lubricating oil analysis · Wear of engine components	1000h	Viscosity decreased for SME100 Greater wear of pressure valve seating and also higher carbon deposits for SME100	Wander et al. [65]

Continued

Table 13.2 Effects of biodiesel on lubricating oil degradation in CI engines on endurance tests (Duration: 500–1000 h)—cont'd

Fuels	Engine	Type of tribological analysis	Engine operation duration	Significant results	Reference
10% Karanja oil with mineral diesel K10 and mineral diesel	Four-stroke, single cylinder, constant speed, DICI engine	Wear of engine components	512 h	Carbon deposits slightly higher for K10 · Wear of cylinder liner and big end bearings relatively higher for K10	Agarwal et al. [66]
Diesel, waste cooking oil, biodiesel B2, B5	Common rail injection diesel engine with variable speed	Wear of components of engine	500 h	21% and 7% higher carbon deposit on piston for B2 and B5 fuels compared to diesel Oil deterioration is evident in B5 fueled engine Zn, P, and S were low in B5 engine compared to B2 and diesel engine	Ku et al. [67]
Jatropha biodiesel 40 Diesel	Constant speed diesel engine/ SAE15 W 40	Physical measurement of cylinder surfaces Wear metal analysis in lube oil samples	512 h	Low wear debris in B40 engine	Niraj Kumar et al. [68]

Table 13.3 Effects of biodiesel on lubricating oil degradation by on-road tests.

Fuels	Engine	Type of tribological analysis	Engine operation (km)	Significant results	Reference
Rapeseed oil methyl ester (RME) B20, B100, and diesel	Turbocharged, intercooled, direct-injected, in-line six-cylinder diesel engine	Lubricating oil analysis	80,000	Viscosity is 1.8 times that of diesel Injector coking was low with RME B20	Daryl and Peterson [69]
Rapeseed oil methyl ester (RME) B20	Heavy-duty truck with diesel engine	Lubricating oil analysis	161,000	No unusual deterioration of the engine No change in oil composition with RME	Peterson [70]
Refined palm biodiesel B50 and E100	Indirect fuel injection, in-line, four-cylinder diesel engine/SAE 15 W40	Lubricating oil analysis	200,000	Amount of wear was not higher when compared to diesel Soot level decreased	Raadnui and Meenak [71]
Soybean biodiesel B20	Ford nine-ton cargo van	· Wear analysis	96,560	· Heavy amount of sludge on cylinder heads with B20 · Require injector nozzle replacement with B20	Fraer et al. [72]

Continued

Table 13.3 Effects of biodiesel on lubricating oil degradation by on-road tests—cont'd

Fuels	Engine	Type of tribological analysis	Engine operation (km)	Significant results	Reference
Diesel and B20 biodiesel	Cummins ISM diesel engine (Bus)	Engine durability Lubricating oil analysis	160,934	Indicate no additional wear metals for B20 Similar results as diesel in terms of TBN, viscosity, and fuel dilution	Kenneth Proc et al. [73]
Soybean biodiesel B20	Diesel engine (Bus)	Wear analysis	161,000	Similar wear observed compared to that of diesel	Mazzoleni et al. [74]
Rapeseed biodiesel B100 and mineral diesel	Four-stroke, turbocharged, DI common rail diesel engine	Lubricating oil analysis	30,000	Wear of engine components is lower for biodiesel Carbon deposits found to be lesser for biodiesel	Agarwal et al. [75]

data concerning how the fuel might interact with lubricating oil additives, either in the bulk of the lubricant or in thin film on metal surfaces influencing the lubricating oil performance changes. The possibility of incompatibility problems between the crankcase lubricants and the biofuel additives should also further be clarified by studying the complicated physiochemical processes between modern additives such as nanoparticles, antioxidants, etc., with that of engine lube oils and biofuels [77]. It is evident that nonferrous metals suffer from higher wear loss compared with ferrous metals. Therefore, the main reason for decreasing metal wear with different blended fuels is the anticorrosion effect of the additive in the fuel and lubricating oil, which controls oxidation as well as corrosion in the lubricating oil.

Zinc is used as a constituent of antiwear compounds in automobile lubricants. These compounds react with metal surfaces to make a protective low-friction film. The protective layer, one to several molecules thick, is absorbed on the metal surface at low temperatures. As the temperature increases, the additive undergoes chemical absorption [78]. This produces a chemical bond that is far stronger than the physical attraction of the absorption. Even at high temperatures, a polymeric or oligometric film is formed. Under high-speed collisions, the organic structure of the film decomposes. The inorganic compounds such as phosphides of zinc and iron acts as stable antiwear agents to protect the surface.

13.8 Reformulation of fuels on engine life

Having high viscosity vegetable oils in long-term operation will normally produce gumming, the formation of injector deposits, ring sticking, and incompatibility with conventional lubricating oils [79]. One feasible means of overcoming these problems is to emulsify these fuels with different proportions of water, leading to improved fuel atomization and spray characteristics, possibly through the phenomenon of a microexplosion [80]. The emulsified fuels show lower tendency in NOx formation as there was an increase in the water content. In general, emulsions are characteristically unstable and they will separate into the stable states of dispersed and continuous phase materials. Surfactants are commonly used to maintain the stability of emulsions. Due to the huge droplet sizes of diesel or biodiesel, combustion takes a longer time, thus the available time frame is not sufficient for complete combustion [81]. However, in emulsion fuels, the droplets are relatively smaller due to microexplosions, therefore allowing a complete burning to occur. The microexplosions would cause the smaller fuel particles to be in contact easily with the air for complete combustion, and reduce the generation of NOx emissions without failing the combustion efficiency [82].

It was observed that for emulsified fuels, due to the presence of fatty acid the concentration of iron in the lube oil drops as the percentage of water increases. This indicated that the presence of water in fuel could lower the combustion temperature and hence the wear rate between engine rubbing components. Tribologically, thickened oil may not provide adequate lubrication to critical engine parts and the antiwear agent may also be depleted. This phenomenon may explain the presence of the relatively higher wear debris concentration of various metal elements in the collected lubricating oil samples when using reformulated fuels, particularly emulsified fuels. During the combustion of emulsion fuels, vaporized water reduces the flame temperature and changes the chemical compositions of reactants, resulting in higher OH radical concentration that controls the NO formation rate [83]. The trends of the exhaust temperatures were consistent with the results of NOx emissions. Lower exhaust temperatures for emulsion fuels did reflect lower emissions of NOx. A research study reported that NOx emissions before and after the engine durability test of using both diesel and emulsion fuels showed no destruction on the components of the engine cylinder [84]. The use of emulsion fuels for buses traveling along 50,000 km showed no effect on engine operability and lubricant. The viscosities of the emulsion fuels increased with water percentage, as a larger volume in the dispersed phase would affect the relatively smaller gap between the continuous phase and the dispersed phase. The size of the dispersed water particles may affect the strength of the microexplosion and emissions [85].

13.9 Conclusions

The exploration of different reduction techniques being adopted in automobiles to control NOx emissions resulted in the degradation of engine life in various aspects such as performance and tribological behavior. Through an extensive literature survey, it is noted that NOx control measures implemented in automobiles alter the performance of the engine in its long-term operation. The current study also involved the validation of the review study through experimental results. Some literature studies revealed that the emissions of nitrogen oxides before and after the engine durability test of using various fuels such as biodiesel fuels showed no destruction on the components of the engine cylinder. The NOx reduction technique of EGR affects the quality of the lubricating oil and the durability of the engine. This is because EGR results in more soot formation, which in turn affects the lubricating oil by thickening the oil and increasing the wear debris in lubricating oil. Increased soot level and wear debris in the lubricating oil may adversely affect the piston rings because the piston rings are used to scrape off the excess lubricating oil from the cylinder liner and return it to the oil sump. In light-duty diesel engines applied with a common rail injection system, the late injection events used as a regeneration strategy for diesel particulate filters increased the opportunity for unburned fuel to reach the cylinder walls, enter the lubricating oil, and promote its deterioration. Hence, the research studies revealed that the latest technologies of controlling nitrogen oxide emissions resulted in the degradation of engine life in various aspects of the engine.

References

[1] Stewart KM. Health effects of diesel exhaust. Report from American Lung Association of Pennsylvania; 2001.
[2] Moser FX, Sams T, Cartellieri W. Impact of future exhaust gas emission legislation on the heavy duty truck engine. SAE Technical Paper No. 2001-01-0186; 2001 Mar 5.
[3] Masjuki H, Kalam MA, Maleque M, Kubo A, Nonaka T. Performance, emissions and wear characteristics of an indirect injection diesel engine using coconut oil blended fuel. Proc Inst Mech Eng 2001;215:393–404.
[4] Stumpp G, Banzhaf W. An exhaust gas recirculation system for diesel engines. SAE 1978;780222.
[5] Zheng M, Reader GT, Hawley JG. Diesel engine exhaust gas recirculation – a review on advanced and novel concept. Energy Convers Manag 2004;45:883–900.
[6] Agarwal AK. Experimental investigations of the effect of biodiesel utilization on lubricating oil tribology in diesel engine. P I Mech Eng D-J Aut 2005;219:703–13.
[7] Prateepchaikul G, Apichato T. Palm oil as a fuel for agricultural diesel engines: comparative testing against diesel oil. Songklanakarin J Sci Technol 2003;25(3):317–26.
[8] Agarwal AK, Dhar A. Karanja oil utilization in a direct-injection engine by preheating. Part 2: experimental investigations of engine durability and lubricating oil properties. Proc IMechE Part D: J Automobile Eng 2009;224(1):85–97.
[9] Arumugam S, Sriram G, Ellappan R. Bio-lubricant-biodiesel combination of rapeseed oil: An experimental investigation on engine oil tribology, performance, and emissions of variable compression engine. Energy 2014;72:618–27.
[10] Haseeb ASMA, Fazal MA, Jahirul MI, Masjuki HH. Compatibility of automotive materials in biodiesel: a review. Fuel 2011;90:922–31.
[11] Haseeb ASMA, Sia SY, Fazal MA, Masjuki HH. Effect of temperature on tribological properties of palm biodiesel. Energy 2010;35:1460–4.
[12] Fazal MA, Haseeb ASMA, Masjuki HH. A critical review on the tribological compatibility of automotive materials in palm biodiesel. Energy Convers Manag 2014;79:180–6.
[13] Fontana MG. Corrosion engineering. Tata McGraw-Hill Education; 2005.
[14] Nadkarni RK. Elemental analysis. Manual 37. In: Totten GE, editor. Fuels and lubricants handbook. ASTM International; 2019. p. 707–16 [chapter 26].

[15] Fazal MA, Haseeb AS, Masjuki HH. Degradation of automotive materials in palm biodiesel. Energy 2012;40(1):76–83.
[16] Fazal MA, Haseeb AS, Masjuki HH. Corrosion mechanism of copper in palm biodiesel. Corros Sci 2013;67:50–9.
[17] Fazal MA, Haseeb AS, Masjuki HH. Comparative corrosive characteristics of petroleum diesel and palm biodiesel for automotive materials. Fuel Process Technol 2010;91(10):1308–15.
[18] Agarwal AK. Lubricating oil tribology of a biodiesel-fuelled compression ignition engine. In: ASME 2003 Internal combustion engine division spring technical conference. American Society of Mechanical Engineers; 2003. p. 751–65.
[19] Booser ER. Tribology data handbook: An excellent friction, lubrication, and wear resource. CRC Press; 1997.
[20] Kwon OK, Kong HS, Kim CH, Oh PK. Condition monitoring techniques for an internal combustion engine. Tribol Int 1987;20(3):153–9.
[21] Booser ER. CRC handbook of lubrication. Theory and practice of tribology: Volume II: Theory and design. CRC Press; 1984.
[22] Booser ER. CRC handbook of lubrication and tribology. Volume III: Monitoring, materials, synthetic lubricants, and applications. CRC Press; 1993.
[23] Jones MH. Ferrography applied to diesel engine oil analysis. Wear 1979;56:93–103.
[24] Pocock G, Courtney SJ. Some quantitative aspects of ferrography. Wear 1981;67:287–301.
[25] Hofman MV, Johnson JH. The development of ferrography as a laboratory wear measurement method for the study of engine operating conditions on diesel engine wear. Wear 1977;44:183–99.
[26] Ekanem EJ, Lori JA, Thomas SA. The determination of wear metals in used lubricating oils by flame atomic absorption spectrometry using sulphanilic acid as ashing agent. Talanta 1977;44:2103–8.
[27] Reis BF, Knochen M, Pignalosa G, Cabrera N, Giglio J. A multicommuted flow system for the determination of copper, chromium, iron and lead in lubricating oils with detection by flame AAS. Talanta 2004;64:1220–5.
[28] Miller JA, Bowman CT. Mechanism and modeling of nitrogen chemistry in combustion. Prog Energy Combust Sci 1989;15:287–338.
[29] Agarwal AK, Das LM. Biodiesel development and characterization for use as a fuel in a compression ignition engine. ASME J Eng Gas Turbines Power 2000;123:440–7.
[30] Agarwal AK. Performance evaluation and emission characteristics of a compression ignition engine using esterified biodiesel. M.Tech. thesis, Center for Energy Studies, Indian Institute of Technology; 1996. p. 82.
[31] Williamson AM, Badr O. Assessing the viability of using rape methyl Ester (RME) as an alternative to mineral diesel fuel for powering road vehicle in the UK. Appl Energy 1998;59:187–214.
[32] Plee SL, Ahmad T, Myers JP, Faeth GM. Diesel NOx emissions – a simple correlation technique for intake air effects. In: 19th Int symp on combustion. The Combustion Institute; 1982. p. 1495–502.
[33] Surisetty VR, Dalai AK, Kozinski J. Alcohols as alternative fuels: an overview. Appl Catal A Gen 2011;404:1–11.
[34] Najafi G, Ghobadian B, Tavakoli T, Buttsworth D, Yusaf T, Faizollahnejad M. Performance and exhaust emissions of a gasoline engine with ethanol blended gasoline fuels using artificial neural network. Appl Energy 2009;86:630–9.
[35] Al-Hasan M. Effect of ethanol–unleaded gasoline blends on engine performance and exhaust emission. Energy Convers Manag 2003;44:1547–61.
[36] Rice R, Sanyal A, Elrod A, Bata R. Exhaust gas emissions of butanol, ethanol, and methanol-gasoline blends. J Eng Gas Turbines Power 1991;113:377–81.

[37] De Silva PR, Priest M, Lee PM, Coy RC, Taylor RI. Tribometer investigation of the frictional response of piston rings with lubricant contaminated with the gasoline engine biofuel ethanol and water. Proc Inst Mech Eng PartJ-J Eng Tribol 2011;225:347–58.
[38] Ferrarese A, Marques G, Tomanik E, Bruno R, Vatavuk J. Piston ring tribological challenges on the next generation of flex-fuel engines. SAE Int J Engines 2010;3:85–91.
[39] Costa HL, Spikes HA. Impact of ethanol on the formation of antiwear tribofilms from engine lubricants. Tribol Int 2016;93:364–76.
[40] Besser C, Schneidhofer C, Dorr N, Novotny-Farkas F, Allmaier G. Investigation of long-term engine oil performance using lab-based artificial ageing illustrated by the impact of ethanol as fuel component. Tribol Int 2012;46:174–82.
[41] Poley J. Tribology data handbook: diesel engine used oil analysis. CRC Press LLC; 1997.
[42] Kinney AJ, Clemente TE. Modifying soybean oil for enhanced performance in biodiesel blends. Fuel Process Technol 2005;86:1137–47.
[43] Fort EF, Blumberg PN. Performance and durability of a turbocharged diesel fueled with cottonseed oil blends. In: J ASAE Publ.; (United States); Conference: International conference on plant and vegetable oils as fuels, Fargo, ND, USA, 2 August; 1982.
[44] Sinha S, Agarwal AK. Experimental investigations of the tribological properties of lubricating oil from biodiesel fuelled medium duty transportation CIDI engine. SAE Int J Fuels Lubr 2008;1(1). 2008-01-1385.
[45] Sinha S, Agarwal AK. Experimental investigation of the effect of biodiesel utilizationon lubricating oil degradation and wear of a transportation CIDI engine. ASME J Eng Gas Turbines Power 2010;132:042801. https://doi.org/10.1115/1.3077659.
[46] Fazal MA, Haseeb ASMA, Masjuki HH. Investigation of friction and wear characteristics of palm biodiesel. Energy Convers Manag 2013;67:251–6.
[47] Lang X, Dalai AK. Preparation and evaluation of vegetable oil derived biodiesel esters as lubricity additives. Tribotest J 2001;8(2):131–50.
[48] Pradeep V, Sharma RP. Use of hot EGR for NOx control in a compression ignition engine fuelled with bio-diesel from Jatropha oil. Renew Energy 2007;32(7):1136–54.
[49] Qiu T, Li XC, Liang H, Liu XH, Lei Y. A method for estimating the temperature downstream of the SCR (selective catalytic reduction) catalyst in diesel engines. Energy 2014;68:311–7.
[50] Fahd MEA, Yang WM, Lee PS, Chou SK, Yap CR. Experimental investigation of the performance and emission characteristics of direct injection diesel engine by water emulsion diesel under varying engine load condition. Appl Energy 2013;102:1042–9.
[51] Yang WM, An H, Chou SK, Vedharaji S, Vallinagam R, Balaji M, et al. Emulsion fuel with novel nano-organic additives for diesel engine application. Fuel 2013;104:726–31.
[52] Aldajah A, Ajayi OO, Fenske GR, Goldblatt IL. Effect of exhaust gas recirculation (EGR) contamination of diesel engine oil on wear. Wear 2007;263:93–8.
[53] Van Gerpen J. Biodiesel processing and production. Fuel Process Technol 2005;86:1097–107.
[54] Clark SJ, Wagner L, Schrock MD, Piennaar PG. Methyl and ethyl soybean esters as renewable fuels for diesel engines. JACOS 1984;61(10):1632–43.
[55] Kalam MA, Masjuki HH. Biodiesel from palm oil—an analysis of its properties and potential. Biomass Bioenergy 2002;23:471–9.
[56] Ramdhas AS, Jayaraj S, Muraleedharan C. Characterization and effect of using rubber seed oil as fuel in the compression ignition engines. Renew Energy 2005;30:795–803.
[57] Pandey A, Nandgonkar M. Experimental investigation of the effect of esterified Karanja oil biodiesel on lubricating oil and wear of a 780 hp military CIDI engine. SAE International Journal of Fuels and Lubricants 2010;3(2):273–9.
[58] Celik I, Aydin O. Effects of B100 biodiesel on injector and pump piston. Tribol Trans 2011;54:424–31.

[59] Dhar A, Agarwal AK. Effect of Karanja biodiesel blend on engine wear in a diesel engine. Fuel 2014;134:81–9.

[60] Liaquat AM, Masjuki HH, Kalam MA, Fattah IMR. Impact of biodiesel blend on injector deposit formation. Energy 2014;72:813–23.

[61] Nantha Gopal K, Thundil Karuppa Raj R. Effect of pongamia oil methyl ester–diesel blend on lubricating oil degradation of di compression ignition engine. Fuel 2016;165:105–14.

[62] Agarwal AK, Bijwe J, Das LM. Effect of biodiesel utilization of wear of vital parts in compression ignition engine. J Eng Gas Turbines Power 2003;125(2):604–11.

[63] Agarwal AK, Bijwe J, Das LM. Wear assessment in a biodiesel fueled compression ignition engine. J Eng Gas Turbines Power 2003;125(3):820–6.

[64] Krahl J, Munack A, Schroder O, Ruschel Y. 500 Hours endurance test on a biodiesel running a EURO IV engine. SAE Int J Fuels Lubr 2010;3(2). 2010-01-2270.

[65] Wander PR, Altafini CR, Colombo CL, Perera SC. Durability studies of monocylinder compression ignition engines operating with diesel, soy and castor oil methyl esters. Energy 2011;36:3917–23.

[66] Agarwal AK, Dhar A. Wear, durability and lubricating oil performance of a straight vegetable oil (Karanja) blend fueled direct injection compression ignition engine. J Renew Sustain Energy 2012;4, 063138.

[67] Ku Y, Lin K, Chen Y, Liao C. The impact upon durability of heavy-duty diesel engine using 5 percentage biodiesel. SAE Int J Fuels Lubr 2013;6(1).

[68] Kumar N, Varun, Chauhan SR. Evaluation of endurance characteristics for a modified diesel engine runs on jatropha biodiesel. Appl Energy 2015;155:253–69.

[69] Daryl LR, Peterson C. Biodiesel testing in two on-road pickups. In: SAE conference proceedings. SOC Automotive Engineers Inc.; 1995. p. 171–80.

[70] Peterson CL, Thomson JC, Taberski JS, Reece DL, Fleischman G. Long-range on-road test with twenty percent rapeseed biodiesel. J Appl Eng Agric 1999;15(2):91–101.

[71] Raadnui S, Meenak A. Effects of refined palm oil (RPO) fuel on wear of diesel engine components. Wear 2003;254:1281–8.

[72] Fraer R, Dinh H, Kenneth Proc, McCormick RL, Chandler K, Buchholz B. Operating experience and teardown analysis for engines operated on biodiesel blends (B20). SAE Technical Paper 2005; No. 2005-01-3641.

[73] Kenneth P, Barnitt R, Hayes RR, Ratcliff M, McCormick RL, Lou Ha. 100,000-Mile evaluation of transit buses operated on biodiesel blends (B20). SAE Technical Paper 2006; No. 2006-01-3253.

[74] Mazzoleni C, Kuhns HD, Moosmüller H, Witt J, Nussbaum NJ, Chang MCO. A case study of real-world tailpipe emissions for school buses using a 20% biodiesel blend. Sci Total Environ 2007;385:146–59.

[75] Agarwal AK, Srivastava DK, Dwivedi D, Kawatra G, Kumer MR, Bhardwaj OP. Field trial of biodiesel (B100) and diesel fuelled common rail direct injection Euro-III compliant sports utility vehicles in India condition. SAE Technical Paper 2008; No. 2008-28-77.

[76] Thornton MJ, Alleman TL, Luecke J, McCormic RL. Impacts of biodiesel fuel blends oil dilution on light-duty diesel engine operation. In: SAE International Powertrains, Fuels, and Lubricants Meeting; 2009.

[77] Watson AG, Wong VW. The effects of fuel dilution with biodiesel on lubricant acidity, oxidation and corrosion—a study with CJ-4 and CI-4 PLUS lubricants. In: Diesel engine-efficiency and emissions research (DEER) conference; 2008.

[78] Shimokoji D, Okuyama Y. Analysis of engine oil deterioration under bio diesel fuel use. SAE Technical Paper 2009; N0.2009-01-1872.

[79] Ryan TW, Dodge LG, Callahan TJ. The effects of vegetable oil properties on injection and combustion in two different diesel engines. JAOCS 1984;61:10.

[80] Ziejewski M, Kaufman KR, Schwab AW, Pryde EH. Diesel engine evaluation of a nonionic sunflower oil - aqueous ethanol microemulsion. JAOCS 1984;61:10.
[81] Ahmad MI, Hirofumi N, Hasannuddin AK, Jazair WY. An overview of utilizing water-in-diesel emulsion fuel in diesel engine and its potential research study. J Energy Inst 2014;10:1016.
[82] Dryer FL. Water addition to practical combustion system concepts and applications. Symp Int 1977;16:279–95.
[83] Niko S, Kegl B, Dibble RW. Numerical and experimental study of water/oil emulsified fuel combustion in a diesel engine. Fuel 2002;81:2035–44.
[84] Kweonha P, Inseok K. The effect of water emulsified fuel on a motorway-bus diesel engine. KSME Int J 2000;18:2049–57.
[85] Mura E, Josset C, Loubar K, Huchet G, Bellettre J. Effect of dispersed water droplet size in microexplosion phenomenon for water in oil emulsion. Sprays 2010;20:791–9.

CHAPTER 14

NO_X reduction through various low temperature combustion technologies

Pajarla Saiteja, B. Ashok, Pemmareddy Saiteja, and R. Vignesh
Engine Testing Laboratory, School of Mechanical Engineering, Vellore Institute of Technology, Vellore, Tamil Nadu, India

14.1 Introduction

Low-temperature combustion (LTC) is an advanced, promising strategy to reduce NO_x nitrogen oxides and particulate matter at the same time as it has a significant effect on specific fuel consumption. The lowered specific fuel consumption of exhaust nitrogen oxide emissions results in great improvement of IC engines [1]. The LTC engine mode operates at a lower temperature than conventional diesel engine modes. The working temperature in LTC is reduced by the engine operational mode with a high exhaust gas recirculation, or the engine works with an excess amount of air ratio. In conventional diesel combustion, the fuel is oxidized with high-temperature air in a stoichiometric condition. This results in more nitrogen oxides and also reduces the availability of oxygen closer to the fuel injector periphery, resulting in higher emissions. In general, the vaporization and mixing problems require high-pressure fuel injection. The interruption of wall collisions caused from extended nozzle tip spreading is a tough task in the case of increased injection pressure. Different techniques are applied to extent the ID allowing for more mixing of the A/F, such as reduced compression ratio, high cooled EGR level, and variable value timing control. Even with a longer ignition time, the formation of a diluted cylinder charge with harmony before the SOC is difficult to achieve. To reduce the peak cylinder gas temperature, high EGR rates are required, but before an increase in fuel consumption, the operation is narrow. However, the combustion process is deteriorating at the high EGR rate and thermal efficiency is decreased. The attainment of LTC at various loads is a major challenge in realistic conditions. Higher portions of the exhaust gas recirculation supply may not be feasible in manufactured engines to ensure depletion of the air supply and

NOx Emission Control Technologies in Stationary and Automotive Internal Combustion Engines
https://doi.org/10.1016/B978-0-12-823955-1.00014-0

to maintain high brake thermal efficiency. When the engine is operated on medium exhaust gas recirculation and feeder accelerators, the pressure on the in-cylinder is high and the engine loads are more prominent [2]. The rich mixture (φ) is accumulated in the combustion chamber by improper mixing of air and fuel, which affects the combustion process and it becomes more difficult while changing the engine operating conditions. Moreover, twin fuel manifold injection and negative valve overlapping systems to achieve LTC are employed on modern diesel engines, but this is expensive and complicated. The main objective of these strategies is to promote the premeasured LTC through the supervision of fuel-rich zones. The high direct injection swirl ratio enhances the mixture between air and fuel and the required cooled EGR rate can be assigned to the necessary SOC phase. It is possible to reduce NOx emissions by 85% and particulate matter emissions by 95% in LTC mode compared to conventional diesel engine mode.

The evolution of the diesel combustion process has been significant since the introduction of diesel emission standards that have forced the introduction of NOx and diesel particulate matter after-treatment systems. In addition to these emission levels, the advanced combustion strategy has sought to find an in-cylinder approach that avoids using aftercare to reduce the demand for aftercare systems [3]. Although the focus of the development of combustion systems was on lower NOx emissions, lower PM emissions are also significant. Following treatment technologies, the exhaust emissions of an IC engine are better controlled, but the rear pressure produces poor performance. Further operations and maintenance costs are also achieved by the integration of different after-treatment strategies. However, it was more important to achieve excellent performance and emission characteristics for emission control strategies such as advanced engines in combustion mode.

Advanced combustion engines (with LTC strategies) have the capacity to meet future emission standards with improved power, including HCCI, PCCI, and RCCI. They are also able to run biofuel using homogenous load preparation technologies with distinct physicochemical characteristics [4]. The LTC combustion mode can be achieved by early fuel injection strategies with lean mixture conditions, through air and fuel will get sufficient time to form homogeneous charge. Therefore, the combustion of lean homogeneous charge obtains low in-cylinder temperature. The charge then auto-ignites because no spark or other means of forced inflammation is used to heat the compressed gases, thus resulting in lower NOx and HC emissions, as depicted in Fig. 14.1. However, it is difficult to achieve LTC mode engines in every driving country by fulfilling future emission standards.

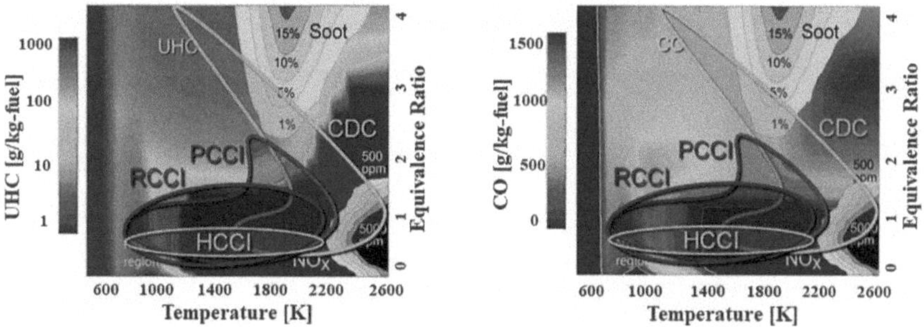

Fig. 14.1 Comparative emission characteristic features of conventional and LTC mode engines.

Efforts must therefore be made to further enhance LTC strategies in existing vehicles.

In recent years, many researchers focused on examining the individual operating conditions and blends in LTC (LTC) mode engines instead of a crucial and exhaustive approach to LTC strategy. For this reason, insight into the techniques of achievement and the influence of operational parameters in the current context are necessary in the LTC engine with enhanced characteristic features. In addition, only some articles that are communicated to the scientific world are essential to address the restrictions and nonlinearities in LTC mode engines [5]. This type of work will help to better understand the correlation between LTC technology with various operational parameters and show how better engine characteristics can be achieved. The current book chapter aimed to explore the various charge preparation techniques to obtain LTC combustion modes and their significance in exhaust NOx reduction. As well distinguished the all LTC mode engines with respective to the fuel type and combustion controlling strategies.

14.2 Homogeneous charge compression ignition engine

The use of an HCCI engine is an excellent way to reduce nitrogen oxide emissions while improving performance. This engine generates a homogeneous charge like a spark ignition engine and burns like a compression ignition engine. Before combustion, air and fuel are gently mixed together using homogeneous charge preparation techniques. By avoiding rapid combustion, the obtained homogeneous charge during the power stroke ensures a longer combustion duration. As a result, LTC is followed by a longer combustion duration, resulting in lower NOx emissions. As a result, while the

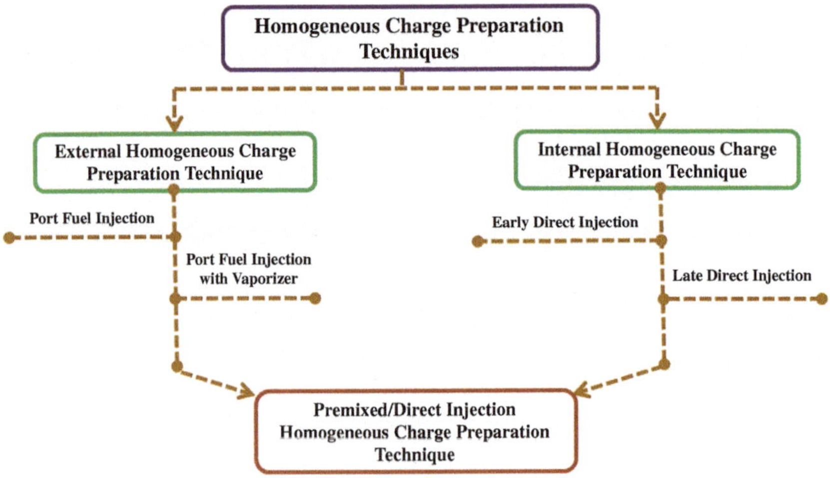

Fig. 14.2 Various homogeneous charge preparation techniques.

preparation of the homogeneous charge is a major concern in the development of the HCCI engine, it can be accomplished through a variety of charge preparation techniques, as listed in Fig. 14.2. One feature shared by HCCI and many other LTC concepts that have evolved from it is that either all or a significant amount of fuel is premixed with air prior to ignition. The chemical kinetics of the mixture direct the combustion rate and ignition timing of such premixed LTC concepts [6]. This greatly complicates combustion process control and makes it sensitive to fuel properties and in-cylinder conditions. Some premixed LTC concepts benefit from fuels with low cetane numbers and volatility characteristics similar to gasoline. Early work with HCCI demonstrated that engine-out NOx and PM emissions could be reduced to 1%–10% of the available diesel engine technology at the time. This raised the prospect of eliminating or simplifying the need for after-treatment devices to meet regulated emission limits.

14.2.1 Significance of external homogeneous charge preparation (EHCP) techniques in NO_X reduction

EHCP is a significant and simple charge preparation method for reducing NOx and particulate matter emissions. In this technique, the injector is located at the intake port, so the fuel arrives along with the air during the suction stroke. As a result, the fuel and air have sufficient tendency to develop a homogeneous charge, which aids in homogeneous combustion.

The EHCP technique is best suited to low viscosity, high -volatility fuels such as alcohols and gasoline. As a result, the addition of a gas vaporizer to the injector nozzle aids in achieving a uniform charge for higher-viscosity fuels. However, meeting emission standards without sacrificing performance requirements in all driving styles is difficult. As a result, innovative HCCI techniques for improving engine output features have been developed. Table 14.1 describes the HCCI characteristic features with respect to various operating techniques for alternative fuels.

14.2.1.1 Influence of port fuel injection (PFI) strategy on NOx emission

PFI is the most basic and straightforward EHCP technique for reducing NOx emissions. Port fuel injectors are located closer to the intake valve, which is located at the engine's intake port or manifold. The injection of fuel starts during the suction cycle and is permitted with air in the engine cylinder, causing disorder in the engine cylinders and helping to form a charge [8]. However, the poor mixing of high and low unstable fuels leads to high emissions of UHC and PM. This makes it recommended to use port fuel injection for low-viscosity and highly unstable fuel such as alcohols and petrol. Higher in-cylinder or high intake charge temperature can improve fuel vaporization through prewarming of low unpredictable fuel and high thick fuel, yet just with a part load. The rich air-fuel mixture is recommended to obtain required output power at higher operating load conditions. Therefore it is difficult to form homogeneous charge with low injection pressures at higher operating load conditions. It brings out inflated charge consumption and fundamentally builds outflow levels [11]. Broad fuel infusion procedures are important to defeat the impediments of the HCCI engine utilizing PFI techniques.

14.2.1.2 Influence of port fuel injection with vaporizer (PFIV) on NOx emissions

In order to homogenize higher-viscosity fuels with inflated request torque, PFIV was developed to reduce NOx-PM emissions. A vaporizer is a piece of equipment that atomizes high-viscosity fuel by means of an external heat source from a liquid to a vapor phase. The vaporizer is further inserted above the port fuel injection system and transfers the evaporated fuel to the intake concentrator. The vaporized charge in the combustion chamber during the suction stroke can be stratified with PFI and PFIV [17, 18]. The limitations of port fuel injection strategy like fuel wall wetting, fuel impingements on piston surface etc. are can be overcome by using port fuel injection with the

Table 14.1 Confined literature on NOx reduction/influencing strategies and its response on HCCI engine.

Ref	Year	Test fuel	Test condition	NOx reduction strategy	NOx response
[7]	2021	Fusel oil/diethyl ether	Speed: 800-1800rpm, λ: 1.35–4, ICT: 80°C, IP: 2.5 bar, DEE: 40%–80%	Port fuel injection (PFI) with LTC	• NOx↓ as λ↑ and blend ↓ • Minimal NOx at 1100rpm, λ:2.2 and 40% DEE
[8]	2021	Alcohols	Equivalence ratio (Φ): 0.2–0.5	Port Fuel Injection with Vaporizer (PFIV)	• NOx↑ as Φ↑ • NOx for sec-butanol and propanol ˃ ethanol and iso-butanol
[9]	2020	Tamanu methyl ester	IP: 220 bar, speed: 1500 rpm, ICT: 80–100 °C and EGR: 5%–15%	Lean mixture combustion with PFIV	• NOx↑ as ICT↑ and EGR↓ • NOx for diesel < Tamanu methyl ester
[10]	2020	Hydrogen enriched kerosene dimethyl ester	Intake air temperature: 248–398 K, speed: 1000–2000 rpm	Lean operating condition ($\Phi = 0.9$)	• NOx↑ as ICT↑ • NOx↑ as Speed↓
[11]	2019	Hydrogen with diesel and Biodiesel (Honge & Cotton seed)	Hydrogen ratio: 0%–45% and EGR: 0%–15%	Low combustion temperature through retarding the combustion process	• NOx↓ as both EGR and Hydrogen↑ • NOx for diesel with biodiesel < pure diesel • NOx for cotton seed oil < Honge oil < diesel at 40% hydrogen energy ratio
[12]	2019	Ethanol and polyoxymethylene dimethyl ether	Ethanol blend ratio (0%–50%) and DOC	Compound combustion	• NOx↓ as Ethanol Ratio↑ and BMEP↑ • NO_2 ↓↓ with DOC than without DOC

Table 14.1 Confined literature on NOx reduction/influencing strategies and its response on HCCI engine—cont'd

Ref	Year	Test fuel	Test condition	NOx reduction strategy	NOx response
[13]	2019	n-butanol	Intake boost pressure (1.8–2.25 bar) Injection timing (300–360° CA)	Leaner cylinder charge combustion with pilot injection	• NOx↓ as Intake boost pressure↑ • Peak NOx at injection timing CA 330°
[14]	2019	n-butanol with distilled water and H_2O_2	λ: 2–4 H_2O_2	Low Temperature Combustion	• NOx↓ as λ↑ and H_2O_2↑ • NOx with n-butanol + H_2O_2 < n-butanol + distilled water
[15]	2016	Polyoxymethylene dimethyl ether	Φ: 0.2–0.5 EGR: 0%–50%	Lean combustion	• NOx↓ as Φ↓ and EGR↑
[16]	2016	Biogas/n-heptane	Energy mean effective pressure: 14.9–18.2, n-heptane: 0%–30% ICT: 300–360°	Compound combustion	• NOx↓ as n-heptane↑ • NOx↓ as energy mean effective pressure↓

vaporizer (fumigating unit). However, improved uniform combustion helps promote extreme temps in cylinders that can result in high nitrogen oxides. As a result, engine variables such as charge intake temperature, EGRn and reformed gases (RG) are adjusted in a way that reduces NOx emissions while maintaining performance. In addition, a new homogeneous charge progressive combustion (HCPC) concept has been developed to prepare a homogenous prepressed load outside. In addition, it is possible to reduce NOx and PM emission levels in the combustion chamber while enhancing thermal efficiency. However, the HCCI engine's PFI and PFIV techniques have limitations such as cold start-ups, combustion controls, and limited operating ranges. As a result, the internal homogenous process of load preparation has been developed to tackle the obstacles linked to PFI and PFIV technology in the HCCI engine.

14.2.2 Significance of internal homogeneous charge preparation techniques in NO_X reduction

In the combustion process, internal homogenous load preparation is carried out, eliminating constraints for preparing external homogeneous loads. At the beginning of the compression stroke, the injection of the cylinder starts, giving sufficient time to form a homogenous blend [7]. In addition, the delayed start of combustion leads to better control of the late injection strategy in the internal homogenous loading technique. Moreover, it accomplished ultralow emissions of NOx and PM with improved proficiency and burning [10]. In any case, the late infusing system can't diminish hydrocarbons like PFI because of the irrelevant homogeneous weight. Insignificant changes in the circumstance of infusion lead to better engine execution. Premixed lean diesel ignition lessens fuel and builds ignition; multistage diesel burning improves power over the consuming stage [16]. The early execution of every strategy is in this manner discussed in the next sections.

14.2.2.1 Influence of early direct injection (EDI) strategy on NOx emissions

EDI, which injects fuel into the combustion process directly, is the most common internal uniform charging technique in the homogeneous charge engine. The EDI method produces high injection cylinders, which help to vaporize the fuel that has been injected early. A more uniform charge with a longer ignition delay period is therefore obtained. The EDI procedure has been utilized under popular choke conditions with a fuel divider mounting because of its limited fuel entrance and oil weakening in the ignition chamber [1–3]. It can be decreased by utilizing an all-around planned CRDI framework. The inadequacy of the controlled ignition stage, which is overwhelmed by using new infusion procedures, is another hindrance of early direct infusion. The most productive early direct infusion procedures are premixed lean diesel combustion, multiple stage diesel combustion, uniform bulky combustion system, PCI, multiple pulse injection with bump combustion chamber, NADI, and homogeneous charge intelligent multiple injection combustion system.

14.2.2.2 Influence of late direct injection strategy on NOx emissions

A late direct injection system for HCCI engines fueled by diesel was developed to address the problem of combustion phasing. Combustion is smarter and emissions are lower than previous techniques for direct and port fuel injection. For preexisting homogeneous load, a lengthy ignition delay is required to prevent advanced combustion, among the other characteristics of combustion [2, 4]. The start deferral is drawn out by postponing the infusion or permitting EGR to lessen the centralization of oxygen. Similarly, quick blending is accomplished by utilizing high whirl outspread bowl math. Subsequently, the ignition period of the HCCI engine is constrained by the infusion time delay. It prompts late direct infusion at medium speed and a 33% evaluated force, bringing about critical outcomes in decreases in nitrogen oxide and ignition commotion with negligible PM emanations. The primary driver for higher HC and CO discharge lies in fuel impressions in the cylinder bowl and cold engine conditions. Therefore, different procedures for improving engine properties for the late immediate infusion technique have been created. The EGR, outfitted with CRDI, improves the proficiency and outflow qualities at full burden and most extreme speed.

14.2.2.3 Influence of premixed/direct injection homogeneous charge technique on NOx

The uniform load preparation technique preimplanted/direct injection can be created through the combination of external and internal loading processes. It can also overcome workable constraints such as combustion phase control and ignition time. Significant amounts of fuel are injected at the entrance port to increase the air-and-fuel time during the homogenous load preparation process. In addition, little fuel is injected directly into the engine combustion chamber to ignite the uniform charge [17]. The secondary fuel injection fuel (direct injection) shall increase the amount of fuel required for full load torque condition; its typical injection timings are depicted in

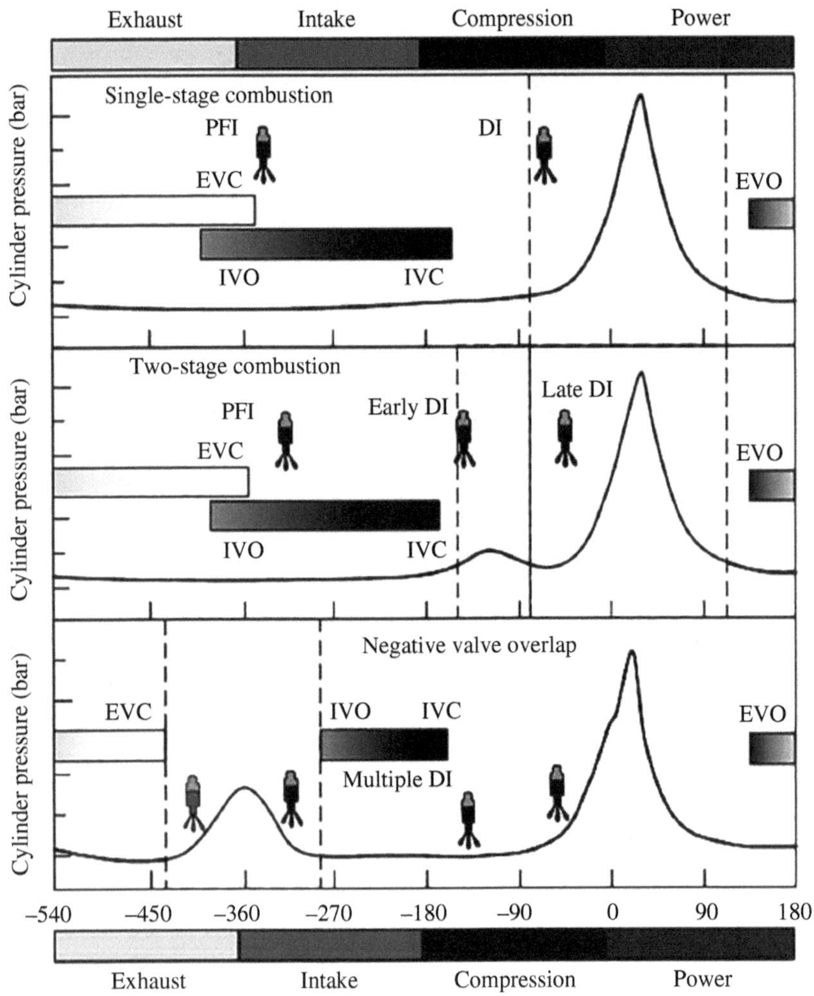

Fig. 14.3 Fuel injection strategies for controlling the HCCI engine ignition timing and combustion phase controlling.

Fig. 14.3. Perhaps the HCCI engine features of premature/direct injection homogeneous technologies are superior to traditional diesel engines due to the timing of injection somewhere between internal and external load preparation techniques, which significantly controls the ignition timing. On the intake collector, a low-pressure pump atomizer is used to vaporize fuel between ignition injections. Also, a high-pressure pump directly removes rapid fuel evaporation by reducing the charge temperature while ultralow smoke and NOx emissions are prevented in various operating conditions

[1–4]. Emission standards are increasing, however, and fossil fuel emissions have increased substantially. Biofuels have therefore drawn attention to the reduced levels of HC and CO emissions, which in their physical and chemical characteristics are similar to conventional fuel. As a consequence, different biofuels, such as biodiesels, alcohols, alkanes, etc., are examined by different parameters in the engine operations of HCCI. It shows important results in engine characteristics, which in the following sections are holistically reviewed and described in detail.

14.2.3 Influence of fuel properties and blends on HCCI engine NO_X emissions

Fuel physicochemical properties, the adoption of alternative fuels, and varying the blend ratio significantly influence NOx emissions. Each fuel property has its influence on NOx emissions while powered with an HCCI engine [1, 17, 18]. Among the fuel properties, the cetane number plays a momentous role in the combustion process. High cetane fuels advance the combustion process and vice versa, which magnifies the HCCI engine start of combustion due to its early start of injection strategies. So, early combustion takes place in various locations due to the ignition of a low cetane numbered homogeneous air-fuel blend. The charge achieves its chemical activation energy, which promotes spontaneous combustion by avoiding the flame front propagation or diffusion flame. It abolishes the high-temperature flame front and secures low in-cylinder temperatures. Accordingly, high cetane fuels with HCCI engines obtain near-zero NOx emissions but also increase difficulties in combustion phase control. Furthermore, the calorific value (CV) is another significant NOx-influencing fuel property, which has a linear relation with HCCI NOx emissions. Early combustion of high calorific value fuel emits greater energy, which notably promotes rapid combustion [17]. The increases in combustion rate provoke high in-cylinder temperatures, which result in exceptional NOx emissions. Other than the cetane number and calorific value, the viscosity and density of the fuel play vital roles in the HCCI exhaust NOx emissions. The higher-viscosity fuels have minimal leakage during fuel injection but also have poor atomization characteristics. So, higher-viscosity fuels are recommended with a higher injection pressure and advanced start of injection, which tremendously increases the fuel injection mass rate and NOx emissions. The increased injected higher-viscosity fuel mass has reduced spray cone angles, larger fuel droplet sizes, and higher spray penetration, which promotes high in-cylinder temperatures by rapid combustion. Therefore, these are the

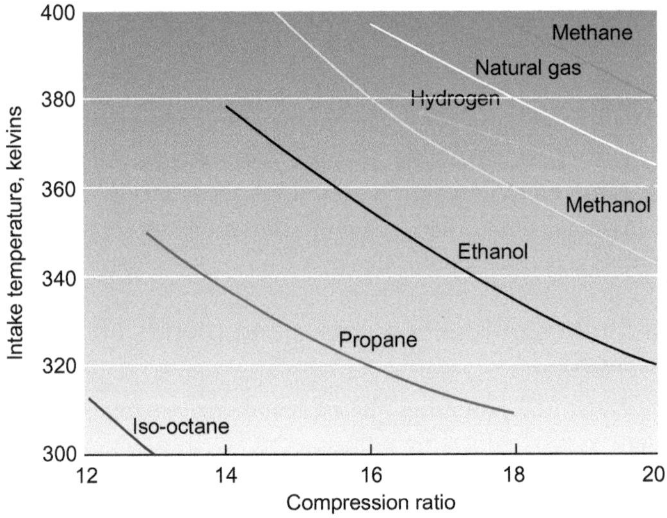

Fig. 14.4 Required intake charge temperature and compression ratio for distinct alternative fuels in HCCI engine.

main reasons for emitting higher NOx emissions through high-viscosity fuels powering the HCCI engine. However, this can be overcome by preheating the fuel or diluting the fuel with other substantial alternative fuels, as mentioned in Fig. 14.4. Biodiesel, alcohols, ethers, and alkanes are the best alternatives to conventional diesel in HCCI engines. Biodiesel is an abundantly available diesel alternative processed by the transesterification of waste cooking, vegetable oil, or agricultural waste as feedstock. Most promotional biodiesel feedstock consists of unsaturated fats, which can result in higher nitrogen oxides compared to HCCI engines powered by petroleum products. In addition, the amount of free radicals that quickly promote nitrogen oxide leads to higher dual bonds of biofuel production. To reduce NOx emissions through lean and low-temperature burning, dual-bonded biofuels with early direct injection techniques of the HCCI engine are necessary. Pure methyl ester emission characteristics in HCCI engines showed a decrease in C12 and C16 saturated NOx compared to diesel gasoline while C18 monounsaturated emissions were slightly increased.

From the explanatory viewpoints above, it is obvious that HCCI combustion can simultaneously reduce the emissions of NOx and soot. However, the HCCI engine is essentially separated from the fuel injector and the sparkplug. This implies that the HCCI engine does not have a vehicle control mechanism and the corresponding combustion phasing and its

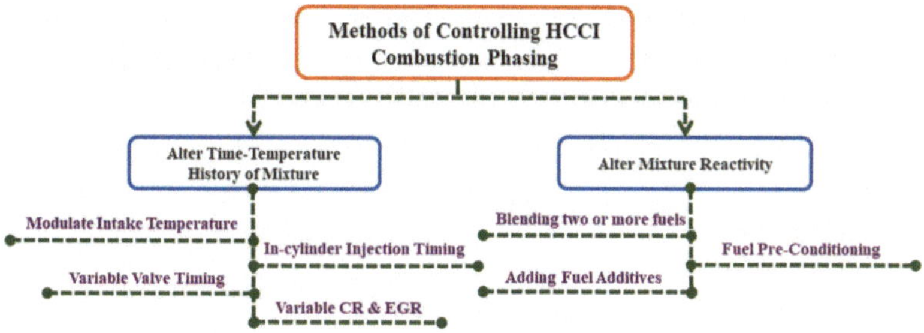

Fig. 14.5 Combustion phase controlling methods.

remedies depicted in Fig. 14.5. The HCCI engine is restricted in a slight range of operation because of the problems of cold start, the high intensity and noise of vaporization, and even knocking oxidation at high charge, without effective strategies for time control of the auto-ignition schedule according to operating conditions.

14.3 Premixed charge compression ignition engine

Combustion is carried out in the standard diesel engine after the PCCI stage and then in the combustion phase. During the air-fuel mixing at a particular phase, oxidation reactions are shown to be abundant quicker than the fuel distribution rate into the charge. The largest percentage of fuel ignites during the diffusion process, where fires occur in closely stoichiometric regions in the lean but regionally inhomogeneous combination. The flame temperature is high and large quantities of NOx are therefore formed. Moreover, the diffusion stage is mainly necessary for soot generation because of the rich mixture locally that might not be oxidized by unsatisfactory air/fuel mixing during the later combustion. The nitrogen oxides and particulate matters are therefore largely discharged during the combustion-controlled stage and therefore can be reduced with the A/F mixture by increasing the precombustion phases, which secures better thermal efficiencies and emission characteristics as depicted in Fig. 14.6. In PCCI combustion mode, ignition delay significantly improved by varying early fuel injection timings (bTDC). Increasing in ignition delay period improves the fuel/blend combustion by enhancing the homogeneity of charge; however which advances the start of combustion. EGR incorporation assisted to retard the combustion process which results in better combustion and emission characteristics. This eliminates rich mixture pockets and high-temp areas within the cylinders, increasing the precombustion phase and simultaneously reducing NOx

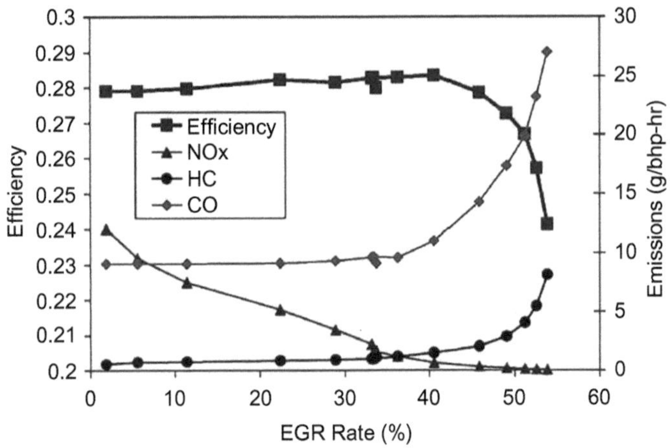

Fig. 14.6 EGR influence on PCCI engine efficiency and emission.

and soot emissions. PCCI combustion offers both DICI and HCCI combustion methods with lesser tailpipe emissions, greater efficiency, and significantly quicker combustion processes than the CI and SI drive systems. In PCCI combustion, a completely homogeneous mixture such as HCCI does not occur and can operate in a broad range of fuels such as, ethanol, gasoline, biodiesel, etc.

HCCI has a small range of functions due to high temperature growth rates and problems in regulating combustion variables such as state of charge, which are dependent on many considerations such as the fuel quality, quick ignition features, and the similarity of oxidation, cylinder pressure, and temperature. These encounters have led researchers to advance another type of LTC, known as PCCI, which has been developed through HCCI combustion to control SOC and the burn length of time. Premixed compression ignition (PCCI) is a further improved steam strategy that reduces nitrogen oxides and lowers soot tailpipe emissions at the same time, similar to the concept of HCCI with enhanced engine efficiency. The combustion of PCCI leads to higher emissions of nitrogen oxides and lower soot emissions, and lower emissions of CO, hydrocarbons compared to HCCIs.

14.3.1 Significance of premixed charge preparation technique in NOx reduction

The PCCI combustion approach injects fuel into the ignition chamber via three approaches: PFI, advance DI, and delayed DI. Although the first two approaches are closely related to HCCI strategy, which primarily reduces

NOx emissions through homogeneous low-temperature oxidation but suffers from restricted operating range and combustion phase control, late injection overcomes the restricted operating range and is used to regulate combustion switching. Therefore, including these three injecting strategies such as adopting a multiple injection strategy is a supreme LTC technique to deal with HCCI engine difficulties with lower nitrogen oxide and soot tailpipe emissions [18]. With LTC, the nitrogen oxide emission is primarily dependent on the rate of EGR rather than the amount and volatility of fuel cetane. EGR cooled is a collective strategy for achieving lower-temperature combustion by dropping the temperature of combustion and extending the ignition time period. The lower temperature of combustion eliminates the formation of NOx and PM. A long ignition delay provides further time for combustion of air/fuel, reducing fueled areas, and eliminating soot formation. Delaying the injection period is a unique operative tactic for prolonging the delay period to LTC operating conditions in addition to EGR. Enhanced PCCI precombustion increases the amount of peak pressure increases, making it difficult to control the combustion phasing, which can be overcome by using suitable alternative fuel for the PCCI.

14.3.2 Role of distinct premixed conventional and alternative fuels on PCCI engine NOx emissions

The section below describes the role of separate, preimplemented alternate fuels in the emissions of PCCI engine NOx based on their typical behavior.

14.3.2.1 Influence of diesel fuel on PCCI NOx emissions

PCCI diesel combustion is based on an enhanced injection strategy for fuel and EGR to reduce the emissions of knockout and NOx and PM [1–3]. Advancing the start-up of the main injection (SoMI) tends to increase the time available for combustion between fuel and air and precipitated combustion as well as reduces the main combustion phase (also the compression ratio pressure and thermal release). The advance of SoPI reduced the cylinder variable due to lower A/F mixing [2, 5]. At perpetual SoPI, the rate of EGR increases due to an extended delay period and lagging at the start of combustion results in a lower cylinder temperature and a low heat rate release speed. The advanced SoMI and SoPI schedules are used to detect advanced combustion and combustion phases and shorter combustion duration. This is due to the better A/F mixing, improved combustion, and increases in the EGR rate; however, it significantly helps to reduce peak knock intensity and noise from combustion with significant NOx emissions.

The emissions of smoke and lower BTE as well as higher BSFC, CO, and HC are also noted with the advance in the injection time and EGR rate [3]. These interpretations are due to increased unwanted column work at more progressive injection times, reduced cylinder parameters, and the chemical activation energy of the blended fuels. Additionally, a decrease in particle diameter and concentration levels is observed as the main injection times advance because the blending time increases, resulting in a more uniform mix. The PCCI tailpipe emission results indicate also that the metals within the cylinders have an increased EGR rate below that of the HCCI combustion mode [1, 3]. The formation of consistent blends and LTC resulted in higher emissions of CO and HC due to increased injection pressure. It favored NOx and smoke reduction. Besides the advanced beginning of the major injection, an increase in fuel injection pressure improves PCCI combustion. It is thus clear that a split injection strategy can be employed by enhancing the injection timing, pressure, and exhaust gas rate for controlling PCCI combustion. In such a way, Table 14.2 describes the PCCI engine characteristic features with respect to varying the operating parameters for distinct alternative fuels.

The PCCI engine's combustion and emission determinants of spray obstruction, injection variables, and EGR rate demonstrate the higher-pressure injection and the greater frequency of heat release when compared with a lower injection pressure and the shortened duration of combustion, as depicted in Fig. 14.7. It also led to improved CO emissions in ITE and IMEP and reduced smoke and HC emissions. This is due to better ignition and atomization at greater injection pressure [19–23]. Injection pressure played a significant role in reduction of NOx emission. Poor homogeneous charge obtained from decreasing the injection pressure, which reduces the combustion temperature there by lower NOx emission is obtained. For decreasing in-cylinder pressure and a high rate of heat release, increasing ignition postponement as well as delayed combustion and EGR are observed.

14.3.2.2 Influence of biodiesel on PCCI NOx emissions

The medium diameter and the penetration by the sauters are more advanced than diesel fuel with greater hydrocarbon emissions by increasing the density, viscosity, and surface stress. Optimizing the injection time in biodiesel PCCI can produce the lowest emissions compared with diesel fuel of NOx, THC, or CO [1, 20]. From the earlier formation of the OH radical in biodiesel fuel, the maximum heat release rates, combustion characteristics, and

Table 14.2 Confined literature on NOx reduction/influencing strategies and its response on PCCI engine.

Ref	Year	Test fuel	Test condition	NOx reduction/ influencing strategy	NOx response
[19]	2021	Wheat germ oil/ bioethanol	Bioethanol energy share 0%–30% Load: 0–5 KW (BP)	Compound combustion	• NOx for diesel ˃ NOx for wheat germ oil • NOx↓ as ↑bioethanol% and ↑load
[20]	2021	Waste cooking oil/diesel	Waste cooking oil: 20% and 40% Load: 0%–100%, premixed ratio: 10%–30%	Low temperature combustion through PCCI-DI	• NOx↓ as ↓waste cooking oil • NOx↑ as ↑load and ↑premixed ratio
[21]	2020	Diesel/diesel-methanol	BMEP: 1-4 bar PCCI and RCCI	Low temperature combustion	• NOx↓ as ↑BMEP • NOx in RCCI < PCCI
[22]	2020	Diesel/light hydrocarbons	Load: 0%–100% Conventional mode and light hydrocarbons (LHC)-PCCI mode	Low temperature combustion	• NOx↑ as ↑load • 28.2% and 23.5% NOx reduction with LHC-PCCI mode at 75 and 100% loads respectively
[23]	2020	Diesel/CNG	CNG: 0%–100%, diesel/ PCCI/HCCI mode	Dual fuel combustion	• Conventional mode NOx ˃˃ PCCI ≈ DF PCCI ˃ HCCI. • NOx↑ as ↓CNG
[24]	2019	Diesel/fumigated ethanol	Fumigated ethanol (0%–10%), BMEP: 0–4 bar	Low temperature combustion by fumigated ethanol	• Up to 4bar BMEP NOx↓ as ↑ethanol% • At 4bar BMEP NOx↑ as ↑ethanol%
[25]	2019	Diesel/CNG	CNG: 0%–100% EGR: 40% and 50% and IMEP: 0.3–0.4 MPa	Dual fuel combustion	• NOx↑ as ↑IMEP and ↓EGR • NOx↑ as ↓CNG

Continued

Table 14.2 Confined literature on NOx reduction/influencing strategies and its response on PCCI engine—cont'd

Ref	Year	Test fuel	Test condition	NOx reduction/ influencing strategy	NOx response
[26]	2019	Diesel	Cetane number (30, 40, 51) and pilot fuel mass ratio (0%–40%)	Low rate of combustion	• NOx↓ as ↑Cetane number • NOx↓ as ↑Pilot fuel mass ratio
[27]	2019	Diesel/CNG	CNG: 30%–70% EGR: 0%–50%	Dual fuel combustion	• NOx↓ as ↑EGR • NOx↑ as ↓CNG
[28]	2019	Diesel/water	Water emulsion (3%, 6%, 9%) Load: 0%–100%	Low in-cylinder temperatures	• NOx↑ as ↑Load • NOx↓ as ↑Water emulsion%

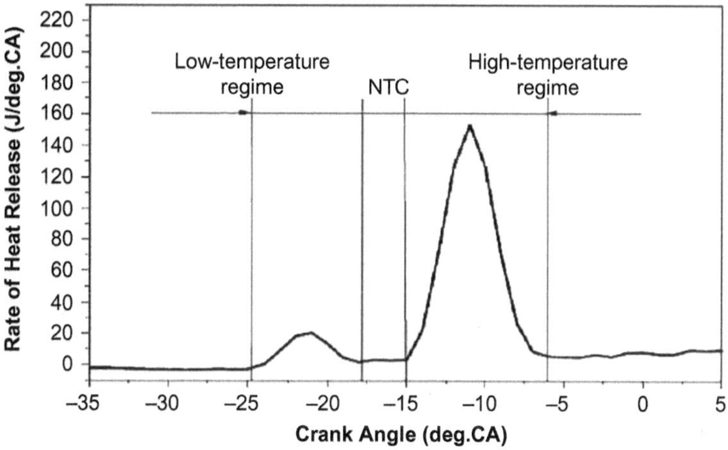

Fig. 14.7 Heat release rate curve for advanced injection timing engines.

cylinder temperature increase have increased. The temperature increases as well as the duration of combustion ultimately decrease NOx emissions.

14.3.2.3 Influence of gaseous fuels on PCCI NOx emissions

In the meantime, one key priority in engine design is to reduce CO2 emissions, which can be reduced by increasing engine efficiency while focusing on reducing NOx and soot emissions from diesel engines. In this respect, the lack of carbon atoms could make alternative fuel such as hydrogen a better choice to eliminate CO2 emissions.

14.3.3 Role of blend/dual fuels on PCCI NOx emissions

The combustion of PCCI occurs mostly through a prior injection of fuel at a significantly lower pressure in the cylinder. This causes long pulling and spray wall impaction, reducing NOx emissions at cylinder temperatures [2–5]. In addition, it does not provide adequate homogeneity of low and high responsive fuel in PCCI combustion (diesel). It is noticed that PCCI combustion with specific restrictions is permitted by the direct injection of fuel. Therefore, with preimplanted isooctane and direct-injected diesel, the PCI dual-fuel concept is advised. That's the only way. It should be noted that the PCI also features advanced levels of exhaust gas recirculation to improve inflammation delays and decrease the burning temperature. But the higher EGR rate (more than 55%) reduces fuel consumption, nitrogen oxides, and soot tailpipe emissions. So, EGR incorporation in the PCCI engine significantly lowers the NOx-PM/soot trade-off by combustion of premixed homogeneous charge at low cylinder temperatures, which increases the fuel reliability. The use of moderate EGR efficiency and poor fuel combustion (diesel mixed with lower fuel cetane such as petrol) are recommended to enhance the ignition time and improve the A/F mixing process before combustion to minimize soot oxides and soot-fuel consumption trade-offs. The literature reports extensively on the double-fuel strategy to regulate the PCCI combustion phase. A wide experimental survey of the timing of self-ignition and the expansion of the PCCI combustion method range of PCCI diesel engines using the dual fuel strategy was conducted with various fuels, such as propane, petrol, alcoholic fuel, dimethyl furan, dimethyl ether, and biodiesel. Below are some of the key studies on the special effects of mixing/dual fuel on the nitrogen oxide emissions from PCCI combustion.

14.3.3.1 Influence of gasoline and diesel blends on PCCI NOx emissions

The low responsive gasoline fuel tends to increase the ignition delay, increasing the premixing ratio between air and fuel. Diesel-gasoline combinations extend the time period that detects fast combustion and thus ensures LTC of ultralow PCCI engine nitrogen oxide emissions [1, 3, 5]. Therefore, increased volatility and high resistance to self-ignition temperatures are the result of diesel fusion/mixing with petrol fuels rather than pure diesel fuel. It also needs to extend the load range by preventing fuel-rich regions and safe combustion with low temperatures, which reduces soot emissions without influencing NOx emissions. In addition, the increased autoignition sensitivity of gasoline increases the precombustion charge (SOC), giving additional freezing.

14.3.3.2 Influence of alcohol and diesel blends on PCCI NOx emissions

Alcohol is one of the most reliable resource for LTC, such as PCCI, due to its favorable physical chemical characteristics such as high autoignition resistance and a long delay time (because of the amount of cetane) that allows adequate time for the A/F mixture and faster sedimentation. Alcohol fuels are oxidized with hydroxyl, which improves the quantity of air during vaporization and therefore decreases exhaust tailpipe emissions from diesel engines, particularly in peak loads. In comparison with diesel, higher vaporization heat in latent applications and a reduced energy content of alcoholic fuels do not substantially increase NOx emissions. High oxygen content and high latent heat of vaporization properties of ethanol allow cooling effect, which results in lower heat release in PCCI engine with early injection strategy. Consequently, the PCCI engine therefore generates lower temperatures of the cylinder, which lowers NOx emissions predominantly.

Correspondingly, low numbered cetane and high temperature-compatible autonomous fuel, such as n-butane, lower NOx emissions by delaying the start of injection via the PCCI engine. Increasing the amount of butane in the mix, however, increases the ignition postponement period and HC emissions respectively due to the robust impact of oxygen and cooling.

14.3.3.3 Influence of biogas and diesel blends on PCCI NOx emissions

Dual low-temperature PCCI fuel combustion mode, together with the intake air and fuel, can be accomplished by incorporating biogas via high-degree injection missions. This leads to a lower rate of high rate of heat release and underdeveloped combustion duration by increasing the injection timing and biogas energy ratio, which contributes to low in-cylinder pressure and reduces NOx emissions [26]. The additional result to the increasing homogeneous charge, which can be reduced by increasing the inlet load temperature, is the higher dominant carbon emission. As well, rapid or uncontrolled combustion can be avoided by increasing the biomass composition percentages, which can lower the cylinder temperatures and results in lower NOx exhaust emissions.

14.3.3.4 Influence of DME and diesel blends on PCCI NOx emissions

Adding dimethyl ether (DME) to the PCCI chamber increases the cylindrical pressure and cylinder temperatures. As the ratio of prepremixed increases, the density of the prepremixed mixture is increased and the heat release rate is lower. Longer burning time and earlier SOC with an increasing DME-fuel premeasured ratio are also observed. Thus, the cylinder temperatures are primarily increased and the pressure leads to increased tailpipe emissions.

14.4 Reactivity controlled compression ignition engine

The RCCI methodology uses a multi-in-cylinder mixing of at least two fuels with separate self-ignition features to regulate ignition and heat generation as well as to help reduce nitrogen oxides and smoke tailpipe emissions by a wide variety of engine workloads below the European Emission IV level [1–4]. The multistage injection approach is used in RCCI mode, which suppresses particle production and results in a better homogenous combination and a longer combustion delay period with maximum EGR [3]. In this method, less reactivity fuel is pumped into the intake valve followed by a tiny supply of higher reactivity fuel delivered directly into the combustion cylinder with two or more discrete injectors that serve as a source of combustion, as depicted in Fig. 14.8 [5]. Implementation of RCCI

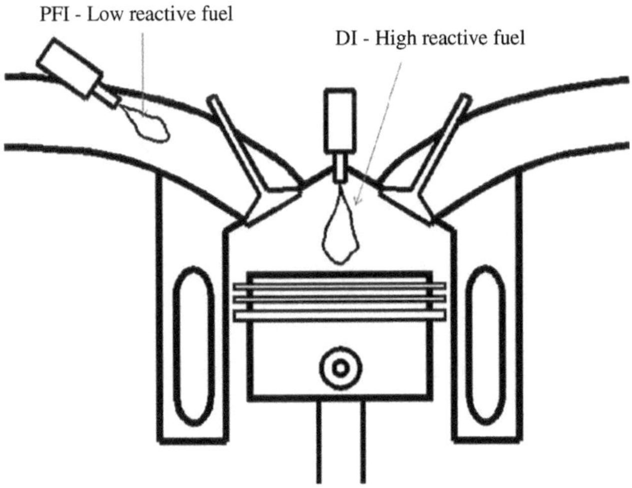

Fig. 14.8 Schematic representation of fuel injection strategy of RCCI engine.

engine with dual fuel approach can lower the peak cylinder pressure rise and improves the combustion duration by altering the injection timings of low and high reactive fuels. Controlling combustion duration is performed by altering the composition amount and/or duration of infusion of the two distinct reactive fuels. As a result, by altering the amount of low reactive fuel in the mixture as well as the length of continuous injection, RCCI provides for elastic actions in an extensive variety of engine working circumstances. Fuel responsiveness adjustments allow for the needed in-cylinder compression ratio and stratified reactivity. However, it has been demonstrated that a bigger fraction of the energy may be produced from lower reactivity fuel to achieve optimum efficiency while dramatically reducing nitrogen oxides and tailpipe emissions. A low reactivity fuel is delivered into the RCCI ignition to increase the blending of the air, fuel, and recycled gases, monitored by a straight combustion chamber infusion of a maximum reactive fuel before combustion of the premixed fuel utilizing single or repeated injection.

14.4.1 Influence of low and high reactive fuel combustion in RCCI engine exhaust NOx emissions

The graphic representation of RCCI processes is shown in Fig. 14.8. Several studies show that in order to achieve greater efficiency with lesser nitrogen oxide and tailpipe emissions, a higher power component (up to 90%) is necessary, mostly from reduced reactivity fuel. The RCCI combustion condition releases extremely little NOx even at full load and with a lower EGR level due to the fuel reaction imbalance within the combustion chamber. As well, the RCCI engine significantly depends upon fuel properties and combustion chamber design, which further plays a role in encouraging A/F mixing within the ignition. The ignition process starts with the self-ignition of strong responsive fuel, monitored by lower sensitive fuel [1]. The in-cylinder reaction stratified by raising the responsiveness variation between the various fuels governs combustion time and duration in this RCCI ignition, allowing for a larger load extended and improved efficiency due to lower thermal energy losses. As a result, this type of ignition process can resolve the problems encountered by other LTC methods such as homogenous combustion and premixed combustion modes. It is also considered to be superior to other LTC modes. In RCCI mode, combustion is ordered and moves from a locally moderate responsiveness fuel zone to a globally low reaction fuel zone. The spray-combustion mechanism of HRF controls the early ignition of RCCI, with the spray peripheral fuel zones having the highest thermal efficiency. This stage of combustion substantially extends the duration of combustion temperature, resulting in high power output, minor pressure increases, and lower tailpipe pollutants, even at advanced loads [2–5]. The primary stages of pure diesel ignition are: (i) cold flame responses initially, (ii) the significant heat release phase (due to premixed combustion, and (iii) a small tail of post vaporization. In RCCI, however, once the diesel is implanted into the natural gas environment, it evaporates and causes nonuniform reaction inequality in the combustion chamber. This results in changes in combustion performance such as: (i) lower surface temp responses and neat flames acting similarly to the clean diesel case encouraged by the diesel implant, and (ii) tracking ability of self-ignition from the extraordinary reaction occurs. The RCCI phase of combustion has been demonstrated to manage ignition more effectively than other strategies such as homogenous charge, premixed charge, dual stage fuel homogenous charge and premixed charge, and single stage fuel PPC, as well as to improve thermal efficiency (60%). In comparison to previous LTC technologies, RCCI ignition produces much more brake power while generating much fewer nitrogen oxides and particulate matter. It also allows for a more even ignition procedure by dropping engine knock and clanging compared to HCCI engines. Several quantitative investigations have also demonstrated that when compared to other LTC strategies, RCCI ignition is a possible LTC mode. However, the RCCI ignition mode has some limitations, including lower fuel economy, particularly at moderate loads; larger capacity enlargement limited by increasing temperature leads to the spread; and increased CO and HC emission levels in crevice zones. This can be minimized by modifying a number of control variables, including injection type, intake heat and acceleration, EGR amount, valve timing, etc. Accordingly, Table 14.3 describes the RCCI engine characteristic features with respect to varying the operating parameters for distinct low and high reactive alternative fuels.

According to the RCCI idea, the degree of reaction of the charge obviously governs the combustion characteristics. As a result, particular conditions must be maintained for the engine to run with various fuel combinations. To fully utilize the RCCI approach, the engines must be able to run on a variety of fuels, ranging from pure gasoline to clean diesel fuel [29]. There are numerous high and low reactive fuel combinations available, such as diesel-gasoline, and gasoline, biodiesel-gasoline, as well as numerous agents to boost the cetane value. Because most studies use diesel fuel, biofuel or its mixtures, and n-heptane fuels as medium responsive sources in RCCI investigations, the current study focuses on low reactive fuels used in RCCI methods, notably gasoline, alcohols, gaseous, and mixed fuels.

14.4.1.1 Influence of low reactive gasoline fuel on RCCI engine exhaust NOx emissions

In RCCI, combustion plays a significant role in enhancing ignition progression, which progressively moves from high reactive to low reactive zones, hence lowering the combustion process. It has been demonstrated that the LRF amount has a key impact in obtaining high efficiency with significant reductions in NOx oxides and pollutants [30]. Over a wide variety of engine rpms and loads, diesel and gasoline fuels can be used in the RCCI process to attain low nitrogen oxides and tailpipe pollutants. The LRF fuel is directly inoculated into the inlet intake by PFI, whereas the HRF fuel is inserted instantly into the ignition cylinder by conventional rail implant. The RCCI engine with gasoline as LRF and diesel as HRF, the exhaust NOx and soot emissions significantly reduced by enhancing both SOI and EGR rates at whole operating ranges. However, lowering EGR increased the HC and CO emissions while lowering the emission levels. Enhanced injection timing lowered CO and HC emissions while improving fuel efficiency. Furthermore, the use of a larger compression ratio necessitates delayed injection duration to reduce unnecessary banging, which leads to undesirable soot tailpipe pollutants at loads greater than 50%. This restriction, however, does not apply to the lesser CR [34–37]. Similarly, minor losses in BTE and increment in peak pressure rise obtained with PODE in gasoline/diesel RCCI combustion. Furthermore, due to features such as higher cetane and

Table 14.3 Confined literature on NOx reduction/influencing strategies and its response on RCCI engine.

Ref	Test fuel: Low reactive	Test fuel: High reactive	Tuning factors	NOx reduction strategy	NOx response
[29]	n-Butanol	n-Heptane	Inlet boosting pressure: 1.75–2.5 bar, Early and Late direct injection	Lean combustion	• NOx↓ as ↑Inlet boosting pressure • NOx for early direct injection < Late direct injection • NOx↑ as ↑Pilot timing
[30]	E85	Diesel	E85 vol (70%–85%)	Retarded combustion	• NOx↓ as ↑E85 Vol%
[31]	Gasoline Gasoline Gasoline	Diesel B20 of TPME B100 of TPME	Gasoline: 0%–50%	LTC with homogeneous lean mixture	• NOx↓ as ↑Gasoline Vol % • NOx for Gasoline with Diesel > B20 > B100
[32]	Wet-ethanol	Diesel	Water in ethanol: 8, 24, 36 vol%; Diesel SOI: 326°–351°; Injection duration: 2–5 ms	Charge cooling effect by wet-ethanol vaporization	• NOx↓ as ↑Water vol% in ethanol • NOx↓ as ↑Diesel SOI • NOx↑ as ↑Wet-ethanol Injection Duration
[21]	Methanol	Diesel	BMEP: 1–4 bar PCCI and RCCI	Low Temperature Combustion	• NOx↓ as ↑BMEP NOx in RCCI < PCCI
[33]	Biogas Syngas	Diesel Diesel	Bio and syngas substitution ratio: 0%–70%	Low in-cylinder temperatures with biogas than syngas	• NOx↑ as ↑Syngas substitution ratio • NOx↓ as ↑Biogas substitution ratio
[34]	Gasoline	Biodiesel	Intake valve closing timing: −112° to −138° EGR: 10%–50%	Charge dilution with EGR	• NOx↓ as ↑EGR • NOx↓ as ↑delay in intake valve closing time

Continued

Table 14.3 Confined literature on NOx reduction/influencing strategies and its response on RCCI engine—cont'd

Ref	Test fuel: Low reactive	Test fuel: High reactive	Tuning factors	NOx reduction strategy	NOx response
[35]	Hexanol	Biodiesel from waste vegetable oil	Hexanol ratio: 20%–50%; Injection pressure: 400, 500, 600 bar; load: medium and rated	Low in-cylinder temperatures though premixed charge combustion	• NOx↓ as ↑Hexanol blend ratio • NOx↑ as ↑Injection Pressure • NOx↑ as ↑Load
[36]	Iso-butanol Gasoline	Diesel Diesel	Premixed Ratio of iso-butanol and gasoline: 40%–60%	Homogeneous charge combustion through extended ignition delay	• NOx↓ as ↑Premixed ratio of both iso-butanol and gasoline • NOx for Iso-butanol < Gasoline
[37]	Natural Gas and Hydrogen/ Syngas	Diesel	Hydrogen/syngas energy share: 0%–50%	Low Temperature Combustion	• NOx↓ as ↑Hydrogen/Syngas energy share at appropriate EGR • NOx for Natural Gas < Hydrogen/ Syngas
[38]	Methanol Methanol	Diesel PODE	Premixed methanol mass: 70%–85% EGR: 0%–26%	Lean and homogeneous premixed methanol-air mixture	• NOx↓ as ↑Premixed Methanol Mass% for both diesel and PODE • NOx↓ as ↑EGR at both combinations
[39]	Gasoline	Diesel	Gasoline mass: 6–28 mg/cycle; Diesel SOI: 10–35° bTDC	Low in-cylinder combustion temperatures	• NOx↓ as ↑Gasoline mass (mg/cycle) • NOx↓ as ↑Diesel SOI (Advancing the SOI)

oxygen, PODE with gasoline ignition releases more oxides and fewer smoke and CO emissions than diesel-gasoline.

In addition, the stepped and immersed piston bowl shapes are predictable to decrease hydrocarbons and carbon monoxide deposition in the splash zones as well as thermal performance losses in the RCCI process while using gasoline as a low responsive fuel. This is accomplished by altering the squish circulation and decreasing the piston contact zone to reduce thermal energy losses. According to the manufacturer, the stock piston provides better charge movement, resulting in faster blending and quicker self-ignition or state of charge than the cylinder and stepped designs. Both the cylinder and stepping pistons are methods to accommodate low nitrogen oxide emissions (0.4 g/kWh) and better fuel efficiency to fulfill European emission standards. The stepwise piston, on the other hand, emits more hydrocarbons and carbon monoxide than some other designs. At moderate load, the lesser region to volume relation of the cylinder piston lowered the thermal energy loss, promoting a higher peak combustion process, a significant decrease in ignition failure and fuel usage, and lesser nitrogen oxide emissions under EURO VI standards [38]. Finally, at maximum load, the stepping piston outperformed the extra two designs in relation to fuel efficiency and nitrogen oxide tailpipe emissions, with somewhat greater smoke density. Gasoline as the LRF also extends the capability of RCCI ignition throughout the entire spectrum of a light-duty engine and solves the primary issues of RCCI engines (curbing hydrocarbons and carbon monoxide pollutants at light loads and fuel usage consequence).

14.4.1.2 Influence of low-reactive alcoholic fuels on RCCI exhaust NOx emissions

Numerous researchers collaborated to investigate the applicability of alternative fuels for RCCI ignition. Because of their greater latent heat of evaporation and reduced reactivity, alcoholic fuels such as methanol, ethanol, and butane have been found to produce stable and prolonged RCCI operation. Several kinds of research have found that using sustainable oxygenation fuels, such as ethanol, boosts hydrocarbons and carbon monoxide pollutants while decreasing oxides and particulate matter [39]. According to reports, the thermodynamic features of ethanol influence the ignition behavior with a small temperature range, resulting in inaccuracy and severe knocking at higher and lower loads, separately. In contrast, it has been demonstrated that a small amount of ethanol (10%–20%) can satisfy European emission oxides and soot pollutant emissions with enhanced hydrocarbon

(HC) and carbon monoxide (CO) emission levels from medium to moderate loads.

The NOx emissions of RCCI engine with hydrous ethyl alcohol can be lowered by enhancing the engine operating parameters like enhancing the intake pressure and temperatures, varying the EGR level, fuel injection timing and pressure. At full load, the intake ambient temp, diesel timing of injection, and first diesel injection pulse portion all affect CO and HC emissions. Nitrogen oxides and soot tailpipe emissions are lowered concurrently with the optimization method, which exhibits excellent diesel injection, delayed combustion timing, and a greater diesel proportion at initial pulse implant with maximum pressure that promotes the A/F mixing process and thus more premixed ignition [31, 32]. When the rail pressure is boosted along the optimum path, the stated energy utilization decreases and the combustion time shortens due to delayed ignition phasing (CA50) and improved premixed combustion, accordingly. Furthermore, nitrogen oxide emissions are responsive to all active parameters, including the percentage from first diesel injection quantities, the time of the subsequent diesel pulse injecting, conventional diesel rail level, and input pressure under all loading situations. It is reported that throughout direct injection of n-heptane and PF alcohol, n-butane, and n-amyl alcohol fuels in RCCI engines, the higher heat energy and less CN of ethanol fuel enhanced the delay period associated with the extra two alcohols. Whenever the premixed ratio is raised, the port injection ethanol fuel exhibits reduced peak in-chamber pressure and thermal efficiency amount with delayed ignition timing. This is due to the decrease in the amount of immediately injected n-heptane gasoline, which inhibited the reactive compositions' response activity. The higher heat energy of alcohols and the lower number of n-heptane fuel postponed the CA-10, CA-50, and CA-90 at larger premixed ratios (> 0.7). Greater premixed percentages result in lower oxides (NOx) and soot emissions, but higher hydrocarbon and carbon oxide emission levels due to reducing in-cylinder reactivity and the enhanced freezing impact of alcohol gases, correspondingly.

14.4.1.3 Influence of low-reactive gaseous fuels on RCCI NOx emissions

In terms of alternative fuels, low responsive gas fuel such as ethanol fuels can also be used to produce RCCI ignition. Because gas fuels have a higher octane value (110) than other short reactivity fuels, they could be a viable alternative for RCCI ignition. Because diesel fuel and gasoline have a considerable reactivity difference due to a higher-octane value, they enable superior control over the highest temperature increase rate and maximum cylinder pressure with longer ignition time. For RCCI ignition, several low reactive gas fuels such as fossil fuels, biogas, and compressed natural gas (CNG) are employed. RCCI combustion with gasoline as low reactive fuel and diesel as high reactive fuel can significantly reduces the NOx emission. Furthermore, at the advanced injection period and greater premixed ratios, the total pollution and their size was reported to be quite low. While hydrocarbons and carbon monoxide pollutants from twin fuel ignition are greater than those from normal solo diesel infusion, they are less than those from sophisticated SOI diesel ignition [1]. This is due to the progressive injection time, wherein the fuel is burned near the squish and crevice zones, and the creation of the A/F combination on the outside wall of the ignition cylinder results in improved fuel vaporization and spray absorption. Similarly, in RCCI ignition, fossil fuel (more octane and less reactive fuel) is blended with air intake, and fuel (poor octane and moderate reactive fuel) is delivered directly inside the combustion cylinder using numerous injection techniques. It can be seen that increasing the premixed proportion lengthens the delay period. Reducing diesel fuel SOI lowers the energy release rates, ignition temperature, and ringing duration, resulting in decreased nitrogen oxides and HC pollutants. Increasing the quantity of fuel and the injection rate of the initial diesel pulse enhanced nitrogen oxide (due to higher in-cylinder heat), CO, and HC emission levels.

14.5 Comparative study on LTC mode advanced combustion engines

LTC techniques are used to power advanced combustion engines such as HCCI, PCCI, and RCCI. From Fig. 14.9, the fuel injection duration of the PCCI engine is bounded by the periods of the traditional diesel engine and the PFI technique of the HCCI engine. Furthermore, the dwell period for a PCCI engine is less than that of a homogeneous charge compression ignition engine, resulting in a stratified premixed charge [1, 3, 29]. The stratified load of premixed ignition, on the other hand, increases the combustion time by minimizing the chemical sensitivity, resulting in greater evaporation phasing regulation than HCCI combustion. Moreover, the PCCI engine has superior command over the maximum ICP and energy discharge rate by increasing the combustion period using a combination of premature direct and early injection techniques, as described in Table 14.4. Furthermore, the PCCI engine has restrictions in respect to particular fuel usage

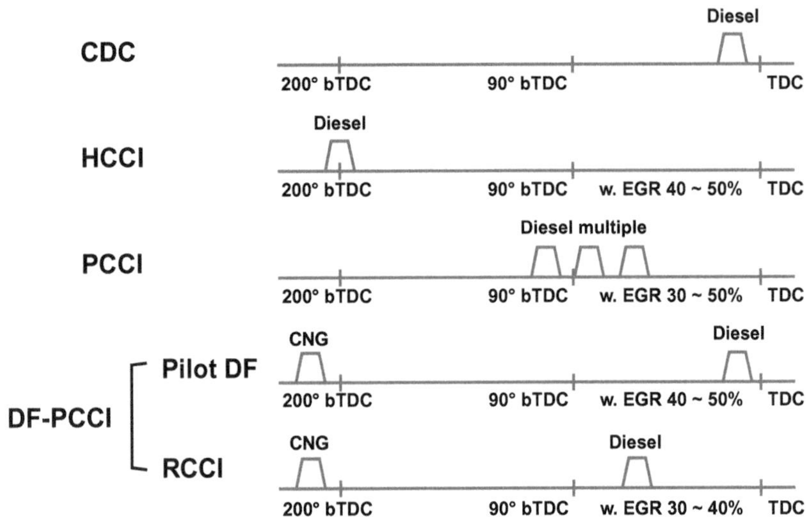

Fig. 14.9 Typical injection timings of conventional and advanced LTC mode.

Table 14.4 Comparison among the low temperature combustion strategies.

Basic characteristics	HCCI	PCCI	RCCI
Fuel requirement	Match with CR	Match with CR	Match with CR
Fuel type	Flexible fuels	Blend of every liquid or	high octane/high cetane fuels
Injection mode	PFI/Early DI/ fumigation	PFI/DI	PFI/DI
Injection pressure	PFI of low pressure and DI of high pressure	PFI of low pressure and DI of high pressure	PFI of low pressure and DI of high pressure
Fuel/air mixture	Premixed	Partially premixed	Partially premixed
Aspiration mode	NA/boosted	Boosted	boosted
Air-fuel ratio	Very lean (depending on fuel)	Air/fuel ratio ($1 \leq \varphi \leq 2$)	Lean (depending on fuel)
Ignition type	Auto-ignition	Compression ignited	Compression ignited
Ignition timing control	Pressure and temperature	Injection timing	Injection timing

Table 14.4 Comparison among the low temperature combustion strategies—cont'd

Basic characteristics	HCCI	PCCI	RCCI
Combustion type	Premixed volumetric combustion, dominated by chemical kinetics	Premixed and diffusion combustion	Premixed and diffusion combustion
Flame front	No flame propagation, homogeneous oxidation	Diffusion flame propagation	Diffusion flame propagation
Throttle	Un-throttled	Un-throttled	Un-throttled
Thermal efficiency	High	High	High
Combustion temperature	Low temperature combustion	Low temperature combustion	Low temperature combustion
Emission characteristics	Extremely low NO_X, soot and CO_2, high HC and CO	Low NO_X, soot and CO_2, high HC and CO	Relatively lower NO_X and PM, high HC and CO

stating that raising the stated maximum cylinder pressure has an impact on the significant use of specific fuel and contributes to increased knocking. As a result, various research works have been carried out using the HCCI and PCCI engines to broaden their working range [17]. The findings demonstrate that a multiple fuel injection technique of a DF-PCCI) engine, also known as a RCCI engine, may achieve greater working loads. Low sensitive fuels such as gas, ethanol, and so forth are permitted in the RCCI modes using a premixed charge compression injection approach, which aims to enhance charge consistency. In addition, significant responsiveness fuels such as diesel, DME, and others are permitted using standard diesel injection strategies, which commences the primary stage of combustion. Generally, low reactive fuels such as gasoline, alcohols, and respective physical and chemical properties has shown greater influence on combustion characteristics of multifuel injection strategy. Furthermore, RCCI technology has been incorporated in the GDI engine, with fuels serving as mild and maximum reactivity fuels. In particular, the injection intensity and injection durations including both sources have been demonstrated to have a stronger impact on ignition sequencing. Furthermore, RCCI may be easily achieved by employing biofuels such as ethanol, natural gas, biofuels, and so on.

Because of their lesser sensitivity and greater heat of vaporization, the types of alcohol significantly improve the working regions of RCCI ignition. Moreover, bioethanol and biodiesel in RCCI oxidation have a significant effect on ignition delay. It claims that as the proportion of ethanol production grew, the combustion duration improved, resulting in a greater suggested average efficient pressure (IMEP) than gasoline-biodiesel RCCI ignition. Furthermore, both PCCI and RCCI ignition have greater combustion features than HCCI, resulting in a loss of efficiency and emission attributes, as described in Table 14.4.

The PCCI and RCCI combustion methods offer slightly better operating parameters than the HCCI engine. The separated charge developed by the PCCI engine's initial injection method improves ignition quality, leading to less knocking and improved power output. Furthermore, PCCI engine working characteristics such as EGR, A/F ratio, CR, injection intensity, and injection duration have been improved to reduce exhaust pollutants such as HC, CO, NOx, and PM. The premixed fuel of the PCCI engine reaches higher in-cylinder heats than the HCCI engine, which promotes oxidation temps and, as a result, fewer hydrocarbon and carbon oxide emissions than the HCCI engine. Consequently, the low heat combustion techniques of the PCCI engine have greater regulation over the in-cylinder temps, resulting in fewer nitrogen oxides and soot pollutants than the HCCI engine. Furthermore, low-pressure gases with too optimum injection strategies of PCCI engines cause fuel wall moisture and coating on the piston, which minimizes NOx emissions through low temp oxidation, which is a result of insufficient fuel vaporization. Additionally, the absence of oxidation temperatures raises smoke, HC, and CO pollutants. The early injection approach, which implies nearly TDC with wide cone inclination injectors, can reduce overall soot and NOx. Furthermore, the PCCI engine's premixed load has a single phase of ignition, whereas the HCCI engine has two phases of ignition, such as high and LTC, which helps to emit fewer NOx than PCCI ignition. However, in comparison to the PCCI engine, these ignition phases produce more HC and CO pollutants.

The dual fuel injection technique (RCCI) has been applied in this manner to improve performance attributes while lowering emission levels at greater working loads, as illustrated in Fig. 14.10. Limited responsiveness fuels have been put into the entrance intake of an RCCI engine to improve charge uniformity through a prolonged delay period as well as to encourage fewer hydrocarbon pollutants through the combustion process. Similarly, injecting enhanced reactivity fuel near the TDC accelerates ignition, resulting in higher ignition and thermal efficiency [1]. Furthermore, the injection

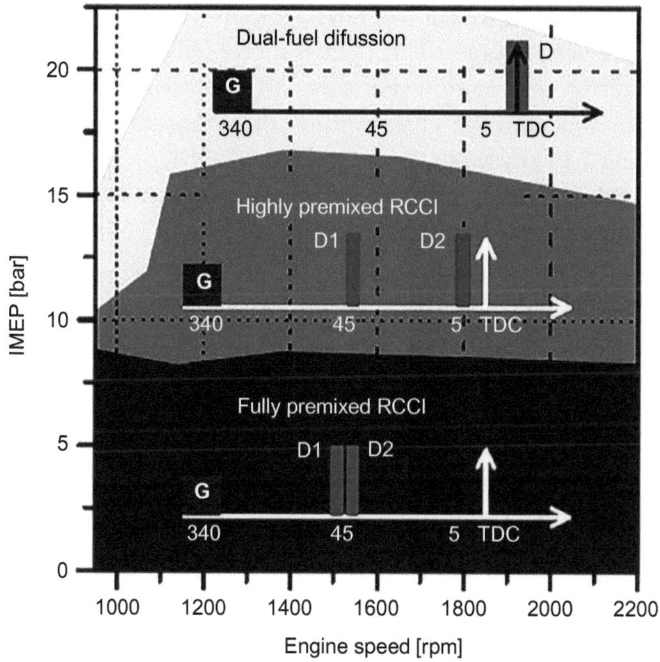

Fig. 14.10 RCCI combustion control strategies according to gasoline and diesel fuel injection.

of enhanced responsiveness fuel improves the oxidation of toxic gases, resulting in fewer HC and CO pollutants but an increase in NOx. By admitting alternate gases into the combustion process, the RCCI engine can reduce emission levels. Alternative energy sources, such as ethanol, biodiesels, fossil fuels, and so on, have a larger oxygen concentration, which helps to improve RCCI ignition. Early injection of bioethanol into RCCI combustion improves homogenization and increases the delay period, resulting in lower HC emissions through the full combustion process and lower NOx by maintaining low in-cylinder pressures. Collectively, both PCCI and RCCI engines enhance efficiency while also lowering NOx and PM pollutants slightly more than the HCCI engine. However, a premixed charge of combustion and twin fuel combustion result in much greater HC and CO pollutants than an HCCI engine.

14.6 Conclusion

To meet existing and future pollution limits for diesel engines below Euro VI standards, particularly NOx and PM emissions, efforts have been concentrated on enhanced in-cylinder/ignition strategies known as LTC over the last several decades. This chapter thoroughly detailed the influence of various charge preparation strategies and biofuels on the performance, combustion, and emission characteristics of advanced combustion mode engines using LTC methods, notably HCCI, PCCI, and RCCI. LTC via EGR and higher premixing improves fuel economy while cutting emissions by removing excessive heat and rich zones of fuel during the ignition development phase. As a consequence of the above-detailed analysis, it is obvious that the fundamental problem of NOx/PM export in a standard diesel engine can be resolved by LTC. The LTC approach has the ability to reduce NOx and PM pollutants from diesel engines to incredible levels while significantly enhancing the efficiency of traditional fuel engines. This LTC mode eliminates the need for after-treatment technologies, which frequently increase the expense and difficulty of automobiles. Furthermore, employing environmentally friendly biofuels in LTC minimizes CO, NOx, and other emissions simultaneously. Despite significant advancements in LTC technology, the implementation of LTC engines has been delayed by a limited working region, control over combustion start at full load, and higher HC and CO pollutants. Though the LTC concept has been demonstrated to be effective in low-load conditions, even in high engines, its application in high-load situations, even in light-duty engines, remains limited.

- In HCCI, both the equivalency ratio and the charge temperature influence the self-ignition properties of the fuels. HCCI can simultaneously reduce nitrogen oxides and particulate matter while boosting thermal efficiency by utilizing initial injections and then repeat injections to prolong the working range. It does, however, emit more CO and HC pollutants.
- The PCCI mode evolved from HCCI and is recommended for mitigating the drawbacks of the HCCI combustion mode (restricted working range, knocking and regulatory combustion constraints). The fundamental distinction between these two combustion processes is the degree of air-fuel mixing.
- RCCI is the twin fuel ignition strategy that uses two fuel mixtures with various self-ignition features (low and high responsive fuel) to regulate ignition going to phase and combustion temperature (the main challenges in HCCI and PCCI modes) while emitting less. Low responsive fuel, which can be used up to 90% of the time, has a key influence in the engine's efficiency, ignition, and emission parameters.

References

[1] Pachiannan T, Zhong W, Rajkumar S, He Z, Leng X, Wang Q. A literature review of fuel effects on performance and emission characteristics of low-temperature combustion strategies. Appl Energy 2019;251:113380.

[2] Krishnamoorthi M, Malayalamurthi R, He Z, Kandasamy S. A review on low temperature combustion engines: performance, combustion and emission characteristics. Renew Sust Energ Rev 2019;116:109404.

[3] Azad AK, Rasul MG, Khan MM, Sharma SC, Bhuiya MM. Recent development of biodiesel combustion strategies and modelling for compression ignition engines. Renew Sust Energ Rev 2016;56:1068–86.

[4] Agarwal AK, Singh AP, Maurya RK. Evolution, challenges and path forward for low temperature combustion engines. Prog Energy Combust Sci 2017;61:1–56.

[5] Saxena S, Bedoya ID. Fundamental phenomena affecting low temperature combustion and HCCI engines, high load limits and strategies for extending these limits. Prog Energy Combust Sci 2013;39(5):457–88.

[6] Gan S, Ng HK, Pang KM. Homogeneous charge compression ignition (HCCI) combustion: implementation and effects on pollutants in direct injection diesel engines. Appl Energy 2011;88(3):559–67.

[7] Ardebili SM, Solmaz H, Calam A, İpci D. Modelling of performance, emission, and combustion of an HCCI engine fueled with fusel oil-diethylether fuel blends as a renewable fuel. Fuel 2021;290:120017.

[8] Gainey B, Yan Z, Lawler B. Autoignition characterization of methanol, ethanol, propanol, and butanol over a wide range of operating conditions in LTC/HCCI. Fuel 2021;287:119495.

[9] Parthasarathy M, Ramkumar S, Lalvani JI, Elumalai PV, Dhinesh B, Krishnamoorthy R, Thiyagarajan S. Performance analysis of HCCI engine powered by tamanu methyl ester with various inlet air temperature and exhaust gas recirculation ratios. Fuel 2020;282:118833.

[10] Yontar AA, Zhou M, Ahmad S. Influence of intake air temperature control on characteristics of a homogeneous charge compression ignition engine for hydrogen-enriched kerosene-dimethyl ether usage. Int J Hydrog Energy 2020;45(41):22019–31.

[11] Khandal SV, Banapurmath NR, Gaitonde VN. Performance studies on homogeneous charge compression ignition (HCCI) engine powered with alternative fuels. Renew Energy 2019;132:683–93.

[12] Liu J, Yang J, Sun P, Gao W, Yang C, Fang J. Compound combustion and pollutant emissions characteristics of a common-rail engine with ethanol homogeneous charge and polyoxymethylene dimethyl ethers injection. Appl Energy 2019;239:1154–62.

[13] Yu S, Li L, Zheng M. N-butanol reactivity modulation with early fuel injections for low NOx compression ignition combustion. Fuel 2019;242:243–54.

[14] Zhou A, Zhang C, Li Y, Li S, Yin P. Effect of hydrogen peroxide additive on the combustion and emission characteristics of an n-butanol homogeneous charge compression ignition engine. Energy 2019;169:572–9.

[15] Wang Z, Liu H, Ma X, Wang J, Shuai S, Reitz RD. Homogeneous charge compression ignition (HCCI) combustion of polyoxymethylene dimethyl ethers (PODE). Fuel 2016;183:206–13.

[16] Kozarac D, Taritas I, Vuilleumier D, Saxena S, Dibble RW. Experimental and numerical analysis of the performance and exhaust gas emissions of a biogas/n-heptane fueled HCCI engine. Energy 2016;115:180–93.

[17] Saiteja P, Ashok B. A critical insight review on homogeneous charge compression ignition engine characteristics powered by biofuels. Fuel 2021;285:119202.

[18] Duan X, Lai MC, Jansons M, Guo G, Liu J. A review of controlling strategies of the ignition timing and combustion phase in homogeneous charge compression ignition (HCCI) engine. Fuel 2021;285:119142.
[19] Nibin M, Raj JB, Geo VE. Experimental studies to improve the performance, emission and combustion characteristics of wheat germ oil fuelled CI engine using bioethanol injection in PCCI mode. Fuel 2021;285:119196.
[20] Elkelawy M, Bastawissi HA, El Shenawy EA, Shams MM, Panchal H, Sadasivuni KK, Choudhary AK. Influence of lean premixed ratio of PCCI-DI engine fueled by diesel/biodiesel blends on combustion, performance, and emission attributes; a comparison study. Energy Convers Manag 2020;X:100066.
[21] Singh AP, Kumar V, Agarwal AK. Evaluation of comparative engine combustion, performance and emission characteristics of low temperature combustion (PCCI and RCCI) modes. Appl Energy 2020;278:115644.
[22] Chen H, Wang X, Pan Z. A comprehensive study of fuel composition, combustion and soot nanostructure characteristics of a diesel/light hydrocarbons premixed charge compression ignition engine. Fuel 2020;274:117858.
[23] Shim E, Park H, Bae C. Comparisons of advanced combustion technologies (HCCI, PCCI, and dual-fuel PCCI) on engine performance and emission characteristics in a heavy-duty diesel engine. Fuel 2020;262:116436.
[24] Pandey S, Bhurat S, Chintala V. Combustion and emissions behaviour assessment of a partially premixed charge compression ignition (PCCI) engine with diesel and fumigated ethanol. Energy Procedia 2019;160:590–6.
[25] Park H, Shim E, Bae C. Expansion of low-load operating range by mixture stratification in a natural gas-diesel dual-fuel premixed charge compression ignition engine. Energy Convers Manag 2019;194:186–98.
[26] Qian Y, Wu Z, Guo J, Li Z, Jiang C, Lu X. Experimental studies on the key parameters controlling the combustion and emission in premixed charge compression ignition concept based on diesel surrogates. Appl Energy 2019;235:233–46.
[27] Park H, Shim E, Bae C. Improvement of combustion and emissions with exhaust gas recirculation in a natural gas-diesel dual-fuel premixed charge compression ignition engine at low load operations. Fuel 2019;235:763–74.
[28] El Shenawy EA, Elkelawy M, Bastawissi HA, Shams MM, Panchal H, Sadasivuni K, Thakar N. Investigation and performance analysis of water-diesel emulsion for improvement of performance and emission characteristics of partially premixed charge compression ignition (PPCCI) diesel engines. Sustain Energy Technol Assess 2019;36:100546.
[29] Han J, Bao H, Somers LM. Experimental investigation of reactivity controlled compression ignition with n-butanol/n-heptane in a heavy-duty diesel engine. Appl Energy 2021;282:116164.
[30] Willems R, Willems F, Deen N, Somers B. Heat release rate shaping for optimal gross indicated efficiency in a heavy-duty RCCI engine fueled with E85 and diesel. Fuel 2021;288:119656.
[31] Harari PA, Banapurmath NR, Yaliwal VS, Khan TY, Soudagar ME, Sajjan AM. Experimental studies on performance and emission characteristics of reactivity controlled compression ignition (RCCI) engine operated with gasoline and Thevetia Peruviana biodiesel. Renew Energy 2020;160:865–75.
[32] Rosa JS, Martins ME, Telli GD, Altafini CR, Wander PR, Rocha LA. Exploring the effects of diesel start of injection and water-in-ethanol concentration on a reactivity controlled compression ignition engine. Fuel 2020;281:118751.
[33] Mahmoodi R, Yari M, Ghafouri J, Poorghasemi K. Effect of reformed biogas as a low reactivity fuel on performance and emissions of a RCCI engine with reformed biogas/diesel dual-fuel combustion. Int J Hydrog Energy 2020;46(30):16494–512.

[34] Li J, Yu X, Xie J, Yang W. Mitigation of high pressure rise rate by varying IVC timing and EGR rate in an RCCI engine with high premixed fuel ratio. Energy 2020;192:116659.

[35] Thomas JJ, Sabu VR, Nagarajan G, Kumar S, Basrin G. Influence of waste vegetable oil biodiesel and hexanol on a reactivity controlled compression ignition engine combustion and emissions. Energy 2020;206:118199.

[36] Pan S, Liu X, Cai K, Li X, Han W, Li B. Experimental study on combustion and emission characteristics of iso-butanol/diesel and gasoline/diesel RCCI in a heavy-duty engine under low loads. Fuel 2020;261:116434.

[37] Rahimi HM, Jazayeri SA, Ebrahimi M. Hydrogen energy share enhancement in a heavy duty diesel engine under RCCI combustion fueled with natural gas and diesel oil. Int J Hydrog Energy 2020;45(35):17975–91.

[38] Duraisamy G, Rangasamy M, Govindan N. A comparative study on methanol/diesel and methanol/PODE dual fuel RCCI combustion in an automotive diesel engine. Renew Energy 2020;145:542–56.

[39] Saxena MR, Maurya RK. Influence of direct injection timing and mass of port injected gasoline on unregulated and nano-particle emissions from RCCI engine. Fuel 2020;282:118815.

Index

Note: Page numbers followed by *f* indicate figures and *t* indicate tables.

D

E

F

O

P

FSC
www.fsc.org
MIX
Papier | Fördert
gute Waldnutzung
FSC® C083411

Kolibri 360 GmbH
Ferdinand-Jühlke-Straße 7
99095 Erfurt, Deutschland
produktsicherheit@kolibri360.de